Spatio-Temporal Heterogeneity

Concepts and Analyses

- Do earthquakes occur in specific locations at regular intervals in time?
- How do the mean value and the variance of tree variables change over time?
- Are air and soil temperatures correlated at particular scales spatially and temporally?

Attempts by ecologists to establish models for predicting the growth of a population or the fluctuations of a natural resource can be confounded by environmental heterogeneity. *Spatio-Temporal Heterogeneity* explores a range of available statistical methods to help ecologists in the attempt to unravel complexities, demonstrating how to place these changes into an understandable statistical framework.

It addresses several key questions, including how to interpret the parameters of statistical models in relation to the biological and environmental realities; how to design a study to collect the best sample data; and how to avoid pitfalls in modeling, design, statistical assessment, and interpretation. Dutilleul uses a variety of examples to facilitate understanding, from plant ecology, earth and atmospheric sciences, animal biology, forestry, and limnology. The accompanying CD-ROM contains MATLAB® and SAS codes to aid analyses.

PIERRE R. L. DUTILLEUL is Professor, Department of Plant Science at McGill University, Associate Member, McGill School of Environment, and Associate Member, Department of Mathematics and Statistics at McGill University, Canada.

ECOLOGY, BIODIVERSITY AND CONSERVATION

The world's biological diversity faces unprecedented threats. The urgent challenge facing the concerned biologist is to understand ecological processes well enough to maintain their functioning in the face of the pressures resulting from human population growth. Those concerned with the conservation of biodiversity and with restoration also need to be acquainted with the political, social, historical, economic and legal frameworks within which ecological and conservation practice must be developed. The new Ecology, Biodiversity and Conservation series will present balanced, comprehensive, up-to-date, and critical reviews of selected topics within the sciences of ecology and conservation biology, both botanical and zoological, and both 'pure' and 'applied'. It is aimed at advanced final-year undergraduates, graduate students, researchers, and university teachers, as well as ecologists and conservationists in industry, government and the voluntary sectors. The series encompasses a wide range of approaches and scales (spatial, temporal, and taxonomic), including quantitative, theoretical, population, community, ecosystem, landscape, historical, experimental, behavioural and evolutionary studies. The emphasis is on science related to the real world of plants and animals rather than on purely theoretical abstractions and mathematical models. Books in this series will, wherever possible, consider issues from a broad perspective. Some books will challenge existing paradigms and present new ecological concepts, empirical or theoretical models, and testable hypotheses. Other books will explore new approaches and present syntheses on topics of ecological importance.

Ecology and Control of Introduced Plants
Judith H. Myers and Dawn Bazely

Invertebrate Conservation and Agricultural Ecoystems
T. R. New

Risks and Decisions for Conservation and Environmental Management
Mark Burgman

Ecology of Populations
Esa Ranta, Per Lundberg, and Veijo Kaitala

Nonequilibrium Ecology
Klaus Rohde

The Ecology of Phytoplankton
C. S. Reynolds

Systematic Conservation Planning
Chris Margules and Sahotra Sarkar

Large Scale Landscape Experiments: Lessons from Tumut
David B. Lindenmayer

Assessing the Conservation Value of Freshwaters: An international perspective
Phil Boon and Cathy Pringle

Insect Species Conservation
T. R. New

Bird Conservation and Agriculture
Jeremy D. Wilson, Andrew D. Evans, and Philip V. Grice

Cave Biology: Life in darkness
Aldemaro Romero

Biodiversity in Environmental Assessment: Enhancing ecosystem services for human well-being
Roel Slootweg, Asha Rajvanshi, Vinod B. Mathur, and Arend Kolhoff

Mapping Species Distributions
Janet Franklin

Decline and Recovery of the Island Fox: A case study for population recovery
Timothy J. Coonan, Catherin A. Schwemm, and David K. Garcelon

Ecosystem Functioning
Kurt Jax

Spatio-Temporal Heterogeneity

Concepts and Analyses

PIERRE R. L. DUTILLEUL
McGill University, Canada

CAMBRIDGE UNIVERSITY PRESS
Cambridge, New York, Melbourne, Madrid, Cape Town,
Singapore, São Paulo, Delhi, Tokyo, Mexico City

Cambridge University Press
The Edinburgh Building, Cambridge CB2 8RU, UK

Published in the United States of America by Cambridge University Press, New York

www.cambridge.org
Information on this title: www.cambridge.org/9780521791274

First published 2011

Printed in the United Kingdom at the University Press, Cambridge

A catalog record for this publication is available from the British Library

Library of Congress Cataloging in Publication data
Dutilleul, Pierre, 1961–
Spatio-temporal heterogeneity : concepts and analyses / Pierre Dutilleul.
p. cm. – (Ecology, biodiversity, and conservation)
Includes bibliographical references and index.
ISBN 978-0-521-79127-4 (hardback) – ISBN 978-1-107-40035-1 (paperback)
1. Spatial ecology – Mathematical models. 2. Population biology – Mathematical models.
3. Ecological heterogeneity. I. Title.
QH541.15.S62D88 2011
577.01′5195 – dc22 2010053579

ISBN 978-0-521-79127-4 Hardback
ISBN 978-1-107-40035-1 Paperback

To Marie-Pierre, with all my love

Contents

Foreword

It has been said that, when preparing a Foreword, one should read the book first. I did so, eagerly, and was in no way disappointed. Pierre Dutilleul, who is an active and prolific academic at McGill University in Montréal, has written a book that has a friendly style, includes pertinent examples, as well as explanations and motivations, and further includes important value judgments. I know of no other book quite like it. I recommend the book highly to students, teachers, and researchers in fields including biology, ecology, environmental science, as well as in mainstream and applied statistics.

The book is well organized. The title refers to "Spatio-Temporal Heterogeneity," while the subtitle is "Concepts and Analyses." These are exactly what the book presents, from the definition of models to the interpretation of data analyses. The coverage and structure are complete. In part, this is due to the author's use of a "Space-Time Response Cube." This is a $2 \times 2 \times 2$ array with the three factors having respective levels: (point, surface), (space, time), (deterministic, random). Laying out these eight combinations has assured the author of presenting his material in a complete orderly unified manner.

Besides the chapters (of which there are ten), there are "Key notes" that pop up from time to time, such as "*With irregular sampling grids, equal-frequency distance classes are recommended over equally spaced distance classes, because the former then provide a more even precision and power in the estimation and testing of all autocorrelation coefficients.*" There are also "Summaries." The one following a key question list is "*Spatio-temporal heterogeneity can take different forms, and certain statistical methods and parameters of statistical models are more appropriate than others to measure spatio-temporal heterogeneity, depending on its form.*" Some chapters also have appendices presenting the more technical material. This allows the various concepts to be provided via words instead of formulas. Formulas are set down later. This makes the book accessible to a broad class of less technically trained users. These persons will be pleased to learn that the book has an accompanying

CD-ROM holding computer commands to carry out analyses of the type highlighted in the book.

In a literature search, Dutilleul found that the term "heterogeneity" came up often in the biological and environmental fields via titles and key word lists. He noted that its frequency of occurrence was increasing with time. The heterogeneity studied in this book refers to that measured by an experimenter, as opposed to functional. Many collections of measurements are analyzed. These include data from earthquakes, sheep, mosquito, temperatures, carbon dioxide, trees, lakes, pink bollworm, plankton, and plants.

The book will be helpful for many levels of reader, starting with final-year undergraduates and moving on to theoretical statisticians. It may prove especially useful to persons seeking to learn and use contemporary random process tools. It may prove inspirational to persons seeking to generalize and synthesize methods important in practice.

It is interesting that the book's material begins with the case of point processes. This makes sense because they are often the basic building blocks of phenomena. The material ends with the important chapter titled "Sampling and study design aspects in heterogeneity analysis of surface patterns."

The book is a topical and timely addition to the scientific literature. Many practitioners will be motivated to employ the tools that Dutilleul makes available in it in clear heuristic manners.

David R. Brillinger
University of California, Berkeley

Preface

The world in which we live is one of continuing change, which affects our environment, and the flora and fauna in it. Under the umbrella of "heterogeneity analysis," the understanding of changes locally and globally at short and longer terms, through accurate and precise quantification based on sound modeling, provides the basis for reliable predictions in space and time. Accordingly, I have written this book first of all as a dissertation on the theme of spatio-temporal heterogeneity. Loosely speaking, that includes "recognizing heterogeneity when it passes by" and being aware of the statistical approaches and tools available for its assessment, to be an informed user in due time. Definitively, I have not written a methods book. Instead, spatial or temporal heterogeneity comes first and the relevant analytical methods follow. Therefore, the concepts are introduced and discussed in words in the body of the text, the corresponding formulas and equations being distributed between the body of the text and appendices or tables.

Since Peter Whittle's (1951) *Hypothesis Testing in Time Series Analysis* and Bertil Matérn's (1960) *Spatial Variation*, numerous books on related topics have been published and each of them has its "raison d'être." One of the particularities of *Spatio-Temporal Heterogeneity: Concepts and Analyses* is that the treatment of heterogeneity analysis performed in it involves space and time taken separately or jointly, which ensures uniformity in notation and terminology throughout regardless of the framework. Two other consequences of this unifying approach are that (i) the material covered here replaces, in a sense, a number of sections in a number of books written for one or the other framework, and (ii) thanks to the CD-ROM with MATLAB® and SAS codes, analytical tools usually offered in different packages are made accessible in one place, together with computer programs not available elsewhere.

Biologists and environmentalists are often challenged on the ground of statistics, not due to lack of information or effort, but because matching a statistical method to the non-statistical question of interest can be a subtle task. Things become even more complicated when the words and concepts used in statistics do not carry the same meanings for non-statisticians. This is the case in particular for the concepts of statistical heterogeneity and ecological heterogeneity, as emphasized in "Spatial heterogeneity against heteroscedasticity: an ecological paradigm vs. a statistical concept," by Pierre Dutilleul and Pierre Legendre, a Forum article published in *Oikos* in 1993. This is why one of my goals in writing this book was to marry the two points of view, by explaining to biologists and environmentalists what the parameters of statistical models exactly mean when they are used to measure heterogeneity in space and time. I also wanted to describe how to estimate these statistical parameters and assess the significance of estimates correctly.

Spatio-Temporal Heterogeneity: Concepts and Analyses responds to the need that biologists and environmentalists have to know more about the statistical analysis of spatio-temporal heterogeneity, but no book on this topic had been written for their audience. From the statistical perspective, the response, or stochastic process, can be modeled with the same fundamental equation or a variant, whether it is purely spatial, purely temporal, or spatio-temporal, without or with replication. This general model is applicable whether space and time are represented by quantitative indices or by levels of qualitative factors. From the biological and environmental perspectives, the model is directly related to the nature of the data, the sampling or experimental design, and the questions of spatial or temporal heterogeneity addressed in the study. Links are facilitated by the use of a cube with eight cells, defined by (i) the pattern serving as support for the heterogeneity analysis, (ii) the axis defining the domain of observation, and (iii) the potentially heterogeneous component of the stochastic process. The questions of interest in a given study will be linked to one or another cell of the cube, and will be assessed by one or several appropriate statistical methods. This way, questions regarding ecological heterogeneity will be defined in a sound statistical framework, and the ecologist will be brought to the statistics *a priori* instead of the statistics being brought to the ecology *a posteriori*.

With a volume providing a compendium that merges effectively the two perspectives on spatio-temporal heterogeneity, non-statisticians will find here a sufficient amount of information about the statistical analysis both in theory and in practice. The volume could serve as a primer, should more specialized statistical literature prove necessary. For their part, statisticians will see how some of their models for heterogeneity analysis in space and time may apply in the biological and environmental contexts. From the sample data collected in those fields, they will be able to judge whether the assumptions of the analytical methods are satisfied or not. In *Spatio-Temporal Heterogeneity: Concepts and Analyses*, the reader will find a great variety of examples, in plant ecology and the earth and atmospheric sciences as well as in animal biology, forestry, and limnology, to name a few. These examples illustrate not only the application of statistical methods, with tables and figures, but also the biological or environmental motivation for the heterogeneity analysis, and finish with the interpretation of results. Focus is on clarity and precision – the two are not contradictory. Focus is also on simplicity because it is sometimes thought that, in the statistical sciences, it is easy to be complicated and difficult to be simple . . .

This book is intended to meet the needs of everyone planning to analyze some spatial or temporal aspect of ecological heterogeneity. It could be used by advanced undergraduate students in the life sciences and environmental engineering and by graduate students in ecology and environmetrics as well as in applied statistics. All that it takes to understand the main part of the material, with the exception of a few sections where a deeper treatment of the topic is performed, is a good introductory statistics course or book with a clear presentation of the concepts of mean, variance, and correlation, and a minimum mathematical background. The book could serve as a reference for researchers in the biological and environmental sciences and for statistical consultants, interested in learning the different types of spatio-temporal heterogeneity and the statistical models and methods appropriate for the analysis of each type. The book could be a source of inspiration for applied statisticians who may find in it ideas for the development of new methods of time series and spatial data analysis. Finally, parametric statisticians with interests in the analysis of spatial and temporal, univariate and multivariate stochastic processes will find useful material in specific chapters.

Pierre Dutilleul
Montréal

Acknowledgments

The author would like to thank a number of people without whom this book project would not have been completed the way it was. These people comprise scientific collaborators, owners of data used in examples, and relatives.

The "seed" of this book was planted together with Martin Lechowicz (McGill University) in the middle of the 1990s. Then, Martin kindly let me take care alone of the seedling and make it grow until the tree became mature . . . Thank you for everything, Martin, including our early meetings in which you shared bits of your tremendous experience in research in ecology with me and the editing of a draft of the first chapters.

Warm thanks go to David Brillinger (UC Berkeley) for his foreword and his inspiration over the years.

Many thanks go to Michael Usher, Editor of the EBC Series at Cambridge University Press, for having been there since 2007. Your comments, advice, and support from beginning to end were greatly appreciated.

Thank you to Philip Dixon (Iowa State University), Bernard Pelletier (McGill University), and Guillaume Larocque (McGill University) for their reading of different chapters. Your comments were appreciated. Bernard and Guillaume, I would like to thank you in particular for our exploration of and discoveries in the geostatistical world in the last years. I am grateful to you for all the questions you raised and all your answers to my "questions à cinq francs belges"! And thank you very much for the Matlab codes, on the website http://environmetricslab.mcgill.ca and on the CD-ROM.

Thank you to Liwen Han (McGill University) for the beautiful Figure 10.2 that he produced from wood CT scan data.

A big thank you to:

Yves Carrière (The University of Arizona), for the pink bollworm data (Chapter 7);

Armand Deswysen (defunct, formerly Université Catholique de Louvain) and Vivian Fischer (Universidade Federal de Pelotas), for the sheep chewing data (Chapter 4);

Louise Filion (Université Laval), for the larch sawfly outbreak data (Chapter 4);

Jim Fyles (McGill University), for the Molson Ecological Reserve data (Chapter 7);

Jean Guillard (INRA, Thonon-les-bains), on behalf of the Station d'Hydrobiologie Lacustre, for the Lac Léman data (Chapter 7);

Tim Haltigin (Agence Spatiale Canadienne / Canadian Space Agency) and Wayne Pollard (McGill University), for the Arctic terrestrial and Martian spatial point pattern data (Chapter 5);

Marc Herman (formerly Université Catholique de Louvain), for the forestry and wood science data (Chapters 6 and 9) – Marc, thank you very much for all that I learned about tree growth rate and its implications for wood properties, through our many interactions;

Guillaume Larocque (McGill University), for the multivariate spatial point pattern data in forest ecology (Chapter 3);

Martin Lechowicz (McGill University), Director of the Gault Nature Reserve in Mont-Saint-Hilaire (Québec, Canada), and collaborators, for the *Arisaema* data (Chapters 3–5) and the MSH air and soil temperature data (Chapters 6–8);

Duyen Nguyen and Manfred Rau (formerly McGill University), for the mosquito larva development data (Chapter 4);

Isabelle Schelstraete and Marc Weyers (formerly Université Catholique de Louvain), for the data on maternal behavior of Wistar rat (Chapter 6);

Jason Stockwell and Gary Sprules (University of Toronto), for the Lake Erie data (Chapter 7); and

Marko Tosic and Robert Bonnell (formerly McGill University), for the Barbados environmental data (Chapter 9).

Thank you very much to Laurence Madden (Ohio State University) for the BBD software and the SAS code for fitting the beta-binomial distribution. The data for the example of California earthquakes in space, time, and space-time (Chapters 3–5 and 10) were retrieved from Web catalogs of the Northern California

Earthquake Data Center; for Chapter 10 in particular, the catalog http://www.ncedc.org/ncedc/catalog-search.html was used. Similarly, the atmospheric CO_2 concentration datasets (Chapter 6) were retrieved from http://cdiac.ornl.gov/trends/co2/sio-keel.htm, and the yearly mean sunspot numbers (Chapter 6) were retrieved from the website http://www.ngdc.noaa.gov/stp/spaceweather.html of the National Geophysical Data Center (NGDC) in Boulder (Colorado, USA).

My most sincere thanks go to my family members, who have supported me (in the two senses of the term . . .) in the course of this project. In particular, thank you, Quentin, for having matched your professional exams to become an actuary, with the revision of the last chapters of my book! Last but certainly not least, thank you, Marie-Pierre, for all your understanding . . .

1 · *Conceptual introduction*

To help readers find their way through the book and to direct non-statisticians towards the statistical methods available for analyzing the type of spatial or temporal heterogeneity they are interested in, a coherent classification of the types of heterogeneity that may be encountered and those that are actually treated in the book is established in this introductory chapter. The classification begins in general terms in Section 1.1, which ends with the key questions addressed in the book. In Section 1.2, the distinction between types of heterogeneity is refined with the presentation of the Space-Time Response Cube, of which the eight cells will be covered in different sections of the book. The use of "*Key note: One or several sentences.*" to highlight a point of some importance in the course of a development and of "*Summary: One or several sentences.*" to summarize the take-home message at the end of a development will be continued in the rest of the book. An overview of Chapters 2 to 10 is given in Section 1.3.

1.1 The concept of spatio-temporal heterogeneity: views and perspectives

Heterogeneity has been the focus of ecological studies for more than half a century. Originally investigated in space, it became increasingly important during the 1950s and 1960s (McIntosh, 1991). As an indication of the interest of biologists and environmentalists and other researchers in the concept, "heterogeneity" occurred in the title or the abstract or as a key word or a heading subject of an increasing number of publications over the years. The number of occurrences characteristically rose since the early 1990s (Table 1.1).

Among the types of ecological heterogeneity reported, a primary classification refers to measured heterogeneity and functional heterogeneity. The former is a product of the experimenter's observation, whereas the latter follows from the ecological entity's perception of the habitat

Table 1.1. *Number of occurrences of "heterogeneity" as a key word or a subject heading or in the title or the abstract of scientific publications: 20-year survey of four main bibliographic databases*

Year	Current contents/all editions	BIOSIS previews	General science full text (Wilson)	Biological and agricultural index plus (Wilson)
2009(*)	7592	4322	312	357
2008	7276	4204	308	348
2007	6836	3956	303	367
2006	6408	3698	328	350
2005	5993	3580	315	335
2004	5505	3279	317	357
2003	5110	3229	276	297
2002	4895	3138	242	245
2001	4637	3104	233	260
2000	4745	3021	216	236
1999	4505	2682	183	181
1998	4324	2693	219	241
1997	4272	2747	178	179
1996	4117	2704	150	180
1995	4035	2732	144	146
1994	3922	2622	131	139
1993	2392	2456	118	146
1992		2489	39	99
1991		2453	20	72
1990		2313	26	88

(*) Latest update performed on December 31, 2010.

(Kolasa and Rollo, 1991). In the statistical approach, the perspective is that of the experimenter, so this book will deal essentially with measured heterogeneity.

Biologists and environmentalists tend to refer to spatio-temporal heterogeneity in a rather informal way, declaring they observe heterogeneity whenever the response values vary in space, time, or both. Non-statisticians would benefit from the more complete appreciation of heterogeneity that pertains to the statistical view, in particular, in the distinction between mean and variance heterogeneities and the heterogeneity arising from autocorrelation (Dutilleul and Legendre, 1993).

Key note: This distinction, and the resulting model presented in Chapter 2, is the backbone of this book.

Conversely, statisticians would benefit from the biological and environmental views on processes underlying the measured heterogeneity, since Life Processes of Mother Nature are very concrete! To be most effective, statisticians should work closely with biologists and environmentalists to merge their complementary views on spatio-temporal heterogeneity. This book has been written on this premise, by an applied statistician who appreciates the needs of non-statisticians for more than two decades.

Following up on the classification of measured heterogeneities, "heterogeneity" in landscape ecology refers to differences in species composition and abundance among parts of the study field (Shachak and Brand, 1991). A forest is said to be "homogeneous" if, whatever the part of the forest considered, the same number of tree species is observed in the same proportions, or diversity is constant. In an extreme case, a forest can be homogeneous because it has been planted with only one species of trees; diversity is then zero or non-existent. Of course, heterogeneity, not homogeneity, may well be the norm in natural systems. This first view of heterogeneity, which refers to the composition of parts of different kinds, is applicable to soil sciences. Generally heterogeneous when they are collected, soil samples are made homogeneous prior to analysis, by making the respective proportions of sand and clay, for example, the same in all sub-samples. A second view of measured heterogeneity refers to the distribution of "points," such as trees in 2-D space and soil particles in 3-D space, and is called "point-pattern approach."

Key note: The point pattern is one of the two main types of pattern that will be scrutinized in this book.

Point patterns can also be observed in time via the occurrence of events. Birth and death, flooding, and fire are possible events of interest. Their occurrence may correspond to the expansion or the extinction of a population in ecological sense, or may mean drastic changes for the equilibrium of an ecosystem.

Key note: The other main type of pattern that will be studied concerns continuous quantitative responses.

When measuring soil pH in space or tree-ring width in time, heterogeneity will refer to the variability of the response, not simply in its values but more importantly in the statistical parameters (e.g., mean, variance, autocorrelation) of its distribution. For example, the mean and variance of the width of growth rings usually decrease with the age of the tree, while soil pH at one location tends to be correlated with soil pH at neighboring locations. This view is called "surface-pattern approach." The expression "surface pattern" will be used in space and space-time, although "curve pattern" and "volume pattern" in 1-D and 3-D space, and "hyper-surface pattern" or "hyper-volume pattern" when time is involved with 2-D or 3-D space, may be more appropriate; "time series" is generally used in time.

Key note: There are also means and variances and possibly autocorrelation in point patterns, but the statistical parameters are then related to counts of points or distances between points.

Scale is an important, intrinsic characteristic of heterogeneity, whether functional or measured. For functional heterogeneity, this characteristic refers to the size of the biological entity (e.g., bird, fish, insect, mammal, plant), its life expectancy, and the area it occupies relative to the ecosystem. When measuring spatio-temporal heterogeneity, scale is usually defined by a grain, the level of resolution between observations in space or time, and an extent, the span of all the observations in a given study (Allen and Hoekstra, 1991). Heterogeneity can thus be measured at micro- to macro-scales. Scale can create heterogeneity from homogeneity and vice versa (Kolasa and Rollo, 1991), and if sampling is conducted within a reduced extent while keeping the same grain, spatial homogeneity at large scale combined with spatial heterogeneity at small scale may resemble patchy heterogeneity (Dutilleul, 1993a).

Key note: The statistical parameters and their estimates obtained from sample data are likely to be scale-dependent or scale-specific.

The following key questions will be addressed in this book. In general terms:

- How to conceptualize spatio-temporal heterogeneity in an understandable statistical framework? In particular, how to interpret the parameters of statistical models in relation to the biological and environmental realities?
- How to adjust the quantification of spatio-temporal heterogeneity to the type of pattern and the component(s) of heterogeneity of interest?
- In view of the statistical models and methods available, how to design a study or an experiment in order to collect "the best sample data

possible" in order to meet the objectives regarding spatio-temporal heterogeneity?
- How to interpret the results of data analysis in the different situations and draw sound conclusions from the non-statistical perspective?
- Overall, how to avoid pitfalls in modeling, design, statistical assessment, and interpretation when conducting a spatio-temporal heterogeneity analysis?

Summary: Spatio-temporal heterogeneity can take different forms, and certain statistical methods and parameters of statistical models are more appropriate than others to measure spatio-temporal heterogeneity, depending on its form.

1.2 The Space-Time Response Cube

A coherent scheme of classification for spatio-temporal heterogeneity must cover the range of situations in which biologists and environmentalists may seek to measure heterogeneity, and each situation should be related to a given type of heterogeneity in the scheme. The classification presented below and illustrated in Fig. 1.1 is based on (i) the pattern (i.e., point vs. surface) supporting the analysis of heterogeneity, (ii) the axis, or the axes (i.e., space vs. time, or both), along which heterogeneity is to be analyzed, and (iii) the component (i.e., deterministic vs. random) that is of particular interest in the response.

Key note: The response will be called "stochastic process" in statistical terms.

The point and surface patterns have been introduced in Section 1.1. The type of pattern refers to the domain of definition of the values taken by the response, among other things. In simple terms, the values are binary (i.e., 1/0 for presence/absence or occurrence/non-occurrence) for a point pattern and continuous quantitative for a surface pattern. In a point pattern, qualitative or quantitative data can be collected in the form of a "mark" associated with each point (e.g., the health status or height of a plant), in addition to the binary response. In a surface pattern, a qualitative or quantitative variable, such as the predetermined level of a classification factor or the observed value of a potential explanatory variable, can be associated with the continuous quantitative response.

The spatial and temporal axes of heterogeneity will be involved together in any study where spatio-temporal heterogeneity will be analyzed. Recall that time is one-dimensional and ordered, whereas space may be one-, two-, or three-dimensional, and is generally not ordered;

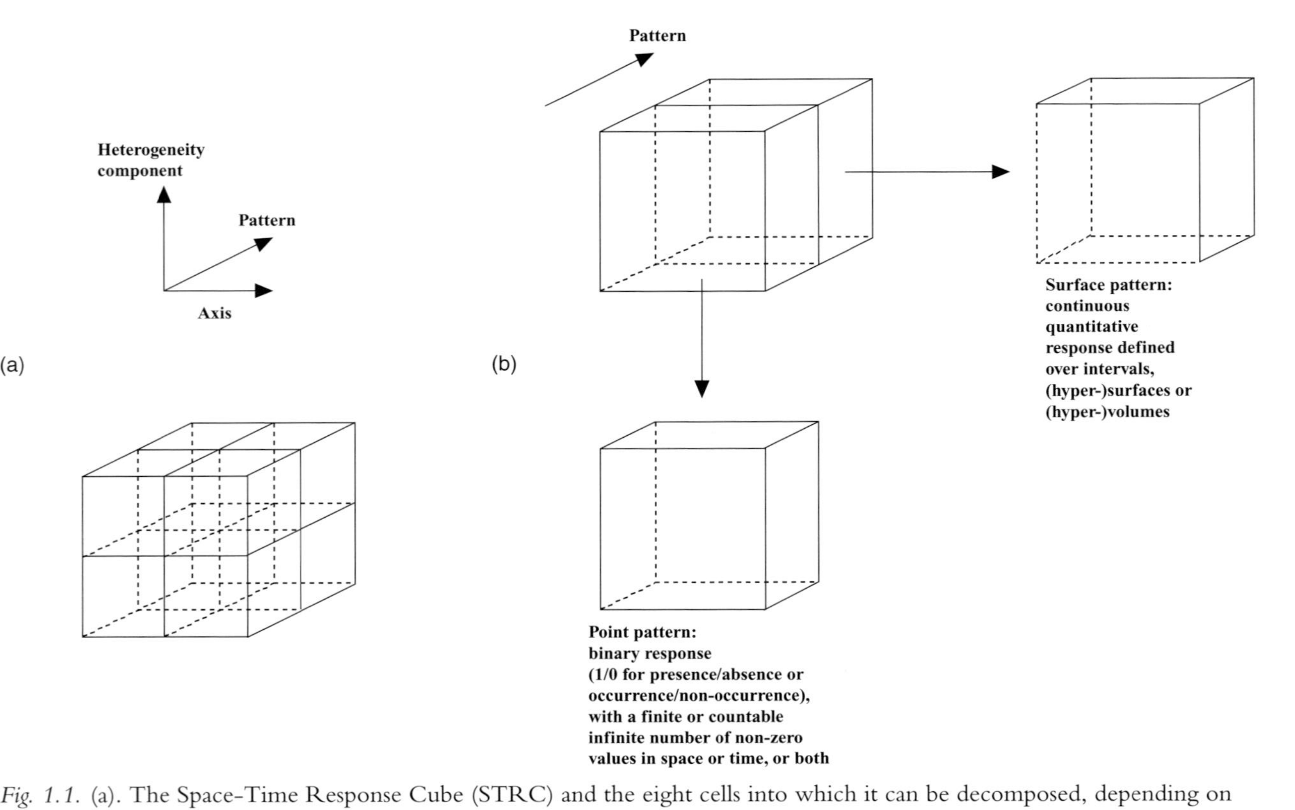

Fig. 1.1. (a). The Space-Time Response Cube (STRC) and the eight cells into which it can be decomposed, depending on (b) the pattern supporting the analysis of heterogeneity, point vs. surface; (c) the axis, or the axes, defining the domain of observation and along which heterogeneity is to be analyzed, space or time, or both; and (d) the component of the response (i.e., the stochastic process) that is likely to be associated with heterogeneity, deterministic vs. random.

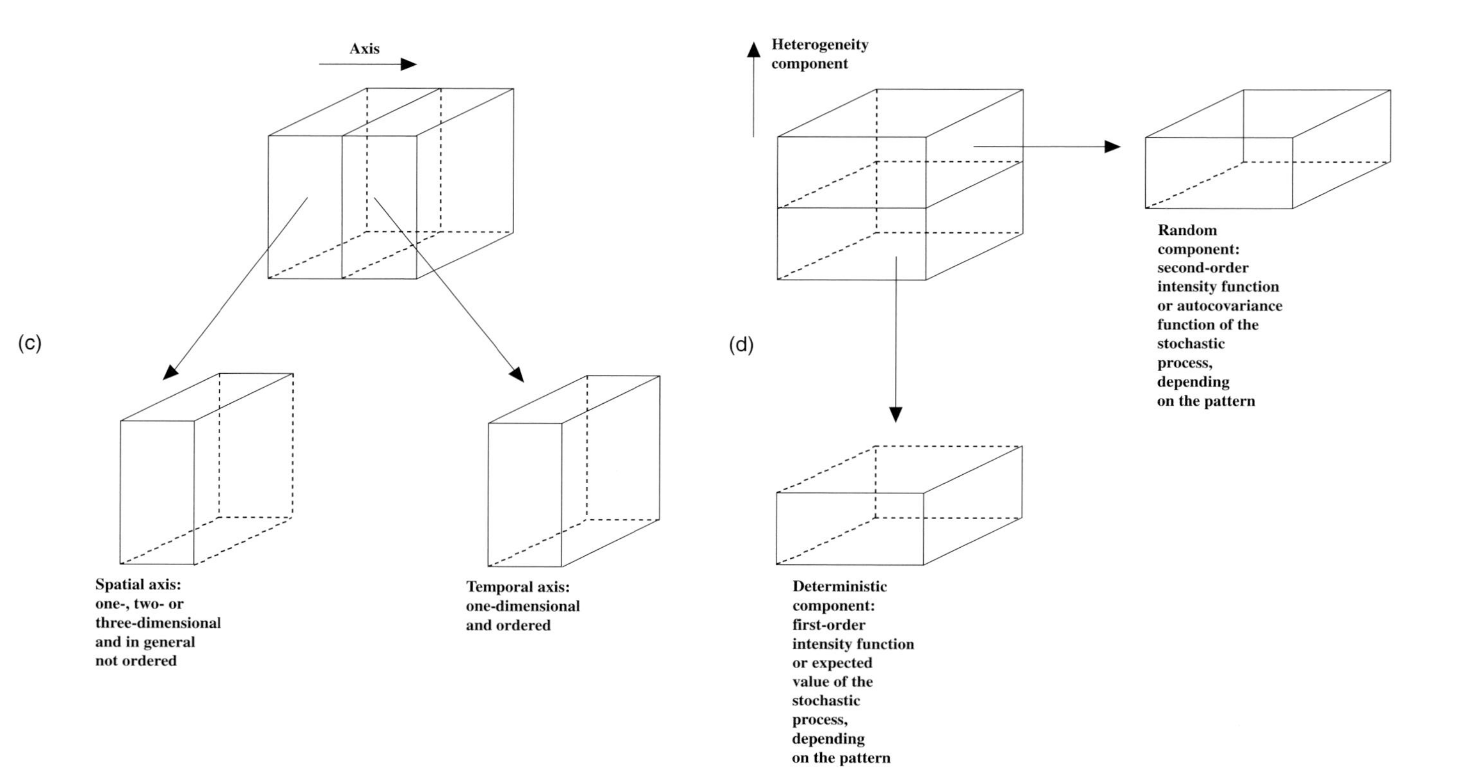
(c)
Axis
Spatial axis:
one-, two- or
three-dimensional
and in general
not ordered
Temporal axis:
one-dimensional
and ordered
(d)
Heterogeneity
component
Random
component:
second-order
intensity function
or autocovariance
function of the
stochastic
process,
depending
on the pattern
Deterministic
component:
first-order
intensity function
or expected
value of the
stochastic
process,
depending
on the pattern

Fig. 1.1. (cont.)

the amount of sediments at the bottom of a river from upstream to downstream provides an example in which the observations would be ordered in space. In a heterogeneity analysis in 3-D space, spatial heterogeneity in the horizontal plane (i.e., perpendicular to the force of gravity) is likely to be different from that along the vertical axis (i.e., parallel to gravity), and some interaction may be anticipated between them: horizontal heterogeneity is expected to change with position on the vertical axis (i.e., altitude).

The deterministic and random components of heterogeneity are fundamentally different. The deterministic component refers to the expected number of points per space or time unit and the associated function (i.e., the first-order intensity function) with point patterns, and to the expected value of the response (i.e., the population mean) at any given location or time with surface patterns. In contrast, the random component is characterized by the variance of, and the covariance between, numbers of points and the associated function (i.e., the second-order intensity function) with point patterns, and by the function defining the variance of and the covariance between responses (i.e., the autocovariance function) with surface patterns.

Summary: Because of the 2 × 2 × 2 combinations of a pattern, an axis and a heterogeneity component in the tripartite classification above, there are basically eight situations in which ecologists can be measuring heterogeneity. These eight situations correspond to the cells of the Space-Time Response Cube (STRC) (Fig. 1.1). The STRC provides a framework to biologists and environmentalists for their analysis of spatio-temporal heterogeneity. It will direct them towards the appropriate analytical method and the relevant parameter in the statistical model for a given type of heterogeneity.

Key note: The STRC must not be confused with the response surface that is fitted in the analysis of experimental data with quantitative treatment factors (e.g., Mead, 1988, Chapter 18).

Five examples are discussed below to illustrate what the eight cells of the STRC are meant for. In the next chapters, these examples will be developed through others, with data analyses.

Example 1

In successive years, a group of plant ecologists surveyed populations of *Arisaema triphyllum* (Fig. 1.2) within a number of plots in the spring

Fig. 1.2. Specimens of *Arisaema triphyllum* (L.) Schott, commonly called "Jack-in-the-pulpit" (photo by Tracy Eades).

and summer seasons. The spatial location of each *A. triphyllum* plant was recorded, and any infection by the systemic fungal pathogen *Uromyces arit-riphylli* was noted when observed. This data collection was motivated by the testing of the following hypotheses. For a given plot, season and year, the spatial distribution of plants is completely random (hypothesis 1), and the distribution of the plants infected by the fungus among all the plants of the plot is completely random (hypothesis 2). Clearly, the observed patterns are 2-D spatial point patterns; 1 mapped *A. triphyllum* plant = 1 point. Under the first hypothesis, complete randomness means (i) *A. triphyllum* plants are not expected to be more numerous in one part of the plot than in another and (ii) the variability of the number of *A. triphyllum* plants is the same in different parts of the plot; in other words, there is constancy of the mean and the variance of the number of points per m^2 over the study area. It also means (iii) absence of correlation between counts in any non-overlapping parts of the plot. It follows that the first hypothesis above refers to two cells of the STRC: point pattern × spatial axis × deterministic component and point pattern × spatial axis × random component. As for the second hypothesis, the presence/absence of infection is a dichotomic variable, which marks the 2-D spatial point process underlying the spatial distribution of *A. triphyllum* plants in a plot. The interpretation of complete randomness made for the first hypothesis applies to this variable if the reasoning is restricted to infected plants and parts of the plot with plants; indeed, the pathogen needs a host to be expressed! In these conditions, the same two cells of the STRC are of interest. Hypotheses concerning spatio-temporal heterogeneity, such as the death of formerly established plants due to unfavorable weather conditions and the emergence of new ones at the same or different locations, could be tested: for a given plot, the spatial distribution of plants in the summer in a given year is independent of that in the previous year (hypothesis 3). The independence or absence of correlation between numbers of plants in the same part of a plot at different times involves the random component of the response. Therefore, the two STRC cells of interest are point pattern × spatial axis × random component and point pattern × temporal axis × random component. This example is inspired by a study conducted at the Gault Nature Reserve of McGill University in Mont-Saint-Hilaire (Québec, Canada) from 1988 to 1991.

Example 2

A seismological station records tectonic activity at the surface of Earth. Earthquakes are part of this activity, and the spatial location of their epicenter, the time of occurrence, and their magnitude are in good place in the recorded information. Such information is available for the State of California for several decades. In addition to hypotheses similar to hypothesis 1 (purely spatial) and hypothesis 3 (spatio-temporal) in Example 1, some hypothesis concerning purely temporal heterogeneity can be tested: the temporal distribution of earthquakes is completely random (hypothesis 4). The observed pattern is a temporal point pattern; regardless of its spatial location, 1 earthquake = 1 point in time. By substituting time intervals to the parts of a plot in hypothesis 1 of Example 1, the interpretation of complete randomness is similar and the two STRC cells of interest are: point pattern × temporal axis × deterministic component and point pattern × temporal axis × random component. Unlike the presence/absence of infection in Example 1, the magnitude is a continuous quantitative mark for the point process associated with the occurrence of earthquakes in space and time, so the magnitude data need to be analyzed differently.

The two preceding examples illustrate situations in which the heterogeneity analysis will be supported by point patterns, where points correspond to biological entities (plants) and may be observed very closely in space (Example 1) and where points represent events with a short "lifetime" (earthquakes) and may be observed closely in space as well as in time (Example 2). A point pattern can also arise from the passage of mobile entities at different locations; this is the case with the hunting expeditions of a predator. The next two examples provide illustrations of surface patterns in space (Example 3) and in time (Example 4).

Example 3

A group of limnologists want to study the horizontal distribution of different zooplankton species in a lake. Therefore, they plan to collect water samples at a number of stations in the lake, and evaluate the densities of zooplankton for each of the water samples collected. The mapping of the observed values of a response (i.e., the density of a given zooplankton species) at the sampling stations will provide a surface with peaks and troughs corresponding to high and low values, that is, a surface pattern. Large-scale gradients (i.e., trends) are expected in the density of some zooplankton species, whereas small-scale spatial variation characterized by zones with high density values (i.e., patches) interspersed with zones with low density values are anticipated for other zooplankton species. Trends are likely to be related to the expected value of the response, whereas patches could be associated with autocorrelation. It follows that the deterministic and random components of heterogeneity are involved in such a study (e.g., Pinel-Alloul *et al.*, 1999), in combination with surface pattern × spatial axis.

Example 4

Small mammals are often used in chronobiological studies, with the ultimate goal of extending results to humans. This is the case for the Wistar rat (Schelstraete *et al.*, 1992, 1993, 1997). In standard conditions of light and temperature, the mother rats and their young respectively possess and develop circadian rhythms of activity, which are characterized by a period of 24 h. Under atypical synchronizers, a lengthening of the rhythm of nursing may be observed in mother rats, in combination with some delay and disorder in the development of drinking and motor rhythms in young rats. The recording of the time spent by a rat in a given activity in successive 15-min intervals during several days provides a time series (i.e., a 1-D particular case of surface pattern); one observed value of the response = the amount of time spent in the activity of interest within one 15-min interval. In the presence of a circadian rhythm, such a curve plotted against time is characterized by peaks every 24 h. The signal, which is related to the expected value of the response, is then easily modeled by trigonometric basis functions. The discrepancies from a circadian rhythm that arise under atypical synchronizers will need a more sophisticated statistical treatment. Autocorrelation will also play a role in the estimation of model parameters in such a temporal heterogeneity analysis. Thus, the two STRC cells of interest are: surface pattern × temporal axis × deterministic component and surface pattern × temporal axis × random component.

Sound experimental designs are crucial to the success of biological and environmental studies, including those aimed at measuring spatio-temporal heterogeneity. The next example is based on an experimental design which is particularly interesting in this context: the repeated measures design.

Key note: When repeated measurement of the response and replication are possible, the repeated measures design allows appropriate testing of the deterministic component of heterogeneity, while taking the heterogeneity in the random component into account (Dutilleul, 1998a, b; see also Chapter 9 here).

Example 5

A forester wants to study the fluctuations from year to year in the width of growth rings made by trees of a given species in different silvicultural conditions (e.g., different frequencies and intensities of thinning). The response is the width of the ring made by a tree at the end of a growing season, so the data are time series or temporal repeated measures. The year of growth ring formation and the silvicultural condition are classification factors. Of primary interest to the forester is the temporal heterogeneity of the mean value of the response: (i) whether the mean ring width is constant over years and (ii) whether differences in the mean ring width between years are the same regardless of the silvicultural condition. These hypotheses refer to the surface pattern × temporal axis × deterministic component cell of the STRC. The surface pattern × temporal axis × random component cell is also involved via possible heterogeneity of the variance (i.e., heteroscedasticity) and correlation between ring widths in successive years. Note that, besides some biological interest, such heteroscedasticity and autocorrelation have a nuisance effect on the statistics because they affect the significance level of the test used to assess the deterministic component of heterogeneity in the statistical model (i.e., an ANOVA model). In addition, the forester may want to include within-tree spatial effects in his study, by collecting wood cores in the stem, at several heights along the north–south and east–west directions. The measured heterogeneity will then be spatio-temporal, and the assessment of position-related effects on tree-ring width, separate from year-related effects or combined with them, will be possible, with two more cells of the STRC covered. This example is inspired from Herman *et al.* (1998).

In Example 5, the heterogeneity of the mean is of primary biological interest, while the heterogeneity of the variance and that due to autocorrelation are rather nuisances for the statistical tests used to assess the heterogeneity of the mean. Nevertheless, heteroscedasticity and autocorrelation may be of primary interest to non-statisticians, as in evolutionary biology where phenotypic plasticity refers to the capacity of a genotype to modulate its response depending on the environment in which it grows. Responses from different genotypes in various environments may be compared in an ANOVA model, where a possibly heteroscedastic and genetically correlated random effect represents the heritable part of plasticity (Dutilleul and Potvin, 1995; Dutilleul and Carrière, 1998) and fixed effects are associated with the "mean plasticity" (Bell and Lechowicz, 1994).

1.3 Overview

The classification scheme introduced in this chapter will be used in the following chapters to organize the "exposé" of statistical methods available for the analysis of heterogeneity in space or time, or both, depending on the cells of the STRC covered by the biological or environmental study. In general, at least two STRC cells will be involved in a heterogeneity analysis.

Chapter 2 is the most formal part of the book, in which the statistical foundations for further developments are established. Presentation is intended to be general in that the text and equations apply to space and time. Simulated data instead of real data are used in this chapter in order to start with known types of heterogeneity, including the stochastic process generating heterogeneity and the equation and parameter values of its model. The two halves of the STRC corresponding to point and surface patterns are investigated in Chapters 3–5 and 6–8, respectively. The analysis of spatio-temporal heterogeneity is developed accordingly. In simple terms, this development is based on counts and distances for point patterns and on sums, sums of squares and cross-products of response values for surface patterns. Chapter 9 is the experimental design chapter of the book; some sampling aspects are covered, too. In this chapter, guidelines are given on how to define the experimental design and the model to analyze the experimental data statistically and to quantify heterogeneity eventually, depending on the situation and the component of heterogeneity of interest or concern. Real data are analyzed in examples presented in Chapters 3 to 9. Heterogeneity may interfere with the assessment of relationships between responses and some of the statistical solutions found to overcome the problem are discussed in Chapters 6 and 7. Concluding remarks and comments on new avenues are made in Chapter 10.

In some chapters, more statistical or theoretical details are presented in an appendix or in parentheses () when left in the body of the text. Depending on the chapter, the space case or the time case is studied first, but the "exposé" is always completed with the space-time case. In the heterogeneity analysis of point patterns (Chapters 3–5), advantage is taken of the graphical features of mapping, by looking where a point is located before recording when it has appeared and incorporating location and time in a spatio-temporal analysis. In the heterogeneity analysis of surface patterns (Chapters 6–8), the time case is studied first because of the simplifications that its one-dimensionality allows in the development of analytical procedures and permits here in their presentation, compared to 2-D or 3-D space.

2 · *Spatio-temporal stochastic processes: definitions and properties*

Patterns observed in space and time can be conceptualized as partial realizations of an underlying "stochastic process," so it is useful to know a number of things about stochastic processes in order to better see the link between the observed pattern and the generating stochastic process in heterogeneity analysis. In this chapter, purely spatial and purely temporal stochastic processes are discussed as particular cases of spatio-temporal stochastic processes. Section 2.1 discusses the links existing between patterns and processes, and illustrates the use of indices to locate data and random variables in space and time. Any underlying stochastic process is defined by an equation, which represents a model with a number of parameters (Section 2.2). Stochastic processes may possess certain properties (Section 2.3), and partial realizations of a given process can be simulated from its model and given parameter values (Section 2.4). Multivariate extensions as well as extensions to be used in the frame of an experimental design with replication are presented in Section 2.5. Recommended readings in spatial and temporal statistics are given in Section 2.6. Important notations are introduced in this chapter, where mathematical development is kept simple and formal definitions of heterogeneity parameters and parameter functions are presented in Appendix A2.

2.1 Observed patterns, indices, and stochastic processes

In very general terms, statistical analyses rely on individual observations, in and outside the spatio-temporal framework. The nature of observations and the type of variable from which they arise structure the statistical analysis. A variable whose value is dependent on the individual on which it is observed, and is not known with certainty before the observation is actually made, is classically referred to as a random variable in statistics; such a variable is denoted U hereafter. The value of U is likely to change with the individual. Because of the variable nature of U, the number of possible values is at least two. When the number of possible values for U

is exactly two, they may be coded 1 and 0 for "success" and "failure," respectively. The health status (i.e., healthy or diseased) and the stage of development (i.e., adult or juvenile) of an individual, the occurrence or non-occurrence of an earthquake in a part of the world and of an insect outbreak in a forest stand, and the presence or absence of a given plant species in a quadrat are examples of binary random variables. At the other extreme, there may be so many possible values for U that they cannot be counted. This is the case when U takes quantitative values in continuous intervals defined by a lower bound and an upper bound, such as soil pH, plankton biomass, the height of a plant, the material density of a wood sample, and the depth of a water table. Unlike the former examples, the numbers of eggs in a bird nest, vegetation types in a region, fruits born by a tree, and parasites contained in a fish, are discrete quantitative random variables characterized by a number of possible values that is generally greater than two and finite. (In theory, some discrete quantitative random variables may take an infinity of different values, but this infinity must be countable; that is the case when possible values are non-negative integers or fractions. Still, in theory, the number of possible values for any continuous quantitative random variable is infinite and uncountable.)

In closer relation to patterns observed in space and time, the spatial distribution of earthquakes of high magnitude is denser at the intersection of tectonic plates, the colonization of a site by plants of a given species follows some environmental gradients, and the rate of occurrence of insect outbreaks in forest stands changes over time. In such patterns, the observations are represented by points, and a "point" may be a location where an event of interest occurs or an individual of interest is present, or a time when an event of interest occurs. As for bird nests, the presence of a nest (i.e., a point) is a prerequisite to the observation of a number of eggs, which is a discrete quantitative mark associated with the spatial point pattern of nests. A similar comment can be made for the number of fruits on a tree and the number of parasites in a fish, whereas health status and development stage are qualitative marks and height is a continuous quantitative mark for the spatial point pattern of plants. By contrast, the measurement of soil pH and water table depth in a field, of plankton biomass in a lake, and of wood density in a tree, provide surface patterns in space, time, or space-time, depending on the experimental situation.

Key note: Of course, the observation of continuous quantitative random variables in the surface patterns above assumes the existence of a field, a lake and a tree, but contrary to marks in point patterns, soil pH, water table depth, plankton biomass and wood density are defined and could be measured anywhere in the field, lake or tree.

The location or the time, or both, at which a measurement is made for a surface pattern represents a "sampling unit," to be distinguished from an observed point, which represents a value of 1 for the random variable in a point pattern. From the examples above, it becomes clear that in the spatio-temporal framework, the nature *and* the indexing of data are dependent on the context in which the observations are collected. In a spatio-temporal point pattern, data are binary; the value of U is 1 where and when a point is observed, and 0 otherwise; and the number of observed points (i.e., values of 1) might be infinite but is always countable. In a spatio-temporal surface pattern, data are continuous quantitative, and with continuous space and discrete time, the number of possible values for U is doubly infinite and uncountable, even in a bounded sampling domain. Therefore, to place the data in their spatial and temporal contexts, the sampling unit in the spatio-temporal framework (i.e., the individual classically) is identified by an index s. Accordingly, a time series is a vector of observations collected repeatedly in time for the same random variable U, an index s specifying the place of each observation in the sequence. A similar indexing applies to data collected in space, except spatial indices may differ in dimensionality to identify linear, areal and volumetric locations. In all cases, each datum, which is one observed value of the random variable $U(s)$, cannot be predetermined from the sole index s. This understanding of the nature and indexing of spatio-temporal data motivates the more formal definition of a stochastic process below.

In simple and general terms, a univariate spatio-temporal stochastic process is a set of indexed random variables $\{U(s) \mid s \in S\}$, where S denotes the set of possible values for the index s and $\in$ is the symbol "belongs to." When $p > 1$ random variables are observed at each location and time, the random variable U is replaced by a $p \times 1$ random vector $\boldsymbol{u}$, and the spatio-temporal stochastic process becomes multivariate and is described by $\{\boldsymbol{u}(s) \mid s \in S\}$. For purely temporal processes, S is the space (in the mathematical sense) of integer numbers $\mathbb{Z}$ if time is discrete or the space of real numbers $\mathbb{R}$ if time is continuous. For purely spatial processes, S may be $\mathbb{Z}$, $\mathbb{Z}^2$ or $\mathbb{Z}^3$ in discrete space or, more generally, it is $\mathbb{R}$, $\mathbb{R}^2$ or $\mathbb{R}^3$ in continuous space, where $\mathbb{Z}^2$, $\mathbb{Z}^3$ and $\mathbb{R}^2$, $\mathbb{R}^3$ denote the Cartesian products $\mathbb{Z} \times \mathbb{Z}$, $\mathbb{Z} \times \mathbb{Z} \times \mathbb{Z}$ and $\mathbb{R} \times \mathbb{R}$, $\mathbb{R} \times \mathbb{R} \times \mathbb{R}$, respectively. It follows that, for spatio-temporal stochastic processes, S

can be one of the combinations of $\mathbb{Z}$, $\mathbb{Z}^2$, $\mathbb{Z}^3$, $\mathbb{R}$, $\mathbb{R}^2$ or $\mathbb{R}^3$ (space) with $\mathbb{Z}$ or $\mathbb{R}$ (time), depending on the situation. This formulation is illustrated in the following examples.

Consider the state transitions from first to fourth instar in developing mosquito larvae. In this example, the value of U is 1, 2, 3, or 4, depending on the stage of development of the mosquito larva, and $s = t$ denotes a continuous time index, so S is a subset of $\mathbb{R}$. Starting with $U(0) = 1$, a mosquito first instar may develop very fast, but it will never transform instantaneously into a fourth instar, so $U(t) = 4$, with $t > 0$. At the other extreme, a mosquito first instar may take a very long time to reach the fourth instar state – $U(t) = 4$ for a large t – or it might never reach the fourth instar state and complete its development – there exists no t such that $U(t) = 4$. In the example of a 2-D spatial point pattern of plants in a field, the value of U is 1 or 0, depending on the presence or absence of a plant; S is a subset of $\mathbb{R}^2$ delimited by the extent of the field (e.g., $[0, a] \times [0, b]$); and $s = (x, y)$ denotes the continuous spatial coordinates of any location in it. If plants are surveyed twice a year (e.g., in the spring and in the summer) in a number of years, this example is extended to space-time, and $s = (x, y, t)$, where t denotes a discrete time index.

As for surface patterns, consider the growth increment of a tree, which is produced through the season in continuous time but provides distinct rings on an annual basis. If $U(s)$ denotes the width of a growth ring made by a tree in a given year s at 1.5 m from the ground in a given radial direction, then S is a subset of $\mathbb{Z}$; $s = t$, where t denotes the discrete time index for the year of tree-ring formation. Downwards, this subset does not contain the integer numbers indexing the first years of life of the tree (e.g., 0, 1) because there is no ring available at 1.5 m from the ground until the tree seedling has reached that height. Upwards, s is bounded by some finite integer number because a tree cannot grow indefinitely in time. If the forester wants to study the heterogeneity in the width of tree growth rings as a joint function of the height on the stem and the year of tree-ring formation, then the height on the stem defines a 1-D spatial axis of analysis and the stochastic process becomes spatio-temporal. Accordingly, $s = (z, t)$ and S is a subset of the Cartesian product $\mathbb{R} \times \mathbb{Z}$; in particular, the values of z are restricted to a continuous interval of finite length (i.e., $[0, H]$ for some upper bound H). If the forester also wants to study heterogeneity in several radial directions, the indexing is extended to $s = (z, \theta, t)$, where θ denotes the angle (in degrees) made by the direction in which tree-ring widths are measured and a pre-defined direction of reference in the horizontal plane. Theoretically, θ can take any value in the interval $[0, 360]$. In that situation, S is a subset of $\mathbb{R}^2 \times \mathbb{Z}$

because of the two continuous spatial indices and one discrete temporal index.

2.2 The fundamental equation

From the definition of a stochastic process as a set of indexed random variables, it follows that:

(i) a "population mean" and a "population variance" are associated with each random variable $U(s)$;
(ii) a population correlation is associated with each pair of random variables $U(s)$ and $U(s')$, with $s \neq s'$ – this particular case of correlation between random variables actually is a "population autocorrelation" because it is the correlation between U and itself for two different values of the index;
(iii) the population mean and the population variance of $U(s)$ may change with the value of s; and
(iv) the population autocorrelation between $U(s)$ and $U(s')$ may change with the values of s and s'.

("Population mean" and "population variance" refer to the theoretical values of the mean and variance of $U(s)$, respectively; the former is also called the expected value of $U(s)$, an intuitively clear expression which was already used in Chapter 1. "Population autocorrelation" refers to the theoretical value of the correlation between $U(s)$ and $U(s')$, to be distinguished from the value of an autocorrelation coefficient estimated from sample data. The population parameters above and more are defined in terms of the expectation operator in Section A2.1 (Appendix A2). Recall that, in statistics, only the normal distribution is entirely characterized by its moments of first and second orders, that is, means, variances and correlations.)

Points (i)–(iv) above are fundamental because they highlight three population parameters from which heterogeneity can arise. More specifically, changes in the value of these population parameters with the index s or the pair of indices (s, s') generate three different types of heterogeneity. In fact, the population mean, variance and autocorrelations are related to different aspects of a stochastic process. The population mean represents a center for the distribution of values of $U(s)$ and the population variance measures the dispersion of the values of $U(s)$ around that center, while a population autocorrelation indicates the sign and strength of the association between any $U(s)$ and $U(s')$ with $s \neq s'$. The three population parameters can be used to represent heterogeneity in a model. Since they have their own meaning, they will appear in different places in the equation. Each also has its own domain of definition of values. The population mean $\mu(s)$ is the expected value of $U(s)$ for a given s (see Section A2.1), so it is expressed in the same unit as $U(s)$, and can be added to or subtracted from $U(s)$ in an equation.

Key note: The set of population means $\{\mu(s) \mid s \in S\}$ *represents the deterministic component of the stochastic process; the value of* $\mu(s)$ *is fixed for a given* s.

The unit of the population variance $\sigma^2(s)$ is obtained by squaring the unit of $U(s)$, as a consequence of its definition [see equation (A2.2)], and any population autocorrelation $\rho(s, s')$, whose value is between –1 and 1 inclusively, is unitless. Accordingly, population means, variances and autocorrelations cannot appear side by side in the same equation. Moreover, to be different from zero, a variance needs to be computed for a random quantity, and a correlation exists only for a pair of random variables.

Key note: Population variances and autocorrelations are characteristics of the random component of the stochastic process, denoted $\{\varepsilon(s) \mid s \in S\}$.

Each random variable $\varepsilon(s)$ is expressed in the same unit as $U(s)$ and has a population mean of zero, because it is the random deviation of $U(s)$ from its expected value, that is, $\varepsilon(s) = U(s) - \mu(s)$.

The fundamental equation below states that any indexed random variable from a stochastic process $\{U(s) \mid s \in S\}$ can be decomposed into a mean parameter $\mu(s)$ plus a random deviation $\varepsilon(s)$ with zero population mean and population variance $\sigma^2(s)$ for a given s:

$$U(s) = \mu(s) + \varepsilon(s). \tag{2.1}$$

This "indexed equation" generalizes the decomposition of a random variable into a population mean plus a random error, and incorporates the cases of spatio-temporal stochastic process and purely spatial and purely temporal stochastic processes. Thus, it provides a general basis for heterogeneity analysis, and is fundamental to a comprehensive understanding of the three main types of heterogeneity:

- **heterogeneity of the mean**, when there are differences between $\mu(s)$ and $\mu(s')$, with $s \neq s'$;
- **heterogeneity of the variance**, when the population variance of $U(s)$, or equivalently of $\varepsilon(s)$, changes with s;

- **heterogeneity due to autocorrelation**, when the population autocorrelations between $U(s)$ and $U(s')$, or equivalently between $\varepsilon(s)$ and $\varepsilon(s')$, are different from zero, with $s \neq s'$.

Key note: The population variances of U(s) *and* ε(s) *are equal.*

In fact, from the definition of population variance [equation (A2.2)], the parameterization in equation (2.1) and the zero population mean of $\varepsilon(s)$, it follows that $\sigma^2(s) = \mathrm{Var}[U(s)] = \mathrm{E}[\{U(s) - \mu(s)\}^2] = \mathrm{E}[\{\varepsilon(s)\}^2] = \mathrm{E}[\{\varepsilon(s) - 0\}^2] = \mathrm{Var}[\varepsilon(s)]$.

Key note: The population autocorrelation between U(s) *and* U(s′) *is equal to that between* ε(s) *and* ε(s′).

This is a consequence of the equalities of $\mathrm{Var}[U(s)]$ and $\mathrm{Var}[\varepsilon(s)]$, of $\mathrm{Var}[U(s')]$ and $\mathrm{Var}[\varepsilon(s')]$, and of the population autocovariances $\mathrm{Cov}[U(s), U(s')]$ and $\mathrm{Cov}[\varepsilon(s), \varepsilon(s')]$. The latter equality follows from equation (A2.3), the parameterization in equation (2.1) and the zero population mean of random deviations, so $\mathrm{Cov}[U(s), U(s')] = \mathrm{E}[\{U(s) - \mu(s)\}\{U(s') - \mu(s')\}] = \mathrm{E}[\varepsilon(s)\varepsilon(s')] = \mathrm{E}[\{\varepsilon(s) - 0\}\{\varepsilon(s') - 0\}] = \mathrm{Cov}[\varepsilon(s), \varepsilon(s')]$.

Clearly, equation (2.1) is general enough to be used to describe heterogeneity in space or time, or both, and it highlights the two components (i.e., deterministic vs. random) of a stochastic process and the possible sources of heterogeneity that they represent. These aspects correspond to the STRC partitions in Fig. 1.1(c) and (d), respectively. Equation (2.1) can also be used with surface patterns and point patterns, following the STRC partition in Fig. 1.1(b). This may be clear for surface patterns and continuous quantitative responses, such as soil pH, plankton biomass, water table depth, and wood density, but may be less obvious for point patterns and their binary responses and associated marks. As explained below, equation (2.1) does apply to point processes, and is applicable to random variables associated with point patterns if these random variables (i.e., the marks) are quantitative in nature or are quantified appropriately if qualitative in nature.

By definition, the indexed random variable $U(s)$ of a point process can take no other value but 1 or 0, depending on whether or not a point of interest (i.e., an individual or event) is observed at s (i.e., a spatial location or a time). As the expected value of $U(s)$, the population mean $\mu(s)$ is then a number between 0 and 1. Such an expected value can be considered a probability: the probability that an individual of a given group is present at s or that a given type of event occurs at s. For example, the germination of a seed depends on the combination of a number of environmental factors, as does the occurrence of an earthquake with respect to geophysical factors, so $\mu(s)$ can be seen as the probability that the environmental conditions are met for the seed to germinate or that the geophysical conditions are satisfied for the earthquake to occur. The population variance of $U(s)$ is then given by $\mu(s)\{1 - \mu(s)\}$ under the Bernouilli distribution model (Kotz and Johnson, 1982, p. 213); this is a very particular case where a variance parameter is a function of the mean parameter.

Equation (2.1) readily applies to quantitative marks associated with point processes, such as the magnitude of earthquakes and the height of plants of a given species in a field. The equation indirectly applies to qualitative marks, such as the health status of an observed plant and the type of habitat in which a bird nest is found; this may be via measures of damaged leaf surface and tree diversity for the area surrounding the bird nest. Recall, however, that the observation of the mark remains conditional on the occurrence of events or the presence of individuals of interest [i.e., $U(s) = 1$ at s]. If no earthquake occurs, there can be no associated magnitude; in the absence of a plant, there can be no associated height or health status. The domain of definition of the index for a mark, in space, time, or space-time, is thus often full of "gaps" corresponding to locations and times where the mark cannot be observed. This is an important difference with a surface pattern!

Counts of events occurring within intervals in time or of individuals present within quadrats in 2-D space, for example, are not restricted to ones and zeroes. Count values are related to the length of intervals or the area of quadrats, and each interval or quadrat can be centered at some s. For each interval or quadrat, there is an expected value of the count and a population variance is associated with each count. A population covariance between counts can also be defined for two intervals or quadrats. Thus, equation (2.1) applies to counts in the analysis of point patterns.

As we will see in greater detail in Chapters 3 and 5, intensity functions are parameter functions for point processes and they are used to define population parameters for quadrat counts. These intensity functions may be deterministic (i.e., their value at each s is the same for all realizations of the point process), or they may be random – the first-order intensity function in particular may itself follow a stochastic process $\Lambda_1(s)$ whose value at each s changes from one realization of the point process to another. An example of the latter situation is given by Cox processes

(Diggle, 2003, pp. 68–71). Equation (2.1) then applies to $U(s)$ as well as to $\Lambda_1(s)$.

The explanations above were intended to clarify the use of equation (2.1) with point patterns and point processes. We can now return to the main stream in this chapter: the fundamental equation and the three types of heterogeneity defined from its parameterization. A number of properties of stochastic processes are defined below. These properties correspond to more or less restricted forms of spatio-temporal heterogeneity that can be analyzed with sample data. The fact of imposing restrictions on population parameters may come as a disappointment, but if there were as many heterogeneity parameters as there were observations, the statistician could not do much: there would be zero degree of freedom!

Key note 1: In practice, the decomposition into a deterministic mean component and a random component is not unique, and equalities that hold in theory may not hold when population parameters are replaced by estimates.

Key note 2: Population means, variances and autocorrelations can themselves be modeled inside the general model defined by equation (2.1) (see, e.g., Section 2.4).

2.3 Properties of spatio-temporal stochastic processes

The knowledge of the possible properties of a spatio-temporal stochastic process in theory is a prerequisite to their assessment, when feasible, on partial realizations in practice. It is important for at least two more reasons:

(1) Some properties represent conditions that are necessary or sufficient to the efficiency and validity of statistical methods; if the spatio-temporal stochastic process studied does not possess the properties in question, then alternative procedures of statistical inference from sample data should be used.
(2) The sources of heterogeneity in a study will be reduced if the stochastic process studied possesses one of the properties defined below; the explanation and prediction of a biological or environmental variable in space and time will then be simplified.

The first properties considered here are different types of stationarity. Basically, stationarity means that the spatio-temporal stochastic process is "behaving the same" in all locations and times. More particularly, stationarity may involve the complete probability distribution of the indexed random variables $U(s)$ or only a part, via one or several of the population parameters (i.e., mean, variance, autocorrelation), but in the spatio-temporal framework, it always refers to constancy in space and time.

Complete stationarity

If the probability distribution of $U(s)$ does not change with s, and the joint probability distribution of $U(s)$ and $U(s')$ does not change with s and s', and that of $U(s)$, $U(s')$ and $U(s'')$ does not change with s, s' and s'', etc., then the spatio-temporal stochastic process $\{U(s) \mid s \in S\}$ is said to be completely stationary in space and time. Complete stationarity is also known as "strong stationarity." It does not imply that all the observed values should be equal to a constant. Instead, complete stationarity means that the probability associated with a given value of $U(s)$ in the discrete quantitative case or with a given interval of values of $U(s)$ in the continuous quantitative case does not change with s, that the probability associated with a given pair of values for $U(s)$ and $U(s')$ or a given pair of intervals of values for $U(s)$ and $U(s')$ depending on the case, does not change with s and s', etc.

The assessment of complete stationarity in space and time would require numerous observations of the stochastic process at each of a number of space-time points. In the absence of "parallel universes," a stochastic process can be observed once at each space-time point and no statistical inference about the distribution of $U(s)$ for a given s and the joint distribution of $U(s)$ and $U(s')$ for given s and s' can be performed from a single observation and a single pair of observations. Anyway, knowing that a spatio-temporal stochastic process is not completely stationary is of limited interest, since this is, in fact, the expectation for most processes. The interest really is in the components of the stochastic process studied that are stationary and those that are not. Such partial types of stationarity are defined below. Recall that the population mean $\mu(s)$ is the first-order moment of $U(s)$, the population variance $\sigma^2(s)$ is the centered second-order moment of $U(s)$, and the population autocorrelation $\rho(s, s')$ is the standardized population autocovariance $C(s, s')$, which itself is the centered second-order joint moment of $U(s)$ and $U(s')$ (see Section A2.1).

Stationarity at order 1

Whatever the value of the population mean in a given location at a given time may be, it is the same for all the other locations and times. Formally, stationarity at order 1 in space and time is defined by $\mu(x, y, z, t) = \mu$ for all (x, y, z, t), a condition rarely satisfied in biological and environmental studies. If $\mu(x, y, z, t) = \mu(t)$ for all (x, y, z) at a given time t, then the population mean does not depend on the location and the spatio-temporal stochastic process is said to be stationary at order 1 in 3-D space. If $\mu(x, y, z, t) = \mu(x, y, z)$ for all t in a given location (x, y, z), then the population mean does not depend on the time and the spatio-temporal stochastic process is said to be stationary at order 1 in time. Intermediate cases are stationary at order 1 in 1-D or 2-D space, combined or not with stationarity at order 1 in time.

Key note: A spatio-temporal stochastic process that is stationary at order 1 cannot generate heterogeneity of the mean in space or time, since $\mu(s)$ *in equation (2.1) is then equal to* μ *for all* s. *In other words, stationarity at order 1 and homogeneity of the mean are very closely related.*

Stationarity at order 2

The population autocovariance $C(s, s')$ depends on the difference $s' - s$ instead of the pair of indices (s, s') themselves; the difference $s' - s$ represents the vector linking the space-time points corresponding to s and s'. Formally, stationarity at order 2 in space and time is characterized by the equality $C(s, s') = C(s' - s)$ for any pair of indices (s, s'). This definition includes the population variance $\sigma^2(s)$ when $s = s'$.

Key note: Stationarity at order 2 implies homogeneity of the variance.

(Proof: By definition of $\sigma^2(s)$ and $C(s, s')$ [see equations (A2.2) and (A2.3)], $\sigma^2(s) = \mathrm{E}[\{U(s) - \mu(s)\}^2] = C(s, s)$. If the spatio-temporal stochastic process is stationary at order 2, then $C(s, s) = C(s - s) = C(0)$. After equaling $C(0)$ to σ^2, it follows that $\sigma^2(s) = \sigma^2$ for all s, which completes the proof.) When $s \neq s'$, it is easy to show that under stationarity at order 2 the population autocorrelation between $U(s)$ and $U(s')$ depends only on the vector $s' - s$: $\rho(s, s') = \frac{C(s, s')}{\sqrt{\sigma^2(s)\,\sigma^2(s')}} = \frac{C(s' - s)}{\sigma^2} = \rho(s' - s)$ [see equation (A2.6) for the definition of $\rho(s, s')$]. Like stationarity at order 1, stationarity at order 2 may be restricted to space or time.

Weak stationarity

This very much used type of stationarity includes stationarity at order 1 and stationarity at order 2 (i.e., up to order 2). It is called "weak" by comparison with complete or strong stationarity. In practice, weak

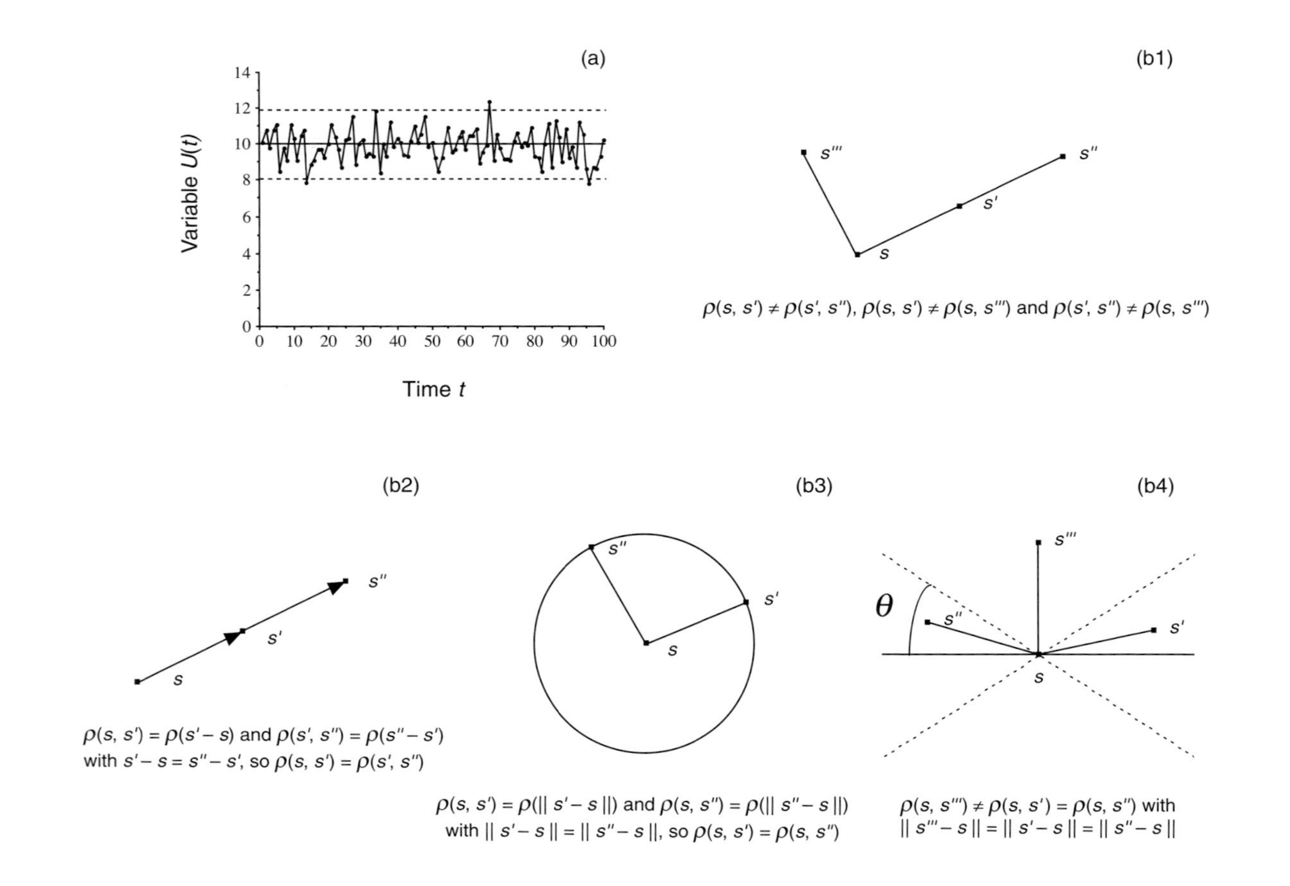
(a)
Variable $U(t)$
Time t
14
12
10
8
6
4
2
0
0
10
20
30
40
50
60
70
80
90
100
(b1)
s'''
s''
s'
s
$\rho(s, s') \neq \rho(s', s'')$, $\rho(s, s') \neq \rho(s, s''')$ and $\rho(s', s'') \neq \rho(s, s''')$
(b2)
s
s'
s''
$\rho(s, s') = \rho(s' - s)$ and $\rho(s', s'') = \rho(s'' - s')$
with $s' - s = s'' - s'$, so $\rho(s, s') = \rho(s', s'')$
(b3)
s''
s'
s
$\rho(s, s') = \rho(\| s' - s \|)$ and $\rho(s, s'') = \rho(\| s'' - s \|)$
with $\| s' - s \| = \| s'' - s \|$, so $\rho(s, s') = \rho(s, s'')$
(b4)
θ
s'''
s''
s'
s
$\rho(s, s''') \neq \rho(s, s') = \rho(s, s'')$ with
$\| s''' - s \| = \| s' - s \| = \| s'' - s \|$

stationarity can be assessed from sample data without replication, but its assessment requires a number of statistical tools that will be introduced later on. In particular, it is a usual condition of application for autocorrelogram and spectral analyses with surface patterns (Chapters 6 and 7).

Formally, a spatio-temporal stochastic process is weakly stationary in space and time if

$$\mathrm{E}[U(s)] = \mu \text{ for all } s, \tag{2.2}$$

$$\mathrm{Var}[U(s)] = \sigma^2 \text{ for all } s, \text{ and} \tag{2.3}$$

$$\mathrm{Corr}[U(s), U(s')] = \rho(s' - s) \text{ for any } s \neq s'. \tag{2.4}$$

As noted above, equations (2.2) and (2.3) imply, respectively, homogeneity of the mean and homogeneity of the variance, but clearly heterogeneity due to autocorrelation may be found in a partial realization of a weakly stationary process because equation (2.4) does not imply that all population autocorrelations are zero. (Since normally distributed random variables are entirely characterized by their moments of first and second orders, weak stationarity is theoretically equivalent to complete stationarity for Gaussian (i.e., normal) stochastic processes.)

Prior to introducing the next type of stationarity, it may be good to discuss further those already defined. The temporal example presented in Fig. 2.1(a) contradicts the misinformed view according to which any changing pattern reflects some heterogeneity. Despite the fluctuations in the observed values of $U(t)$ for $t = 1, \ldots, 100$, there is no heterogeneity to be found in this time series, simply because the temporal stochastic process from which it originates cannot generate heterogeneity of

←

Fig. 2.1. (a). Partial realization of a temporal stochastic process with constant population mean [$\mu(t) = 10$ for all t], constant population variance [$\sigma^2(t) = 1$ for all t] and zero population autocorrelations [$\rho(t, t') = 0$ for any $t \neq t'$]. Autocorrelation in 2-D space: (b1) there is no restriction on population autocorrelations, that is, there are as many values for population autocorrelations as there are pairs of indices (s, s'), (s'', s'''), etc.; (b2) observations at pairs of locations (s, s') and (s', s''), linked by the same vector, present the same autocorrelation (stationarity at second order); (b3) observations at two locations s' and s'' separated from a third location s by the same distance present the same autocorrelation (isotropy); and (b4) observations at two locations s' and s'' separated from a third location s by the same distance and forming with it line segments deviating from a given direction by less than a given angle θ present the same autocorrelation (directional anisotropy).

any type: the population mean is constant, the population variance is constant, and all population autocorrelations are zero. (Weakly stationary stochastic processes like this are called "white noise" in time series analysis.) All population autocorrelations need not be equal to a constant regardless of the length of the vector $s' - s$, for the stochastic process to be stationary at order 2. However, some restrictions will generally need to be placed on the values of the population autocorrelation function, for the following reasons. If there was no restriction, there would be as many values of $\rho(s, s')$ as pairs of indices (s, s'). In the absence of replication (i.e., when only one partial realization of the stochastic process is available for statistical inference), it would also mean a one-to-one match between each pair of observations and one autocorrelation parameter to be estimated, which is not feasible in practice. In the spatial example drawn in Fig. 2.1(b1), there is no restriction on the autocorrelation function: all population correlations between the pairs of indexed random variables in 2-D space are different, although $s' - s = s'' - s'$ and the three vectors $s' - s$, $s'' - s'$ and $s''' - s$ are of same length. Stationarity at order 2 is assumed in Fig. 2.1(b2), where the three spatial locations corresponding to s, s', and s'' are aligned by convenience. In this case, the population correlations between the pairs of indexed random variables, $(U(s), U(s'))$ and $(U(s'), U(s''))$, are assumed to be the same because the corresponding pairs of spatial locations are separated by the same vector in 2-D space. The practical advantage of this is that all the pairs of observations collected at spatial locations separated by the same vector can then be used to estimate the corresponding value of the autocorrelation function, resulting in a higher precision. Note that $\rho(s'' - s)$ will generally be smaller than $\rho(s' - s) = \rho(s'' - s')$ because observations far apart tend to be less dependent than observations close to one another and the length of the vector $s'' - s$ is greater than that of $s' - s = s'' - s'$. Stationarity at order 2 has a similar practical advantage in space-time: for example, $\text{Corr}[U(1, 3, 2, 4), U(4, 2, 3, 1)] = \text{Corr}[U(5, 9, 7, 11), U(8, 8, 8, 8)] = \rho(3, -1, 1, -3)$ under stationarity at order 2.

Intrinsic stationarity

This type of stationarity is about the increments $U(s') - U(s)$ instead of the indexed random variables $U(s)$ and $U(s')$ themselves. In geostatistics, intrinsic stationarity is a usual condition of application for variogram analysis with surface patterns (Chapter 7). Two conditions define intrinsic stationarity. The first one is equivalent to stationarity at order 1, which is simply rewritten to apply to increments; the second condition concerns the population autocovariances, and states that the variance of the increment $U(s') - U(s)$ depends on the vector $s' - s$ instead of the pair of indices (s, s'):

$$\mathrm{E}[U(s') - U(s)] = 0 \text{ for any } (s, s') \text{ and} \tag{2.5}$$

$$\mathrm{Var}[U(s') - U(s)] = 2\gamma(s' - s) \text{ for any } (s, s'), \tag{2.6}$$

where $\gamma(\cdot)$ denotes the population semivariance function. It can be shown that equalities (2.3) and (2.4) imply (2.6), but the reciprocal is not true, so one could say that intrinsic stationarity is even weaker than weak stationarity... Note that the circular variance–covariance structure, which plays a key role in the repeated measures ANOVA (Chapter 9), satisfies (2.6), but not (2.3) in general.

Ergodicity

Contrary to the various types of stationarity defined above and the properties of spatio-temporal stochastic processes introduced hereafter, ergodicity cannot be defined only in terms of population parameters. These are involved, of course, but sample statistics computed from increasingly larger, partial realizations of the process are, too. In simple terms, a stochastic process is said to be ergodic if the sample mean and variance calculated from all the observations of a partial realization are good approximations of the population mean and variance, and approximations are improved when the number of observations increases and become exact when the number of observations is infinitely large (i.e., asymptotically). If one uses the singular form for the population mean and variance, this assumes that they are constant and, by extension, that the stochastic process is weakly stationary. Indeed, ergodicity and weak stationary are related, but a detailed discussion of the relationship (Cressie, 1993, pp. 53–58) falls beyond the scope of this book. Instead, let us use Fig. 2.1(a) again. In this temporal example with simulated data, the observed values of the sample mean and variance are 9.9 and 1.02, while the population mean and variance are constant over time and equal to 10 and 1, respectively. The proximity of values supports the ergodicity of the underlying temporal stochastic process, which is weakly stationary by construction.

Key note: In the presence of autocorrelation, the sample mean and variance are not optimal estimators in general; the former is not efficient, while the latter is biased.

Isotropy

This property is related to the analysis of heterogeneity due to autocorrelation. Accordingly, isotropy does not concern the population mean of a spatio-temporal stochastic process, and does not directly concern the population variance. Instead, it has to do with measures of association, such as population autocorrelations, autocovariances and semivariances for univariate processes. Without loss of generality, the focus is on population autocorrelations in what follows. Note that isotropy may also apply to population cross-correlations and cross-semivariances for multivariate processes (Section 2.5).

Isotropy is also called "stationarity under rotations" by some authors, by comparison with equation (2.4) and the corresponding "stationarity under translations" (Ripley, 1981, pp. 9–10). Basically, it means that the population correlation between $U(s)$ and $U(s')$ depends on the distance between s and s' instead of the vector $s' - s$ or the indices s and s'.

In the spatio-temporal framework, two distances, one in space and one in time, are needed to define isotropy, so points located at the same distance in space and at the same distance in time from a given point of reference are involved in the definition. Formally, the autocorrelation function is said to be isotropic in space-time when

$$\begin{aligned}&\mathrm{Corr}[U(x, y, z, t), U(x', y', z', t')]\\&\quad = \mathrm{Corr}[U(x, y, z, t), U(x'', y'', z'', t'')]\\&\text{if } \|(x' - x, y' - y, z' - z)\| = \|(x'' - x, y'' - y, z'' - z)\|\\&\quad \text{and } |t' - t| = |t'' - t|\end{aligned}$$

or, equivalently,

$$\begin{aligned}&\mathrm{Corr}[U(x, y, z, t), U(x', y', z', t')]\\&\quad = \rho(\|(x' - x, y' - y, z' - z)\|, |t' - t|) \text{ for any } (x', y', z', t'), \qquad (2.7)\end{aligned}$$

where (x, y, z, t) plays the role of point of reference, and $\| \cdot \|$ and $| \cdot |$ denote the Euclidean norm and the absolute value, respectively. Recall that the Euclidean norm of a vector is given by the square root of the sum of squares of its components.

In 2-D and 3-D space, isotropy is defined as follows. For a given s,

$$\mathrm{Corr}[U(s), U(s')] = \mathrm{Corr}[U(s), U(s'')] \text{ if } \|s' - s\| = \|s'' - s\|$$

or, equivalently,

$$\mathrm{Corr}[U(s), U(s')] = \rho(\|s' - s\|) \text{ for any } s'. \qquad (2.8)$$

In this case, the correlation between U and itself at two spatial locations depends on the spatial distance between locations, instead of the specific spatial locations. In other words, all the observations made at sites located at the same Euclidean distance from s – these sites are on a circle in 2-D space [Fig. 2.1(b3)] and on a sphere in 3-D space – can be used to estimate the population correlation with $U(s)$ in all directions.

In time and in 1-D space, isotropy readily follows from stationarity at order 2 because the only rotations are then rotations of 180°, so rotations are equivalent to translations. More specifically, if a temporal stochastic process $\{U(t) \mid t \in S\}$ is stationary at order 2, then

$$\begin{aligned}
\mathrm{Corr}[U(t), U(t-\Delta t)] &= \mathrm{Corr}[U(t-\Delta t), U(t)] = \rho(\Delta t) \text{ and} \\
\mathrm{Corr}[U(t), U(t+\Delta t)] &= \rho(\Delta t) \text{ for any } t \text{ and any } \Delta t > 0, \text{ so} \\
\mathrm{Corr}[U(t), U(t-\Delta t)] &= \mathrm{Corr}[U(t), U(t+\Delta t)] \\
&= \rho(\Delta t) \text{ for any } t \text{ and any } \Delta t > 0.
\end{aligned}$$

It also follows that the population autocorrelation function is symmetrical: $\rho(\Delta t) = \rho(-\Delta t)$ for any lag Δt. *Note:* Sample autocorrelation coefficients will define a symmetrical function, too (see Chapter 6). Similar arguments hold for 1-D spatial stochastic processes.

Consider now a situation in which the random variable of interest is soil pH, which is measured at a given depth at a number of sites on the side of a hill. It is very likely that the correlation between soil pH at a site located at midway and a site located upwards is different from the correlation between soil pH at the first site at midway and another site located at midway, even though the distances between sites are the same. Such spatial stochastic processes are said to be anisotropic, and heterogeneity due to autocorrelation should then be analyzed along preferred directions [Fig. 2.1(b4)].

Separability

This property has its roots in the parsimony principle, which favors models with fewer parameters whenever possible. It is not restricted to the spatio-temporal framework, but in this context as well as in 3-D space it has been mainly considered for the autocovariance function of the stochastic process, so it is potentially related to two of the three types of heterogeneity. Separability attributes distinct variance–covariance structures to space and time in space-time and to the horizontal 2-D space and vertical 1-D space in 3-D space, and assumes that the spatio-temporal variability can be described only by purely spatial and purely temporal autocovariances, and the 3-D spatial variability, only by horizontal and vertical autocovariances. Formally, it is defined as follows:

$$\begin{aligned} \text{in space-time: } C(s, s') &= \mathrm{Cov}[U(x, y, z, t), U(x', y', z', t')] \\ &= C_1((x, y, z), (x', y', z'))C_2(t, t') \\ &\quad \text{for any } (x, y, z, t) \text{ and } (x', y', z', t'); \end{aligned} \tag{2.9}$$

$$\begin{aligned} \text{in 3-D space: } C(s, s') &= \mathrm{Cov}[U(x, y, z), U(x', y', z')] \\ &= C_1((x, y), (x', y'))C_2(z, z') \\ &\quad \text{for any } (x, y, z) \text{ and } (x', y', z'). \end{aligned} \tag{2.10}$$

Summary: The conditions on the autocorrelation and semivariance functions excepted, the various types of stationarity are related to homogeneity instead of heterogeneity. In particular, two of the three conditions defining weak stationarity correspond to homogeneity of the mean and homogeneity of the variance. When a spatio-temporal stochastic process possesses that property, which is the exception in the biological and environmental studies, the estimation of the autocorrelation function and the quantification of heterogeneity due to autocorrelation will be facilitated by the absence of the two other types of heterogeneity. More particularly, isotropy will allow the computation of omnidirectional correlograms and variograms, and the separable variance–covariance structure is a first example of modeling of heterogeneity parameters.

2.4 Illustrations by simulation

The graphical representation of sample data is a prerequisite to estimation and testing as well as a basic procedure in the applied statistical approach. Such "eye training" is particularly important in the context of spatio-temporal heterogeneity analysis. Therefore, simulations are used in this section to become acquainted with the main spatial and temporal heterogeneous patterns that are susceptible to be observed in biological and environmental studies; to show that *different* stochastic processes can generate *similar* heterogeneous patterns; and to propose guidelines for gauging the stochastic processes that may have generated the observed heterogeneous patterns. In general terms, the following set of eight illustrations, referred to as Cases 1–8, provides a first look at the two components (i.e., deterministic, random) of a spatial or temporal stochastic process and the types of heterogeneity (i.e., mean, variance, autocorrelation) that they can generate. More specifically, Cases 1–6 illustrate the differences and common points between heterogeneous point and surface patterns in

2-D space on the one hand and between heterogeneous surface patterns in 2-D space and time on the other hand. Cases 7 and 8 provide an introduction to the multi-scale perspective on spatial heterogeneity.

Simulations are based on particular cases of the fundamental equation (2.1). Equations and population parameter values used in each illustration are given in the legends of Figs. 2.2–2.9. Data were simulated in SAS by using the pseudo-random number generators RANNOR and RANUNI for the normal and uniform distributions, respectively (SAS Institute Inc., 2009).

Case 1: Heterogeneity of the mean vs. heterogeneity due to autocorrelation in spatial point patterns

The 2-D spatial point process of which three partial realizations on the [0, 9] × [0, 9] square are represented in Fig. 2.2(a)–(c) is a heterogeneous Poisson process. Its first-order intensity function is a second-order polynomial in the spatial coordinates x and y, which has its maximum at (0, 9) and its minimum at (9, 0). Thus, the expected number of points per area unit decreases from the top-left corner to the bottom-right corner of the square. Such a deterministic first-order intensity function defines a specific and reproducible form of heterogeneity of the mean for 2-D spatial point patterns. In fact, very similar heterogeneous patterns are observed for the three partial realizations in Fig. 2.2(a)–(c). Because of the steady decrease of the expected number of points per space unit and the deterministic nature of the first-order intensity function, such a form of heterogeneity of the mean in spatial point patterns represents a "true trend."

The other 2-D spatial point process of which three partial realizations are represented in Fig. 2.2(d)–(f) is a Cox process. The main characteristic of a Cox process is that its first-order intensity function is random, so the first-order intensity function defining the expected number of points per area unit for one partial realization is different from that of another. This is why the spatial distribution of points looks so different among partial realizations in Fig. 2.2(d)–(f), although the process fixing the values of first-order intensity functions in the three panels is the same. *Note:* The second process is associated with, but different from, the Cox process.

In this case in particular, the expected number of points in a quadrat of unit side is positively and strongly correlated with the expected number of points in neighboring quadrats, following a 2-D spatial first-order autoregressive [AR(1)] process. Accordingly, it is an example of heterogeneity

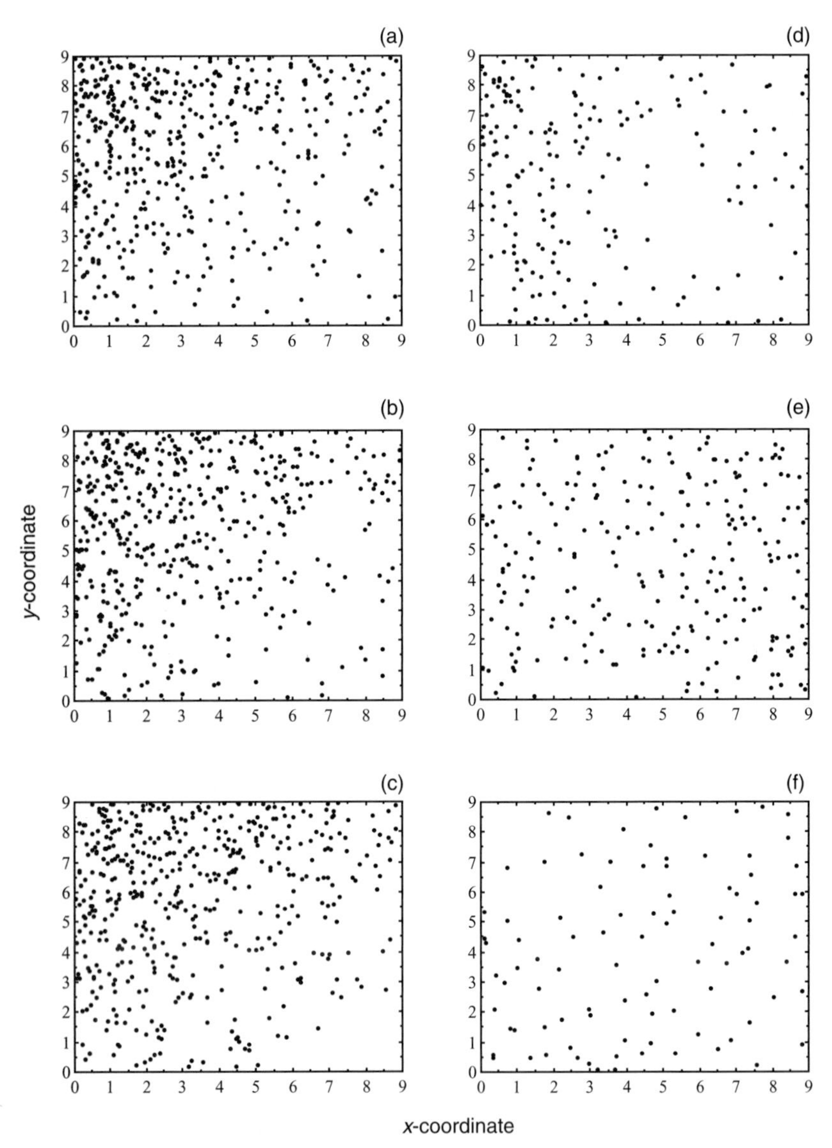

Fig. 2.2. Case 1: 2-D spatial point patterns. (a)–(c) Partial realizations of a heterogeneous Poisson process, for which the expected number of points in a given area is defined by the first-order intensity function $\lambda_1(x, y) = (x^2 + y^2 - xy - 18x + 9x + 81)/12$ for any $(x, y) \in [0, 9] \times [0, 9]$, and (d)–(f) partial realizations of a Cox process, for which the numbers of points observed in the 81 quadrats of unit side follow from one partial realization of a 2-D spatial AR(1) stochastic process with autoregressive parameter $\rho = 0.249$, plus or minus a positive quantity.

due to autocorrelation for 2-D spatial point patterns. Different heterogeneous patterns may then be observed depending on the partial realization: the heterogeneous pattern in Fig. 2.2(d) is similar to that in Fig. 2.2(a)–(c), but the other two are slightly aggregated [Fig. 2.2(e)] and regular [Fig. 2.2(f)]. Due to the lack of similarity among partial realizations combined with an apparent systematic change in point density in some of them, such a form of heterogeneity arising from autocorrelation in spatial point patterns is called a "false trend" occasionally.

To relate to real-life examples, consider a population of specialist plants for which a lakeshore or a hillside represents the habitat of preference, so its density is a function of the distance from the lake or the middle of the hillside; the number of points per area unit is expected to be large near the lake and midway on the hillside, and small far from them. A spatial point process with a deterministic first-order intensity function defining a true trend of mean heterogeneity can be postulated in such situations, assuming the topographical structures and their environment are permanent. By contrast, consider geologic events such as volcanic eruptions beneath the sea level. These can heavily influence the colonization of shallows by a marine plant community. Globally, the colonization process may remain the same, but the spatial distribution of plants on disturbed structures (i.e., portions of the shallows covered by lava) is likely to change with the disturbance (i.e., the lava flow). A spatial point process with a random first-order intensity function characterizing spatial heterogeneity due to autocorrelation will be more appropriate in such a situation, assuming the frequency of volcanic eruptions is sufficient without being excessive. This assumption is required for the trend to be false instead of true and for giving enough time to the marine plants to establish between successive volcanic eruptions.

Case 2 below illustrates true and false trends in temporal surface patterns when the observed time series are short. Case 3 shows how to distinguish a true trend of mean heterogeneity from heterogeneity due to autocorrelation in temporal surface patterns, by extending and replicating the time series. Time being one-dimensional, Case 4 can be seen as an extension, since it illustrates a true trend of mean heterogeneity and heterogeneity due to autocorrelation in 2-D spatial surface patterns.

Case 2: True vs. false trends in short time series

In Fig. 2.3(a) and (b), our eye discerns essentially the same form of temporal heterogeneity, that is, an increase of the values of the continuous

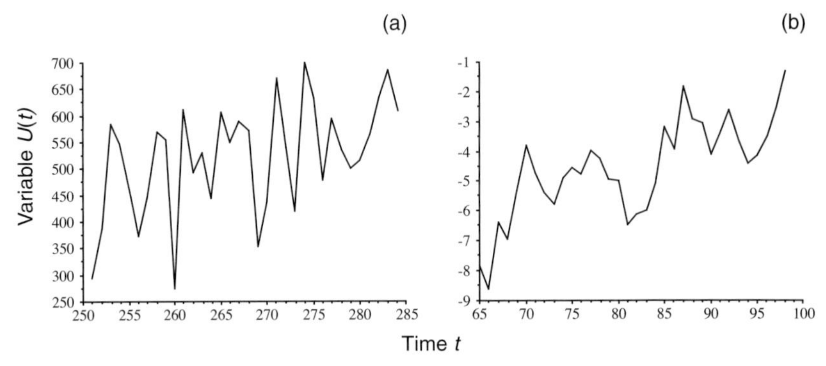

Fig. 2.3. Case 2: short time series. (a) True trend in time [$\mu(t) = 10 + 2t$ for any t] plus a non-autocorrelated noise [$\sigma^2(t) = 100$ for all t; $\rho(t, t') = 0$ for any $t \neq t'$] vs. (b) false trend in time arising from the autocorrelation in one partial realization of a discrete-time AR(1) stochastic process [$\mu(t) = 0$ for all t; $\varepsilon(t) = 0.95\varepsilon(t-1) + \eta(t)$ for any t, with $\mathrm{Var}[\eta(t)] = 1$ for all t, so $\sigma^2(t) = \frac{1}{1-0.95^2}$ for all t and $\rho(t, t') = 0.95^{|t'-t|}$ for any $t \neq t'$]. Thirty-four equally spaced simulated data were used for illustration in each panel.

quantitative random variable over the period of observation. Such trends are not restricted to be increasing; they can be decreasing as well. They are not constrained to be strictly monotonic either; the first differences between successive observations can be positive or negative. However, the overall tendency reflects an increase, a decrease or a particular shape (e.g., a quadratic trend of ∪ shape). The true trend [Fig. 2.3(a)] was simulated by using a deterministic first-degree polynomial as mean function $\mu(t)$. The false trend [Fig. 2.3(b)] is a partial realization of a discrete-time AR(1) process with $\mu(t) = 0$ for all t [i.e., $U(t) = \varepsilon(t)$]. The appellations "true trend" and "false trend" will be clarified and fully justified in Case 3. For now, the two temporal surface patterns in Fig. 2.3 look similar, except that fluctuations in the time series of Fig. 2.3(a) are greater than in that of Fig. 2.3(b). Nevertheless, they were produced by different mechanisms involving the deterministic component of the stochastic process via the population mean values for the former and the random component of the stochastic process via the autocorrelation function for the latter.

How can similar heterogeneous patterns be generated by different stochastic processes? If it is not a general case, how can one explain that two different stochastic processes produced so similar heterogeneous patterns in this case? Perhaps most importantly, how could one identify with confidence the actual mechanism underlying an observed trend in practice? One way to obtain some elements of response is to increase the data available in each of the two situations, by extending the time series. Indeed, the collection of several, longer partial realizations can reveal the true nature of the underlying stochastic process, as illustrated in the next case.

Case 3: True trend of mean heterogeneity vs. heterogeneity due to autocorrelation in time series

The time series of Fig. 2.3 were lengthened by increasing the number of observations (i.e., the extent), while keeping the length of the time interval between successive observations (i.e., the grain) unchanged. Thus, a true trend of mean heterogeneity in time, with the same parameter values as in Case 2, was simulated at large scale for $t = 1, \ldots, 1000$ [Fig. 2.4(a)]. Portions corresponding to intermediate scales of analysis are plotted in Fig. 2.4(b) and (c). In addition, three partial realizations of a discrete-time AR(1) process, with the same parameter values as in Case 2, were simulated for $t = 1, \ldots, 1000$ [Fig. 2.4(d)–(f)]. The original series of Fig. 2.3(b) are framed in Fig. 2.4.

At large scale [Fig. 2.4(a)], the increase in the values of the continuous quantitative random variable over time is maintained and even reinforced, compared to Case 2, for the true trend of mean heterogeneity. To avoid redundancy, only one long time series is presented for it because the increase of values follows the first-degree polynomial defining the underlying mean function $\mu(t)$ very closely. The two portions [Fig. 2.4(b) and (c)] reproduce the same pattern for the same reason. Actually, the true trend looks like a trend, whatever portion of the time series is plotted [Figs. 2.3(a) and 2.4(a)–(c)]. Provided the observations are not too noisy (i.e., the variance of the random deviation is not too high) and the extent is not too small, a true trend will look like a trend at all scales, from small to large, and thus fully justifies its name.

The following aspects concerning long, positively autocorrelated time series are noteworthy. The increasing pattern observed on an AR(1) partial realization at small scale [Fig. 2.3(b)] disappears at large scale [Fig. 2.4(d)]. In fact, tendencies are upward or downward within relatively short intervals of time in all three AR(1) partial realizations in Fig. 2.4(d)–(f). Via the autoregression involved in the discrete-time AR(1) model (Chapter 6), the current value of the response is a function of the previous value in the time series, plus a random noise. At

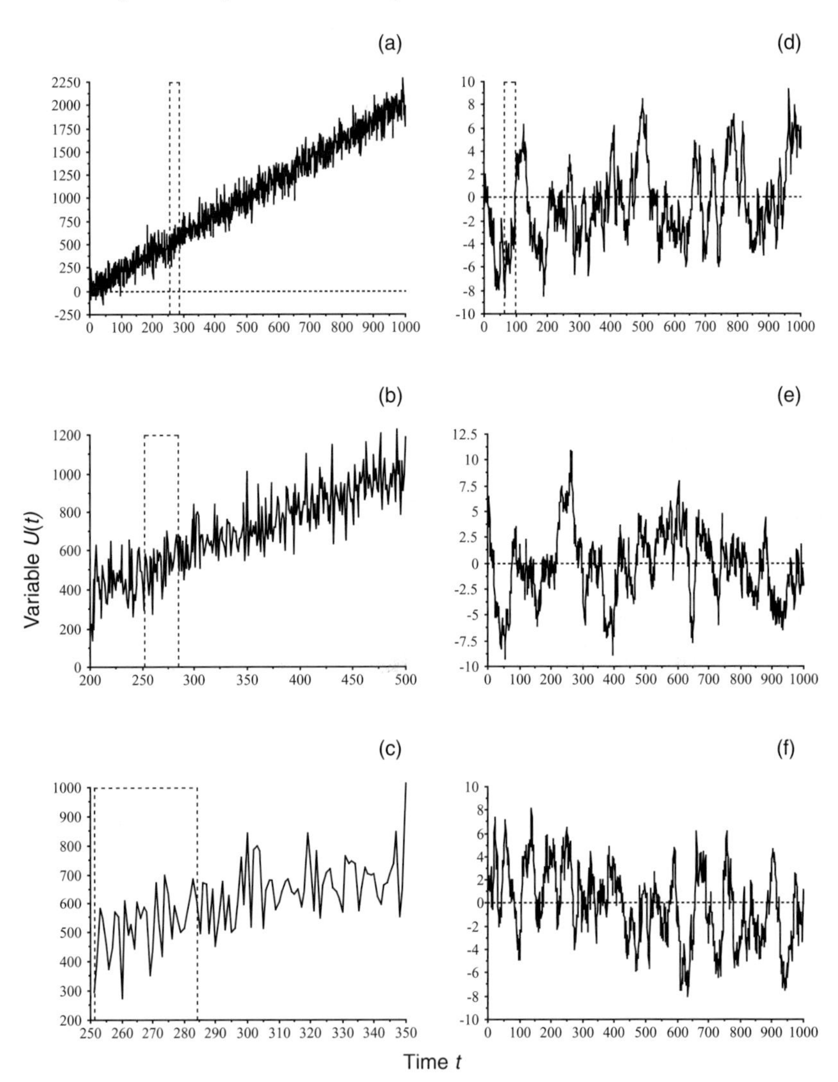

Fig. 2.4. Case 3: time series. (a)–(c) True trend in time plus a non-autocorrelated noise in series of decreasing length ($n = 1000$, 300, and 100), and (d)–(f) three partial realizations of equal length ($n = 1000$) for a discrete-time AR(1) stochastic process. Model parameter values are the same as in Fig. 2.3(a) and (b), respectively. The short time series presented in Fig. 2.3 are located by dashed frames.

some moment, the stochastic nature of the AR(1) process "tips" the response in a new direction, up or down, and the positive and strong autocorrelation sustains the new tendency over some interval thereafter. These up-and-down fluctuations become apparent only in longer series. Whereas a true trend in time characterized by a positive slope cannot do otherwise but be increasing [Figs. 2.3(a) and 2.4(a)–(c)], a false trend in time, although characterized by a positive autoregressive coefficient, can be increasing or decreasing depending on the portion of the series [Fig. 2.4(d)–(f)]. Furthermore, the heterogeneous patterns provided by three partial realizations of the same discrete-time AR(1) process are not consistent [Fig. 2.4(d)–(f)]. Such lack of consistency follows from the changes in tendency over short intervals of time, which randomly occurred at different moments for the three partial realizations because of the stochastic nature of the process. Thus, the appellation "false trend" for temporal heterogeneity arising from positive and strong autocorrelation is fully justified.

Case 4: True trend of mean heterogeneity vs. heterogeneity due to autocorrelation in spatial surface patterns

Three replicates of a true trend in 2-D space were simulated by using a second-degree polynomial in the spatial coordinates x and y (i.e., a "trend surface") as mean function $\mu(x, y)$, plus a non-autocorrelated noise [Fig. 2.5(a)–(c)]. An isotropic exponential variogram model was used to generate spatial heterogeneity due to positive and strong autocorrelation; in simple terms, the autocorrelation generated corresponds to the autocorrelation function of a 2-D spatial AR(1) process with a positive autoregressive coefficient. Three AR(1) partial realizations were simulated on a relatively small grid [Fig. 2.5(d)–(f)].

The 2-D spatial surface patterns in Fig. 2.5(a)–(c) are very similar. All three are characterized by a flat zone of low values for the continuous quantitative response. That zone is located on the bottom-left side and covers about one-third of the area. At the other extremity (top-right corner), a peak or zone of high values is observed. The two zones can be easily joined by a straight line mainly perpendicular to the contour lines in the map. This straight line defines an axis of fastest increase, or steepest descent, which is consistent in the three 2-D spatial surface patterns. Such similarity follows from the deterministic nature of the mechanism that generated this true trend in 2-D space (i.e., the trend surface).

Although the 2-D spatial surface patterns in Fig. 2.5(d) and (e) show two zones of lower and higher values, the global pattern appears to have

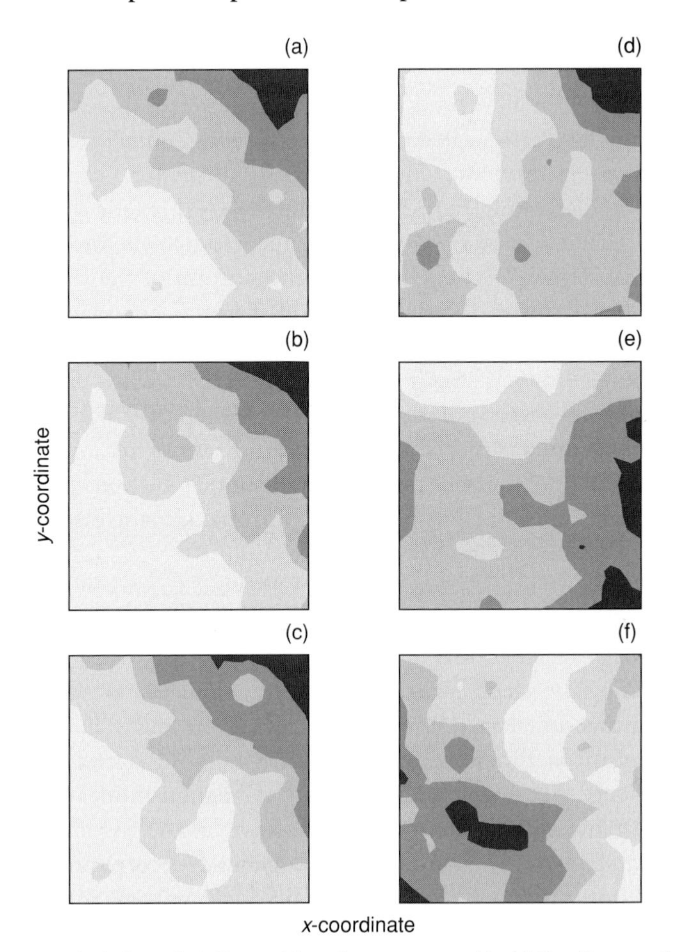

Fig. 2.5. Case 4: 2-D spatial surface patterns. (a)–(c) Replicates of a true trend in 2-D space [$\mu(x, y) = x^2 + y^2 + xy - 2x - 2y + 1$ for any (x, y)] plus a non-autocorrelated noise [$\sigma^2(x, y) = 400$ for all (x, y); $\rho((x, y), (x', y')) = 0$ for any $(x, y) \neq (x', y')$], and (d)–(f) partial realizations of a weakly stationary and isotropic 2-D spatial stochastic process with positive autocorrelation [$\mu(x, y) = 0$ for all (x, y); $\sigma^2(x, y) = 400$ for all (x, y); $\rho((x, y), (x', y')) = \exp(-\frac{3d}{16})$, where d denotes the Euclidean distance between any (x, y) and (x', y')]. The bottom-left corner of a panel is the origin of the coordinate system. Data were simulated on a 9×9 regular grid ($n = 81$) first and then were interpolated to produce maps. Gray tones are in ascending order of the values of $U(x, y)$, from low (light gray) to high (dark gray).

been rotated by 90° from the first to the second AR(1) partial realization. Peaks are approximately of same size in Fig. 2.5(d) and (e), but the axes of fastest increase, which are less clearly defined than in Fig. 2.5(a)–(c), seem to be left to top-right in Fig. 2.5(d) and left to bottom-right in Fig. 2.5(e). The third AR(1) partial realization [Fig. 2.5(f)] contrasts with the two others, by showing a flat zone of low values mostly where peaks were observed in Fig. 2.5(d) and (e). Furthermore, a peak of larger size is observed in Fig. 2.5(f), where a flat zone of intermediate values was observed in Fig. 2.5(d) and (e). As was the case in time in Case 3, the dissimilarities among AR(1) partial realizations in 2-D space are related to the random nature of the underlying mechanism that generates them.

True trends in 2-D space are of special interest in studies of river basins. Indeed, the distance from the river can be used to predict the value of variables such as soil moisture and nitrogen content, vegetation cover, biomass, etc. Assuming the topographical structure and environment of the basin are permanent, replicates [Fig. 2.5(a)–(c)] can be made at the same period, year after year. False trends in 2-D space can fluctuate smoothly at different levels [Fig. 2.5(d) and (e)], and show peaks or troughs in different places [Fig. 2.5(f)]. Therefore, these will be more appropriate when the observed heterogeneous pattern may drastically change among partial realizations, while being generated by the same mechanism. For example, a coast ecosystem can be heavily influenced by climatic events such as hurricanes. Globally, physical and chemical properties of water and soil may remain the same, but locally there may be a complete redistribution of sandy zones after a hurricane, resulting in modifications of soil pH and drainage, preference habitats for crabs and mussels, etc.

The next two cases illustrate further aspects of heterogeneity of the mean and heterogeneity due to autocorrelation in temporal and 2-D spatial surface patterns. More specifically, the heterogeneous forms that partial realizations of a stochastic process [e.g., AR(1)] can take include pseudo-periodicity in time (Case 5) and patchiness in 2-D space (Case 6). In these two cases, the features of interest are of intermediate length and size.

Case 5: Periodic heterogeneity of the mean vs. heterogeneity due to autocorrelation in time series

Three replicates of a periodic signal in time, with a period of 28 units, were simulated by using a trigonometric function as mean function $\mu(t)$ for $t = 1, \ldots, n = 200$, plus a non-autocorrelated noise

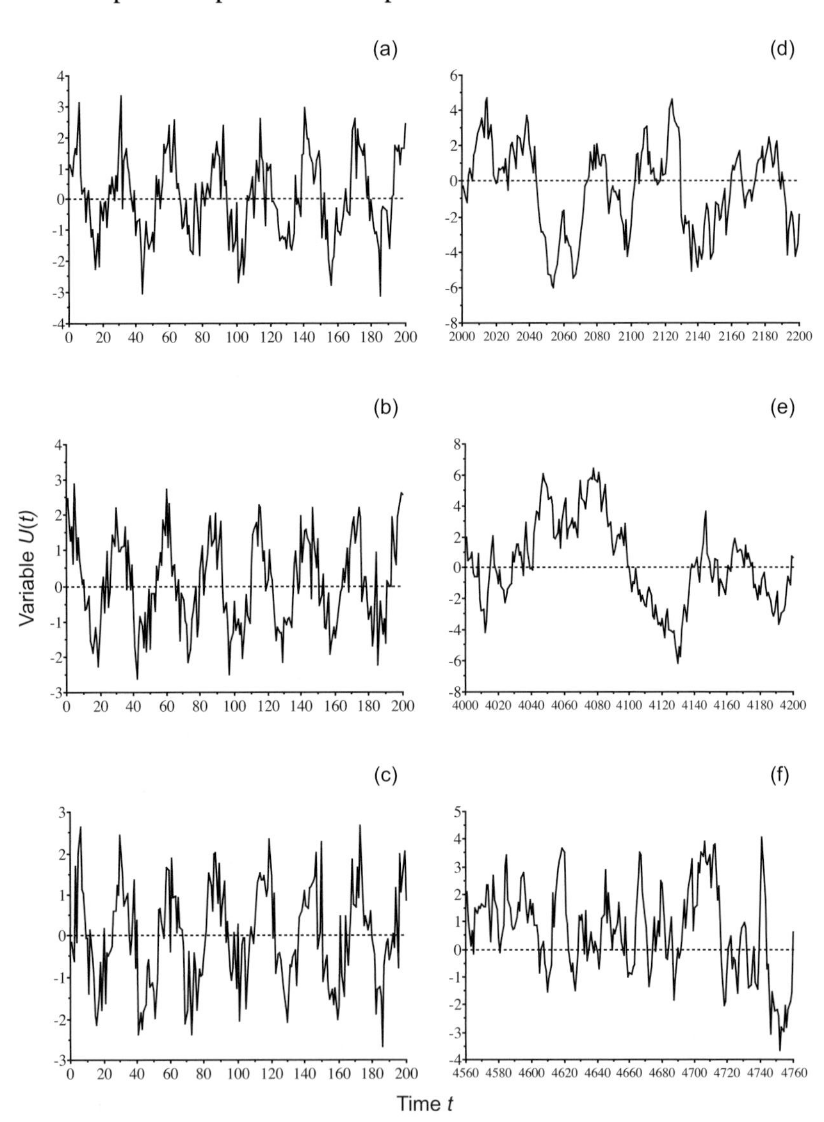

Fig. 2.6. Case 5: time series. (a)–(c) Replicates of a periodic signal in time with a period of 28 units [$\mu(t) = \cos(\frac{2\pi}{28}t) + \sin(\frac{2\pi}{28}t)$ for any t] plus a non-autocorrelated noise [$\sigma^2(t) = 0.49$ for all t; $\rho(t, t') = 0$ for any $t \neq t'$], and (d)–(f) partial realizations of a discrete-time AR(1) stochastic process [$\mu(t) = 0$ for all t; $\varepsilon(t) = 0.9\varepsilon(t-1) + \eta(t)$ for any t, with $\text{Var}[\eta(t)] = 1$ for all t, so $\sigma^2(t) = \frac{1}{1-0.9^2}$ for all t and $\rho(t, t') = 0.9^{|t'-t|}$ for any $t \neq t'$]; $n = 200$ for the six time series.

[Fig. 2.6(a)–(c)]. For comparison purposes, three partial realizations of a temporal AR(1) process with positive and strong autocorrelation were simulated by following a procedure similar to that used in Cases 2 and 3 [Fig. 2.6(d)–(f)].

The three periodic signals mimic the repetition of the lunar cycle with a period of 28 days over 200 days, and show about eight well-delimited peaks and troughs [Fig. 2.6(a)–(c)]. The cosine-wave shape is somewhat hidden in the observed time series because of a moderate signal-to-noise ratio, but the overall pattern is clear and consistent.

In examining the temporal surface pattern in Fig. 2.6(d), our eye may discern some periodicity in this time series, with perhaps eight peaks or intervals of higher values interspersed with eight troughs or intervals of lower values. The periodicity of the signal is not evident, though, and other partial realizations are required to help complete the examination [Fig. 2.6(e) and (f)]. The two supplementary partial realizations provide very different pictures: two major bumps separated by one gap are observed in Fig. 2.6(e); the successive intervals of higher and lower values are very short in Fig. 2.6(f). This is the proof that the mechanism generating these time series, the same for all three, is not deterministic but random!

Case 6: True patches vs. heterogeneity due to autocorrelation in spatial surface patterns

Four true patches in 2-D space, with fixed centers and same size, were simulated three times over a given area by using the same threshold mean function $\mu(x, y)$ plus a non-autocorrelated noise [Fig. 2.7(a)–(c)]. For comparison purposes, three 2-D spatial AR(1) partial realizations with positive and strong autocorrelation [Fig. 2.7(d)–(f)] were simulated by following a procedure similar to that used in Case 4.

If one considers that patchiness in space corresponds to periodicity in time, what does periodicity look like in 2-D space? Answering this question is one of the objectives of Case 6. In the surface-pattern approach, patches are defined as regularly spaced, well-delimited portions of the sampling domain, with higher, or lower, values of the continuous quantitative response. Examples are presented in Fig. 2.7(a)–(c), where four patches with about the same center and size and higher response values shape the three 2-D spatial surface patterns. Many will say that the heterogeneous pattern in Fig. 2.7(d) is patchy, although the patches there are not as clearly defined as in Fig. 2.7(a)–(c). It is true, however, that

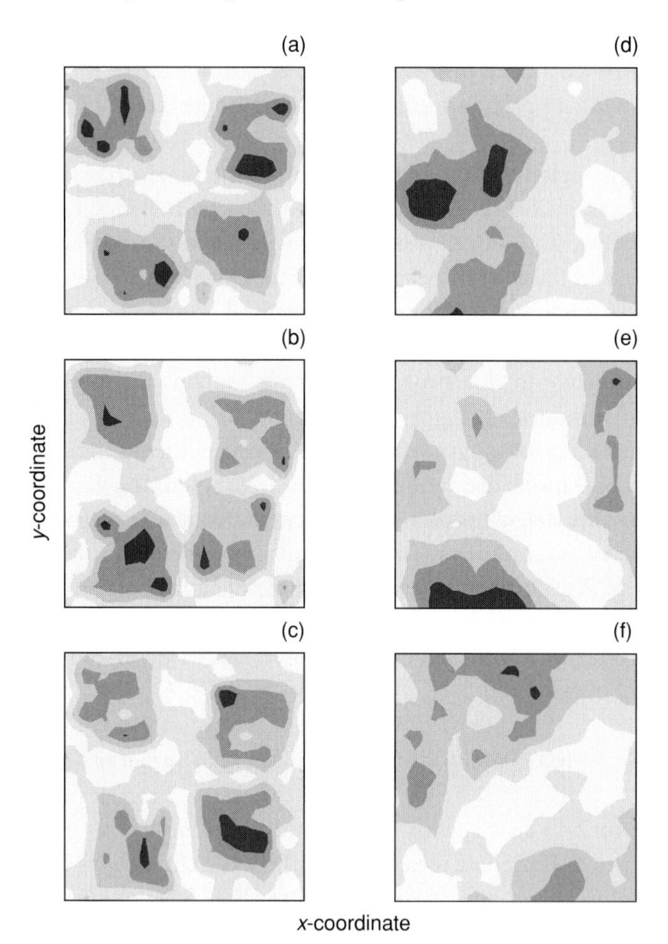

Fig. 2.7. Case 6: 2-D spatial surface patterns. (a)–(c) Replicates of four true patches in 2-D space, with a higher population mean [$\mu(x, y) = 5$ within patches and 0 outside], fixed centers [(3.5, 10.5), (4.5, 3.5), (9.5, 4.5), (10.5, 9.5)] and size 4×4, plus a non-autocorrelated noise [$\sigma^2(x, y) = 3$ for all (x, y); $\rho((x, y), (x', y')) = 0$ for any $(x, y) \neq (x', y')$], and (d)–(f) partial realizations of a weakly stationary and isotropic 2-D spatial stochastic process with positive autocorrelation [$\mu(x, y) = 0$ for all (x, y); $\sigma^2(x, y) = 3$ for all (x, y); $\rho((x, y), (x', y')) = \exp(-\frac{3d}{6})$]. The bottom-left corner of a panel is the origin of the coordinate system. Data were simulated on a 13×13 regular grid ($n = 169$) first and then were interpolated to produce maps. Gray tones are in ascending order of the values of $U(x, y)$, from low (light gray) to high (dark gray).

one predominant peak (bottom-left) and three secondary peaks (top-left and right) are observed in Fig. 2.7(d). Do the two other AR(1) partial realizations also show patchiness at the scale considered? The answer is no... In Fig. 2.7(e), a major trough (in diagonal, top-left to bottom-right) separates a secondary peak (bottom-left) from a large zone of intermediate to high response values (right); the pattern in Fig. 2.7(f) seems to be a 90° rotation of the former. Such differences among patterns are, again, illustrative of a random generating mechanism. The relatively smooth fluctuations, which are maintained over the heterogeneous patterns, are typical of a 2-D spatial stochastic process with positive and strong autocorrelation.

The information learned from Cases 1–6 can be summarized as follows.

- Trends, periodic signals, and patches considered to be "true" are related to spatial and temporal heterogeneity of the mean, whereas false trends are observed occasionally as part of heterogeneity due to autocorrelation (e.g., when autocorrelation is strong and positive and the extent-to-grain ratio is small or intermediate). The generating mechanism is deterministic for the former, and random for the latter.
  Key note: The AR(1) autocorrelation structure was used here for convenience and simplicity. The AR(1) processes exist in time and space, whether discrete or continuous (i.e., in continuous space via the isotropic exponential variogram model). Their model, which is a simple autoregression in discrete time and space, is intuitively clear. Population autocorrelations are not restricted to be positive for AR(1) processes; in time, see Chapter 6.

- Replication, or the collection of more than one partial realization for the same stochastic process, is an asset in the assessment and recognition of heterogeneous patterns, particularly for the distinction between patterns generated by deterministic vs. random mechanisms.
  Key note: In real-life situations, it seems easier to replicate in time than in space; plants and animals commonly provide a basis for replication in growth and chronobiological studies, for example. Replication in space can be performed in essentially two ways: by sampling the domain at comparable times (see the river basin example) or by sampling different portions of the domain at the same period (e.g., by including a sufficient number of apparent patches in each replicate).

- Scale, or the extent-to-grain ratio provided by the number of quadrats of unit side in Case 1 and the value of n in Cases 2–6, influences the form (i.e., *not* the type) of the heterogeneity observed. For example, the first half of a cycle ($n = 14$) in the time series of Fig. 2.6(a)–(c) would look like, and would actually represent, a true nonlinear trend. Depending on scale, partial realizations of an autocorrelated stochastic process are versatile and may provide various heterogeneous patterns [see the AR(1) partial realizations in Cases 2–6]. A multi-scale approach to heterogeneity analysis is thus highly recommended.

- The comments and guidelines above apply to the analysis of point patterns as well as to that of surface patterns. So far, the emphasis has been on the latter, but this was intentional in order to keep some original developments for the next chapters. Via the intensity functions and quadrat counts (i.e., a discrete quantitative response), the resemblance between the analyses of point and surface patterns may eventually be greater than one might think at first sight . . .

In Case 7 below, different grains are considered for the same extent, while autocorrelation is present at different ranges in Case 8. Both cases are 2-D spatial.

Case 7: Alternating homogeneity and heterogeneity in spatial hierarchy
The objective here is to illustrate an important statement made in the opening chapter: "Scale can create heterogeneity from homogeneity and vice versa." To show this by simulation, some adjustments need to be made to the model equation. In particular, the index s in equation (2.1) needs to be adapted to the multi-scale analysis approach. This is why the simulation procedure is exceptionally detailed below instead of in the legend of Fig. 2.8.

Instead of using $s = (x, y)$, the location of 2-D spatial data collected on the 32×32 grid is identified with a four-component index $s = (i, j, k, l)$, where each component refers to a given scale, from large (i) to small (l). Unlike Cases 2–6, in which scale was changed by modifying the extent while fixing the grain, the extent is fixed here, so different scales correspond to different grains or sizes of the sampling unit. In decreasing order, large scale is associated with main plots, large-to-intermediate scale with subplots, intermediate-to-small scale with sub-subplots, and small scale with sampling points within a sub-subplot.

In the simulation model, the mean function $\mu(i, j, k, l)$ is a constant $\mu = 5$ (i.e., U could be soil pH); the random deviation $\varepsilon(i, j, k, l)$ is made of as many terms as there are scales (i.e., four), all random obviously:

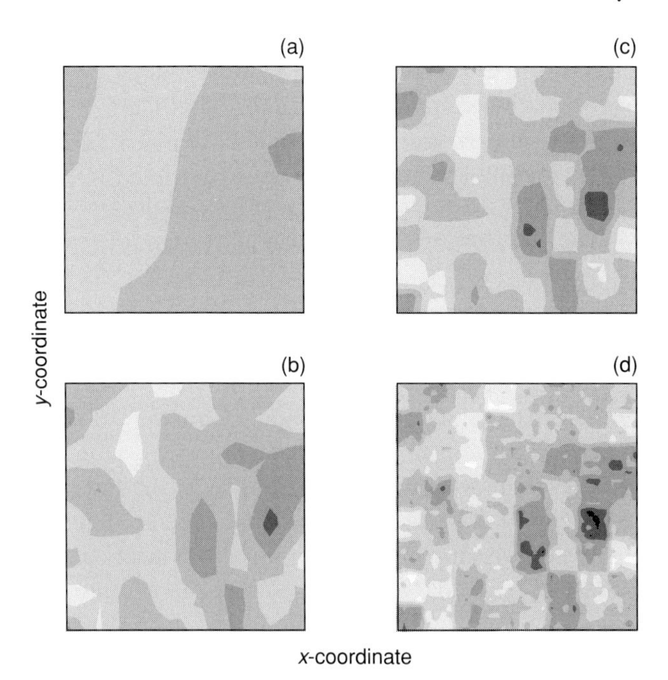

Fig. 2.8. Case 7: 2-D spatial surface patterns. Spatial hierarchy characterized by (a) homogeneity at large scale; (b) a dominant random component of variability at intermediate-to-large scale; (c) ancillary variability at intermediate-to-small scale; and (d) heterogeneity and maximum variability at small scale. See text for details. The same 2-D contours [i.e., from light to dark gray by increments of 1 in ascending order of the values of $U(x, y)$] were used in the four maps.

Main plot(i), *Subplot(j | i)*, *Sub-subplot(k | i, j)*, and $\eta(i, j, k, l)$, assumed to be independent and normally distributed with zero population mean and variance components $\sigma^2_{large\ scale}$, $\sigma^2_{large\text{-}to\text{-}intermediate\ scale}$, $\sigma^2_{intermediate\text{-}to\text{-}small\ scale}$, and $\sigma^2_{small\ scale}$, respectively. Thus, equation (2.1) can finally be rewritten as

$$U(i, j, k, l) = \mu + Main\ plot(i) + Subplot(j \mid i) + Sub\text{-}subplot(k \mid i, j) + \eta(i, j, k, l), \qquad (2.11)$$

where subscript i ranges from 1 to 16 (i.e., the field is divided into 4 × 4 main plots); for a given main plot i, subscript j ranges from 1 to 4 (i.e., each main plot is divided into 2 × 2 subplots); for a given subplot j in main plot i, subscript k ranges from 1 to 4 (i.e., each subplot is divided into 2×2 sub-subplots); and subscript l ranges from 1 to 4 (i.e., there are four sampling points on a 2×2 grid within each sub-subplot).

To initiate the spatial hierarchy, the values of $\sigma^2_{large\ scale}$ and $\sigma^2_{large\text{-}to\text{-}intermediate\ scale}$ were fixed at 0 and 1, respectively. At intermediate-to-small scale, no additional variability was assumed, with $\sigma^2_{intermediate\text{-}to\text{-}small\ scale} = 0$. At the lowest level of this 2-D spatial hierarchy, the value of $\sigma^2_{small\ scale}$ was fixed at 0.25. The way this spatial heterogeneity of the variance among scales was simulated, via a sum of terms for the random component of the stochastic process, recalls the analysis-of-variance models of Dutilleul (1993a) and the author's recommendation of using plot splitting for multi-scale analysis of spatial heterogeneity.

In Fig. 2.8(a), where the sample means computed for the 16 main plots are mapped, there is no apparent spatial heterogeneity, which is in agreement with the constancy of the expected value of the response and the fact that $\sigma^2_{large\ scale} = 0$. The observed pattern is totally different in Fig. 2.8(b), where the 64 subplot sample means are mapped and important fluctuations are observed. No further heterogeneity can really be seen in Fig. 2.8(c), which is in agreement with $\sigma^2_{intermediate\text{-}to\text{-}small\ scale} = 0$. Spatial variability is maximum in Fig. 2.8(d), that is, at the level of sampling points within a sub-subplot, where the population variance of $U(i, j, k, l)$ is equal to $\sigma^2_{large\text{-}to\text{-}intermediate\ scale} + \sigma^2_{small\ scale} = 1.25$. In view of Fig. 2.8(a)–(d), it is clear that working at too large a scale may hide traces of spatial heterogeneity present at smaller scales.

Case 8: Spatial heterogeneity due to autocorrelation and mixtures of random structural components

In Case 7, we learned that the random component of a 2-D spatial stochastic process can be further decomposed into a sum of random terms, each characterized by a variance component associated with a given scale; none of the random terms was spatially autocorrelated in that case. In Case 8, partial realizations of weakly stationary and isotropic 2-D spatial stochastic processes with three random "structural components," some of these being spatially autocorrelated, were simulated. In this context and in simple terms, autocorrelation is characterized by a range (i.e., maximum distance for spatial correlation) and the sign and strength of correlations at distances up to the range (i.e., positive and decreasing with distance here). One partial realization of each of the three structural components and three weighted sums of them, or "mixtures," are mapped in Fig. 2.9(a)–(c) and (d)–(f), respectively.

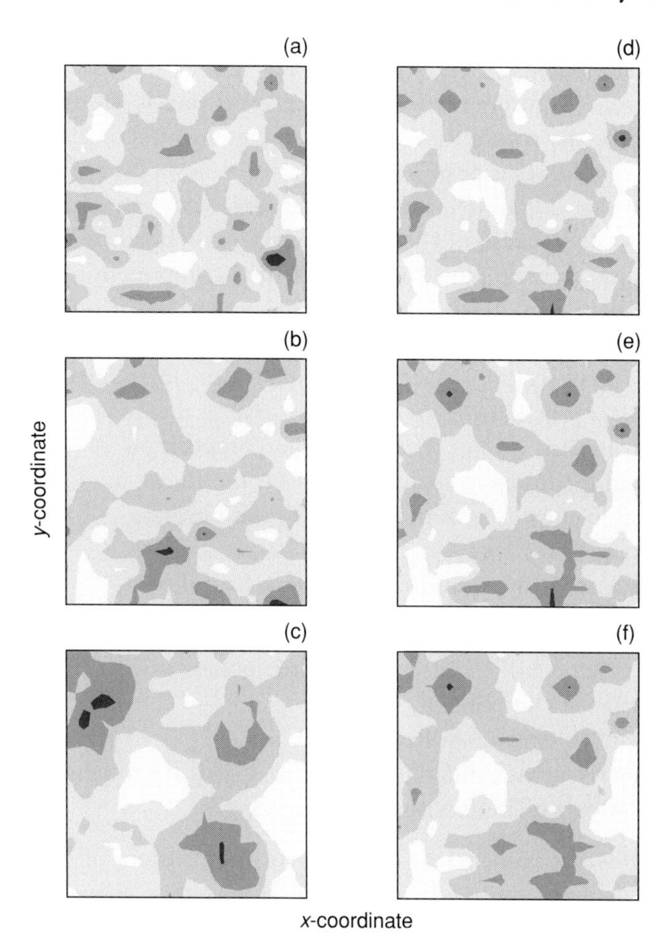

Fig. 2.9. Case 8: 2-D spatial surface patterns. Partial realizations of weakly stationary and isotropic 2-D spatial stochastic processes [$\mu(x, y) = 0$ for all (x, y); $\sigma^2(x, y) = 1$ for all (x, y)]: (a) a purely random process [$\rho((x, y), (x', y')) = 0$ for any $(x, y) \neq (x', y')$]; (b) a positively autocorrelated process with range 2 [$\rho((x, y), (x', y')) = 1.0 - 1.5\frac{d}{2} + 0.5\frac{d^3}{8}$ if $d < 2$ and 0 otherwise]; and (c) a positively autocorrelated process with range 5 [$\rho((x, y), (x', y')) = 1.0 - 1.5\frac{d}{5} + 0.5\frac{d^3}{125}$ if $d < 5$ and 0 otherwise]. Normalized weighted sums of the three partial realizations: (d) uneven weights (0.707, 0.577, 0.408) favoring the purely random process; (e) even weights (0.577, 0.577, 0.577); and (f) uneven weights (0.408, 0.577, 0.707) favoring the most autocorrelated process (i.e., with the longer range). Data were simulated on a 15 × 15 regular grid ($n = 225$) first and then were interpolated to produce maps. The same 2-D contours [i.e., from light gray to dark gray by increments of 1 in ascending order of the values of $U(x, y)$] were used in the six maps.

From the erratic fluctuations observed in Fig. 2.9(a) and the very smooth fluctuations in Fig. 2.9(c), one can easily guess that the first structural component is purely random and the third one is the most autocorrelated, while the second one is only moderately autocorrelated [Fig. 2.9(b)]. To keep the same population variance for all six 2-D spatial surface patterns in Fig. 2.9(a)–(f), the sum of squares of the weights in mixtures was set to 1. Because of that, it is possible, but not necessarily easy, to determine that the predominant structural component is the purely random one in Fig. 2.9(d) and the most autocorrelated one in Fig. 2.9(f). Since the three random structural components are present in each mixture, the observed surface pattern does not differ substantially among mixtures [Fig. 2.9(d)–(f)]. Thus, the graphical approach has limitations when the time comes to identify the composition of mixtures like these, and quantitative analyses based on the linear model of coregionalization (Chapter 7) were developed for this purpose.

From Cases 7 and 8, it is important to remember what follows.

- Different spatial surface patterns are observed at different scales when the population variance of the spatial stochastic process is the sum of scale-specific variance components.
- A spatial stochastic process may have several autocorrelated random structural components with different ranges, so spatial heterogeneity may arise from autocorrelation within different ranges.

Cases 1–8 provided an informative overview of the plethora of heterogeneous patterns that may be observed in space and time as well as a good understanding of the meaning of parameters in equation (2.1) for the definition of models for the underlying stochastic processes. In the next section, this equation is extended to multivariate and replicated responses.

2.5 Extensions of the fundamental equation

Though fundamental for the definition of the three types of heterogeneity (i.e., mean, variance, autocorrelation), equation (2.1) is nonetheless restricted to one variable U, whereas several variables $U_1, \ldots, U_p$ are generally observed at each sampling location or time in biological and environmental studies. In these situations, equation (2.1) can be the basis for heterogeneity analyses performed for each variable U_j ($j = 1, \ldots, p$) separately, but must be extended to allow a multivariate analysis that incorporates the relationships between variables in space and time. Moreover, replication is a practice that is highly recommended in study designs in general and in the analysis of spatio-temporal heterogeneity in particular, as illustrated in Section 2.4. Thus, another extension of equation (2.1) is required if one wants to model spatio-temporal heterogeneity by incorporating replication and any classification factors of the study design into the equation. Such extensions, including the one for multivariate replication (i.e., replication for several variables), are presented below.

When $p > 1$ variables are to be analyzed jointly for spatio-temporal heterogeneity, equation (2.1) extends to a p-variate stochastic process $\{\boldsymbol{u}(s) \mid s \in S\}$ as follows:

$$\boldsymbol{u}(s) = \boldsymbol{\mu}(s) + \boldsymbol{\varepsilon}(s), \tag{2.12}$$

with $\boldsymbol{u}(s)$, the indexed $p \times 1$ random vector of responses; $\boldsymbol{\mu}(s)$, the corresponding population mean vector; and $\boldsymbol{\varepsilon}(s)$, the $p \times 1$ vector of random deviations (see Section A2.2 for more formal definitions). Equation (2.1) applies to each of the p components $U_j(s)$ of $\boldsymbol{u}(s)$, so equation (2.12) represents a system of p equations similar to equation (2.1):

$$U_j(s) = \mu_j(s) + \varepsilon_j(s) \quad (j = 1, \ldots, p). \tag{2.13}$$

In addition to the analyses of spatio-temporal heterogeneity of the mean, of the variance and that due to autocorrelation for each variable separately, equation (2.12) allows the analysis of relationships between variables in space and time. The population cross-correlations $\rho_{jj'}(s, s')$ (see equation A2.15) and the population cross-semivariances $\gamma_{jj'}(s, s')$, which are involved in the linear model of coregionalization used in geostatistics (see Chapter 7), allow such analysis in well-defined conditions. Under the assumption of stationarity at order 2, $\rho_{jj'}(s, s') = \rho_{jj'}(s' - s)$; under the assumption of intrinsic stationarity, $\gamma_{jj'}(s, s') = \gamma_{jj'}(s' - s)$; and under the assumption of isotropy in 2-D and 3-D space, $\rho_{jj'}(s, s') = \rho_{jj'}(\|s' - s\|)$ and $\gamma_{jj'}(s, s') = \gamma_{jj'}(\|s' - s\|)$.

Multivariate stochastic processes may be studied through point patterns as well as surface patterns. In fact, points in space such as plants of different species and events in time such as transitions between behavioral acts of different types correspond to non-zero values for different components of a multivariate point process. The value of $U_j(s)$ is 1 if a plant of species j or a transition of type j is observed at s, and 0 otherwise ($j = 1, \ldots, p$). Two marks associated with the same univariate point process can also be analyzed jointly. For example, the magnitude of an earthquake (first mark) may be correlated with the time elapsed since the previous earthquake (second mark) in the same region; the height of a plant can be related to its status, healthy or infected by a fungus. It is possible that multivariate stochastic processes are studied more directly in the surface-pattern approach. For example, tree diversity may be analyzed in space in relation to a number of soil properties, from vectors of observations collected for all the variables at the same sampling locations; similarly in time, tree-ring width, wood density, and fiber length could be analyzed from triplets of observations made in the same sampling years.

In equation (2.1), it is assumed that one partial realization of a univariate stochastic process is available for heterogeneity analysis. The importance of replication in heterogeneous pattern recognition has been demonstrated graphically, with simulated examples, in Section 2.4. In the context of real-life situations, replicates are obtained in space when plankton biomass is sampled in the epi-, meta- and hypolimnion strata along the vertical axis at a number of stations in a lake, and in time when the width of growth rings in the years following an insect outbreak is measured for a number of trees. The extension of equation (2.1) to such situations or, more generally, when spatial or temporal data are collected in accordance with an experimental design with a number of fixed and random classification factors, is the following:

$$U_{f_1 \dots f_F r_1 \dots r_R k}(s) = \mu_{f_1 \dots f_F r_1 \dots r_R k}(s) + \varepsilon_{f_1 \dots f_F r_1 \dots r_R k}(s), \tag{2.14}$$

where the subscripts $f_1 \dots f_F$ denote the levels of F fixed classification factors, the subscripts $r_1 \dots r_R$ denote the levels of R random classification factors, and subscript k identifies the replicate for a given combination of $f_1 \dots f_F$ and $r_1 \dots r_R$. The population mean $\mu_{f_1 \dots f_F r_1 \dots r_R k}(s)$ includes the main effects of fixed factors and their interactions, while the variance and covariance components associated with the main effects of random factors and the random interactions are present in the population variances and the population autocovariance function of the stochastic process, via $\varepsilon_{f_1 \dots f_F r_1 \dots r_R k}(s)$. Particular cases of equation (2.14) are used in the repeated measures ANOVA to assess the presence of mean and variance heterogeneities, among other things (Chapter 9).

The combination of the two extensions provides the equation in the case of multivariate replication:

$$\boldsymbol{u}_{f_1 \dots f_F r_1 \dots r_R k}(s) = \boldsymbol{\mu}_{f_1 \dots f_F r_1 \dots r_R k}(s) + \boldsymbol{\varepsilon}_{f_1 \dots f_F r_1 \dots r_R k}(s), \tag{2.15}$$

where the notations are similar to those in equations (2.12) and (2.14). A particular case of equation (2.15) is used in the ANOVA of the finite Fourier transform, where replicated time series are analyzed in the frequency domain instead of the time domain (Dutilleul, 1998b).

2.6 Recommended readings

The distinction of three main types of heterogeneity (i.e., mean, variance, autocorrelation) was initiated in the spatial context by Dutilleul and Legendre (1993), who also pointed out a number of common points and differences between the point-pattern and surface-pattern approaches. Ripley (1981) and Cressie (1993) elaborate on more mathematical aspects of the properties of spatial stochastic processes. Note that Cressie (1993, p. 61) suggests that isotropy is a property restricted to spatial processes that are intrinsically stationary, whereas Ripley (1981, pp. 9–10) distinguishes spatial processes that are stationary under translations from isotropic spatial processes which are stationary under rotations; the content of Section 2.3 here is closer to Ripley's view on isotropic 2-D and 3-D spatial processes. More specific aspects of temporal stochastic processes may be found in Brillinger (1981) and Priestley (1981). Griffith (1987) gives a very interesting introduction to the concept of spatial autocorrelation, and so do Isaaks and Srivastava (1989) for the geostatistical view on spatial stochastic processes, called "random functions" by geostatisticians. Diggle (2003) is a key reference for the statistical analysis of spatial point processes, including the basic properties and elements introduced in the previous sections here.

The next three chapters are devoted to the heterogeneity analysis of point patterns in space, time, and space-time.

Appendix A2: Population parameters and the expectation operator

Knowing the calculation procedure for population parameters is important for at least two reasons:

(i) This knowledge is a prerequisite to any modeling of spatio-temporal heterogeneity [see the fundamental equation (2.1) and its extensions (2.12), (2.14) and (2.15)] and to the definition of most properties of spatio-temporal stochastic processes (see Section 2.3).

(ii) When a partial realization of a stochastic process is available for statistical inference (i.e., a sample of data has been drawn from the statistical population), the definition of the sample statistics that will estimate the population parameters (see Chapters 3 to 8) is inspired from that of the relevant population parameters. The operators defining the two sets of quantities are different, but the interpretation of the quantities in terms of heterogeneity (i.e., observed on the partial realization vs. hypothesized for the underlying stochastic process) is basically the same.

A2.1 The univariate case

In this section, population parameters such as the population mean and variance of $U(s)$ and the population autocovariance and autocorrelation between $U(s)$ and $U(s')$ with $s \neq s'$ are defined by using an operator called "expectation" and denoted $\mathrm{E}[\cdot]$. The substitution of $U(s)$ (i.e., any indexed random variable from a univariate stochastic process) for the dot in $\mathrm{E}[\cdot]$ provides $\mathrm{E}[U(s)]$, which is the expected value of $U(s)$ or the population mean $\mu(s)$:

$$\mu(s) = \mathrm{E}[U(s)]. \tag{A2.1}$$

If $U(s)$ is discrete quantitative, then its distribution is defined by a probability function $p(u; s)$ and $\mathrm{E}[U(s)]$ is calculated as $\sum_u up(u; s)$, where $\sum_u$ denotes the summation over all possible values u of $U(s)$ and $p(u; s)$ is the probability associated with a given value u. This is the case in the analysis of point patterns, when $U(s)$ represents the count in a quadrat centered at s. If $U(s)$ is continuous quantitative, then its distribution is defined by a probability density function $f(u; s)$. As a density, only the integral of this function over an interval of values u represents a probability; each value of $f(u; s)$ for a given u does not represent a probability. For example, when the stochastic process is real-valued in the analysis of surface patterns, $\mathrm{E}[U(s)]$ is calculated as $\int uf(u; s)du$, where $\int \cdot\, du$ denotes the integral over all possible values of $U(s)$. In both cases (i.e., discrete quantitative and continuous quantitative), $\mathrm{E}[U(s)]$ is the first moment of $U(s)$. Unlike the second-order moments below, it is not centered. Note that the probability function and the probability density function are unknown in practice, and so is the population mean as well as the population variance, autocovariance, and autocorrelation.

Whereas the population mean $\mu(s)$ represents a center for the probability distribution of $U(s)$, the population variance $\sigma^2(s) = \mathrm{Var}[U(s)]$

represents a measure of dispersion of the possible values of $U(s)$ around $\mu(s)$, in the same way that the moment of inertia in physics is calculated with respect to a reference axis of rotation. Thus, a centering with respect to $\mu(s)$ is performed in the calculation of the population variance, while taking into account the probabilities or densities of probability associated with the different values of $U(s)$. Accordingly, the population variance is the centered second-order moment of $U(s)$:

$$\sigma^2(s) = \mathrm{E}[\{U(s) - \mu(s)\}^2]. \tag{A2.2}$$

The expectation $\mathrm{E}[\{U(s) - \mu(s)\}^2]$ in (A2.2) is calculated as $\sum_u \{u - \mu(s)\}^2 p(u; s)$ in the discrete case and as $\int \{u - \mu(s)\}^2 f(u; s)du$ in the continuous case.

As for the population autocovariance $C(s, s') = \mathrm{Cov}[U(s), U(s')]$ with $s \neq s'$, it is defined as the centered second-order joint moment of $U(s)$ and $U(s')$:

$$C(s, s') = \mathrm{E}[\{U(s) - \mu(s)\}\{U(s') - \mu(s')\}]. \tag{A2.3}$$

The expectation $\mathrm{E}[\{U(s) - \mu(s)\}\{U(s') - \mu(s')\}]$ in (A2.3) is calculated by using the joint probability function $p(u, u'; s, s')$ of the random variables $U(s)$ and $U(s')$ when they are discrete quantitative:

$$\sum_u \sum_{u'} \{u - \mu(s)\}\{u' - \mu(s')\} p(u, u'; s, s'); \tag{A2.4}$$

or by using the joint probability density function $f(u, u'; s, s')$ when the two random variables are continuous quantitative:

$$\int\int \{u - \mu(s)\}\{u' - \mu(s')\} f(u, u'; s, s')du\, du'. \tag{A2.5}$$

The population autocorrelation $\rho(s, s') = \mathrm{Corr}\,[U(s), U(s')]$ with $s \neq s'$ measures the strength and direction of the association between two indexed random variables $U(s)$ and $U(s')$ from the same univariate stochastic process. It is obtained by standardizing the population autocovariance:

$$\rho(s, s') = \frac{C(s, s')}{\sqrt{\sigma^2(s)\, \sigma^2(s')}}. \tag{A2.6}$$

The standardization in (A2.6) ensures that $\rho(s, s')$ is not greater than 1 in absolute value: $-1 \leq \rho(s, s') \leq 1$.

A2.2 The multivariate case

The definition of population parameters in terms of expectation extends as follows to any indexed $p \times 1$ random vector $\boldsymbol{u}(s) = (U_1(s), \ldots, U_p(s))^{\mathrm{T}}$ from a multivariate stochastic process, where ^{T} denotes the transpose operator in matrix algebra.

The population mean of $\boldsymbol{u}(s)$ is the $p \times 1$ vector

$$\boldsymbol{\mu}(s) = (\mu_1(s), \ldots, \mu_p(s))^{\mathrm{T}} \tag{A2.7}$$

and its p components are given by

$$\mu_j(s) = \mathrm{E}[U_j(s)] \quad (j = 1, \ldots, p). \tag{A2.8}$$

The population variances of and population covariances between the p components of $\boldsymbol{u}(s)$ are the entries of the $p \times p$ variance–covariance matrix $\mathrm{Var}[\boldsymbol{u}(s)]$, with on the diagonal:

$$\sigma_j^2(s) = \mathrm{Var}[U_j(s)] = \mathrm{E}[\{U_j(s) - \mu_j(s)\}^2] \quad (j = 1, \ldots, p) \tag{A2.9}$$

and off the diagonal:

$$\begin{aligned} C_{jj'}(s) &= \mathrm{Cov}[U_j(s), U_{j'}(s)] \\ &= \mathrm{E}[\{U_j(s) - \mu_j(s)\}\{U_{j'}(s) - \mu_{j'}(s)\}] \;\; (j, j' = 1, \ldots, p; j \neq j'). \end{aligned} \tag{A2.10}$$

In equation (A2.10), the population covariances $C_{jj'}(s)$ involve two components of the random vector $\boldsymbol{u}(s)$. In space, $C_{jj'}(s)$ is the local covariance between $U_j(s)$ and $U_{j'}(s)$; in time, $C_{jj'}(s)$ is the instantaneous covariance between $U_j(s)$ and $U_{j'}(s)$.

Local or instantaneous correlations between $U_j(s)$ and $U_{j'}(s)$ $(j, j' = 1, \ldots, p; j \neq j')$ are given by the population correlations

$$\rho_{jj'}(s) = \mathrm{Corr}[U_j(s), U_{j'}(s)] = \frac{C_{jj'}(s)}{\sqrt{\sigma_j^2(s)\,\sigma_{j'}^2(s)}}. \tag{A2.11}$$

So far in this section, the definition of population parameters involved only one index value, namely s. Below, two index values $s \neq s'$ are involved:

- in the population autocovariances between $U_j(s)$ and $U_j(s')$ $(j = 1, \ldots, p)$

$$\begin{aligned} C_j(s, s') &= \mathrm{Cov}[U_j(s), U_j(s')] \\ &= \mathrm{E}[\{U_j(s) - \mu_j(s)\}\{U_j(s') - \mu_j(s')\}]; \end{aligned} \tag{A2.12}$$

- in the population autocorrelations between $U_j(s)$ and $U_j(s')$ ($j = 1, \ldots, p$)

$$\rho_j(s, s') = \mathrm{Corr}[U_j(s), U_j(s')] = \frac{C_j(s, s')}{\sqrt{\sigma_j^2(s)\,\sigma_j^2(s')}}; \qquad \text{(A2.13)}$$

and

- in the population cross-covariances and cross-correlations between $U_j(s)$ and $U_{j'}(s')$ ($j, j' = 1, \ldots, p$; $j \neq j'$), which are two functions of covariances and correlations specific to multivariate stochastic processes,

$$C_{jj'}(s, s') = \mathrm{Cov}[U_j(s), U_{j'}(s')]$$
$$= \mathrm{E}[\{U_j(s) - \mu_j(s)\}\{U_{j'}(s') - \mu_{j'}(s')\}]; \qquad \text{(A2.14)}$$
$$\rho_{jj'}(s, s') = \mathrm{Corr}[U_j(s), U_{j'}(s')] = \frac{C_{jj'}(s, s')}{\sqrt{\sigma_j^2(s)\,\sigma_{j'}^2(s')}}. \qquad \text{(A2.15)}$$

Equations (A2.14) and (A2.15) define population covariances and correlations between different components of the random vector of responses at different spatial locations or times. Cross-covariances and cross-correlations are especially useful to study complex systems in which heterogeneity may interfere with the assessment of relationships between variables in space and time (see Chapters 6–8).

3 · *Heterogeneity analysis of spatial point patterns*

Quantification of the heterogeneity contained in an observed point pattern and the eventual modeling of parameter functions of the underlying point process are discussed in this and the following two chapters. As a preamble, the aspects of fundamental equation (2.1) that are specific to point processes are emphasized, and the general background appropriate for the analysis of point patterns is introduced in Section 3.1. Heterogeneity analysis is first undertaken in the spatial framework (this chapter), because space is a natural domain to deal with points. Thereafter, the analytical procedures will be transposed to the temporal and spatio-temporal frameworks (Chapter 4), and last but not least, aspects related to modeling will be presented together with a general summary (Chapter 5). In the three frameworks (i.e., spatial, temporal, and spatio-temporal), quadrat, interval and cell counts on the one hand and distances of various types on the other hand support the two main types of statistical methods available for heterogeneity analysis of point patterns. Marked point patterns (when a qualitative or quantitative characteristic called "mark" is observed at each point), multivariate point patterns (when points of different types are observed), state transitions, and movement will be studied in some of the frameworks. Examples with simulated and real data will be used in each framework. More formal definitions and details about simulation procedures are grouped in Appendix A3, while references for complementary readings on the analysis of point patterns are given at the end of Chapter 5. MATLAB® codes and SAS codes implementing analytical procedures are available on the CD-ROM that accompanies the book.

3.1 Preamble

The questions in plant ecology and the earth sciences below typically involve space and time, and are representative of the kind of questions that are addressed in this and the following two chapters.

- How does a community of plants in a given site occupy space from year to year? How does it increase or decrease in size according to climatic fluctuations or following infection by a rust fungus? Are the dead or infected plants grouped together?
- Do earthquakes occur in specific locations at regular intervals in time? If there has been no earthquake for a long time, will the next earthquake in the region be of greater magnitude? Is there a shift over time in the spatial location of earthquakes?

Stochastic processes associated with the location of individuals in space and the occurrence of events in time (i.e., the distribution of points in space and time) are called "point processes," and their partial realizations "point patterns." As briefly mentioned in Chapter 2, a first-order intensity function and a second-order intensity function characterize each point process. In Chapters 3 to 5, the front half of the STRC in Fig. 1.1(b) is explored in greater detail. First of all, we will see that, basically, the two intensity functions may generate the three possible types of heterogeneity in a point pattern, via moments of the probability distribution of quadrat counts.

In general terms, an indexed random variable $U(s)$ for a given s, from a stochastic process $\{U(s)|s \in S\}$, can be decomposed into a population mean $\mu(s)$ and a random deviation $\varepsilon(s)$ (Section 2.2). In the case of point processes, $\mu(s)$, which is the expected value of $U(s)$, can be considered as a probability, that is, a non-negative quantity that cannot be greater than 1. It is true that $U(s)$ is binary here, since it takes the value 1 if there is a point at s, and 0 otherwise. As a matter of fact, the probability that there is a point exactly at s may be very small, especially if space or time is continuous because there is an infinity of possible locations or times for points to be observed at and the number of observed points is usually finite. If one enlarges s to a "quadrat" A centered at s (e.g., a portion of a plot in 2-D space or of an interval in time), that probability increases, and one can then replace the binary variable $U(s)$ by the number of points found in A, or "quadrat count" $N(A)$. Note that a quadrat was originally a square of 1-m side, used by the Uppsala school of plant ecologists as the basic sampling unit for investigating plant communities in the field (Du Rietz, 1930).

Equation (2.1) applies to quadrat counts in the general sense as follows:

$$N(A) = \mu(A) + \varepsilon(A), \tag{3.1}$$

with $\mu(A) = \mathrm{E}[N(A)]$, the expected value of $N(A)$ or the expected number of points in A; $\mathrm{Var}[N(A)] = \mathrm{Var}[\varepsilon(A)]$; and $\mathrm{Cov}[N(A), N(B)] = \mathrm{Cov}[\varepsilon(A), \varepsilon(B)]$, where B denotes a quadrat centered at s'. The first-order intensity function is used to calculate $\mu(A)$, and the second-order intensity function to calculate $\mathrm{Var}[N(A)]$ and $\mathrm{Cov}[N(A), N(B)]$ (Section A3.1). It follows that there is heterogeneity of the mean when $\mu(A)$ varies with A, and heterogeneity of the variance when $\mathrm{Var}[N(A)]$ is dependent on A, and that there is heterogeneity due to autocorrelation when $\mathrm{Cov}[N(A), N(B)]$ is not zero for all pairs of disjoint quadrats A and B.

Key note: The quadrat count can be used as response to define the three possible types of heterogeneity for a point pattern.

The reasoning above applies to any point pattern, whether spatial, temporal, or spatio-temporal, because equation (3.1) and the corresponding expressions, which involve quadrat counts $N(A)$ and $N(B)$, hold whether A and B are portions of a transect, a plot or a volume in 1-D, 2-D or 3-D space, portions of an interval in time, or combinations of such portions in space-time. Accordingly, essentially the same statistical methods can be used to perform heterogeneity analysis of spatial, temporal and spatio-temporal point patterns. As we will see, these methods are based on quadrat and cell counts on the one hand and relevant distances on the other hand. In particular, cell counts made within quadrats in space (Section 3.4) will be used in the fitting of the beta-binomial distribution, which is a discrete quantitative statistical distribution that has an aggregation index as one of its parameters; the use of distances will assume that the location of observed points is known, and when this is the case in space and time the distance-based approach can be followed for the analysis of spatio-temporal point patterns.

Of course, the hypotheses of interest may depend on whether one works in space, time, or space-time. One of the specific aspects of time compared with space lies in the existence of state transitions in time and their absence in space at a given time. One has no choice: to change state, one needs a temporal component! Heterogeneity analysis of point patterns will be performed in a number of situations that researchers in the biological and the environmental sciences can relate to. Two datasets (i.e., *Arisaema* and California earthquakes; see the introductory questions) will be analyzed in different sections of Chapters 3 to 5, and these two main examples will be completed by others in entomology and forest ecology and in animal science and wildlife animal biology. Real data will be used to refine ideas, cover a range of practical situations, and discuss results in concrete terms given the hypothesis tested. In a very first step, the basic principles and methods available for the analysis of point patterns will be introduced using simulated data.

First, spatial point patterns will be studied in this chapter (Sections 3.2–3.5), and then, temporal and spatio-temporal point patterns will be analyzed in Chapter 4 (Sections 4.1 and 4.2). The primary objective is to present statistical methods of heterogeneity analysis in which the nature of point patterns is taken into account, and the great majority of these methods cannot be used to analyze surface patterns. One or several statistical methods that are appropriate to each situation will be presented as simply, clearly, and precisely as possible, in relation to the type of heterogeneity of interest (i.e., mean, variance, autocorrelation). In the course of doing this, we will see how to assess assumptions, estimate population parameters, quantify heterogeneity, test hypotheses about possible underlying processes, and attempt to come up with a model eventually (Chapter 5). In particular, partial realizations of hypothesized point processes will be simulated and special attention will be paid to edge effects in modeling. A number of more specialized topics will also be approached, such as marked spatial and temporal point patterns, multivariate spatial point patterns, state transitions in time, and movement in space-time, in different sections of Chapters 3 to 5.

3.2 Univariate spatial point patterns

The emergence of biological entities (e.g., germinated seeds) and the occurrence of events (e.g., earthquakes) can be located in 1-D, 2-D, or 3-D space depending on the experimental situation, at a level of precision that is a function of the equipment available. Until the 1980s and the advent of the Global Positioning System (GPS) and geographic information systems (GIS), it was tedious, if not impossible, to record the spatial location of, say, each individual plant of the species of interest in a plot. This was particularly true when the number of points was very large, and may explain why the tendency was then to count points in quadrats of a given size without recording the spatial location of each counted individual. As we will see, starting with spatial point patterns, there are several differences in the approach to the analysis of quadrat counts vs. the analysis of explicit positional data and, as a result, in the way heterogeneity can be quantified and assessed.

Prior to presenting any statistical method of heterogeneity analysis for point patterns, it is necessary to introduce the three main types of point patterns that may be encountered in space as well as in time: (1) completely random, (2) regular, and (3) aggregated. Note that "aggregated" is more general than "clustered" and does not have the connotation that points are crowded because of some biological mechanism (e.g., progeny of a mother plant). In what immediately follows, the philosophy is that of a teacher who simulates a sample of data from the probability distribution under study and graphs the simulated data, for the class to visualize the characteristics of the distribution via the sample; the sample and the distribution are, respectively, the point pattern and the point process here. Details about the simulation procedures are given in Section A3.2. Examples with real data will follow.

3.2.1 Simulated examples in 2-D space

In the partial realization of a simple Poisson process in Fig. 3.1(a), points are close to each other in some portions of the plot, are farther apart in other portions, and tend to be at a constant distance from each other in the remaining portions. Overall, no consistent pattern is observed. Such a completely random point pattern falls between a regular point pattern [Fig. 3.1(b)], in which no point has a very close or very far nearest neighbor, and an aggregated point pattern [Fig. 3.1(c)], in which points are grouped at small distances from each other in separate portions of the plot. These visual observations are made in terms of distances. A number of visual observations can also be made in terms of numbers of points per quadrat (e.g., square of unit side). For the completely random point pattern, the number of points per quadrat varies from 0 (empty quadrat) to 4, and the number of empty quadrats seems important. For the regular point pattern, the number of points per quadrat varies from 0 to 3, and there seems to be more quadrats with 1 point than empty quadrats or quadrats with 2 or 3 points. For the aggregated point pattern, the minimum number of points per quadrat is still 0, but the maximum is 8, with many empty quadrats (i.e., over about 75% of the plot) and 4 quadrats with 8 points (i.e., there are 4 clusters). The use of squares of unit side as quadrats for counting points is arbitrary and has a historical background (Du Rietz, 1930). The use of smaller vs. larger quadrats is discussed in a later subsection.

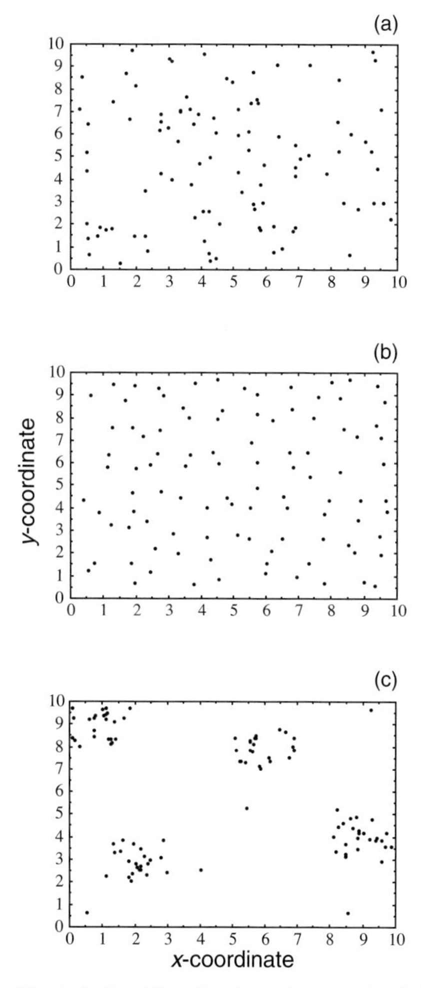

Fig. 3.1. Partial realizations ($n = 100$) of three 2-D spatial point processes: (a) a completely random or simple Poisson process; (b) a simple inhibition process showing regularity; and (c) a Poisson cluster process showing aggregation.

3.2.2 Scale effects

Before proceeding to the quantification of heterogeneity in spatial point patterns, it is important to illustrate some of the effects of scale on the heterogeneity to be measured in this context. The two simulated 2-D spatial point patterns presented in Fig. 3.2 are partial realizations of different Poisson cluster processes. One is characterized by a concentric distribution of points within clusters [Fig. 3.2(a)], whereas the distribution

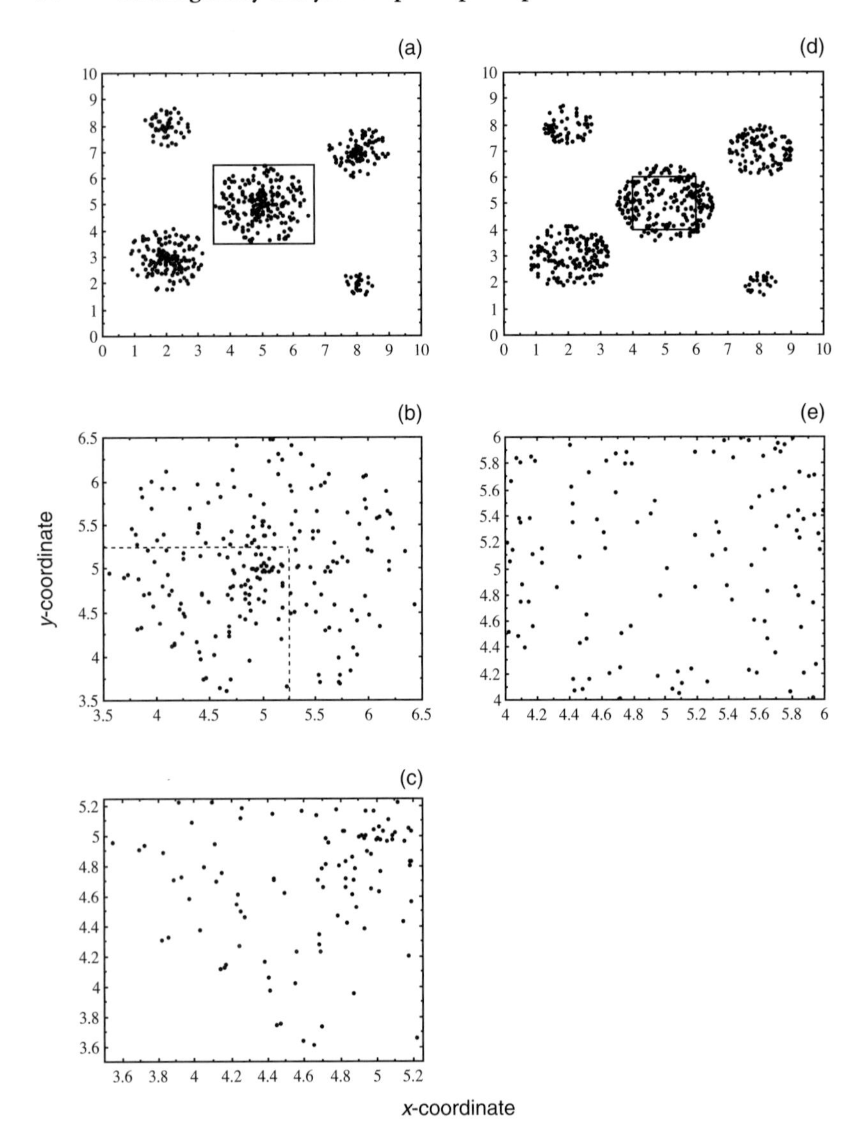

Fig. 3.2. Partial realizations ($n = 1120$) of two 2-D spatial Poisson cluster processes, in which the number of points per cluster is proportional to the area covered by the cluster, with (a) concentric vs. (d) uniform spatial distribution of points within clusters. The largest cluster in (a) is completely displayed in (b) and its bottom left part is amplified in (c), whereas a zoom of the interior of the largest cluster in (d) is shown in (e).

of points within clusters is uniform in the other [Fig. 3.2(d)]. At this extent (i.e., the largest in Fig. 3.2), aggregation is obvious in both point patterns. When the extent is reduced to the interior of the largest cluster in the latter point pattern, the heterogeneity of aggregation is replaced by the homogeneity of complete randomness [Fig. 3.2(e)]. When the focus is on the largest cluster in the former point pattern, a trend of decreasing point density from the center appears [Fig. 3.2(b)]. Furthermore, the bottom-left quarter of the same cluster can be thought of as a partial realization of an inhomogeneous Poisson process characterized by a bottom left-to-top right trend of increasing point density [Fig. 3.2(c)]. As for a biological example, imagine that a forest ecologist has surveyed a tree species specialized in a mountain habitat. In such a situation, Fig. 3.2(a)–(c) could represent survey results for a number of hills, only one of these hills, and a side of this hill, respectively. In many situations, heterogeneity is thus scale-dependent. Therefore, scale is an absolutely necessary component of any heterogeneity analysis of point patterns.

Key note: Scale effects can take multiple forms; they are not specific to point patterns and concern the heterogeneity analysis of surface patterns as well (see Chapters 6 to 8).

3.2.3 The count-based approach

As shown in Section A3.2, the simple Poisson process is characterized, in space as well as in time and space-time, by the equality between the population variance and the population mean of quadrat counts, that is, $\text{Var}[N(A)] = \text{E}[N(A)]$. Naturally, the sample variance S^2 and the sample mean $\bar{X}$ to be calculated from the observed numbers of points in quadrats were used to define the variance-to-mean ratio $\frac{S^2}{\bar{X}}$ as a measure of departure from the homogeneity of complete randomness. Note that the property of ergodicity (Section 2.3) is assumed in doing this. Upton and Fingleton (1985, p. 29) refer to Clapham (1936) as the first published application of the variance-to-mean ratio to the analysis of spatial point patterns. Although the definition of this ratio makes some sense, its use has motivated some questioning in the literature (Hurlbert, 1990). Indeed, the simple Poisson process implies a variance-to-mean ratio value of 1, exactly in theory and approximately in practice (because the sample statistics S^2 and $\bar{X}$ are random), but the reciprocal is not true. A point pattern that produces a variance-to-mean ratio value of 1 is not, therefore, a partial realization of a simple Poisson process. Hurlbert

(1990) gives an example with the spatial distribution of so-called "montane unicorn" populations that show different patterns of aggregation over 10 000 km^2, but all yield a variance-to-mean ratio of 1 when a grid of 1-km^2 quadrats is used. A different example is given below, but some of Hurlbert's statements are first revisited and updated hereafter.

In Hurlbert (1990), the author argues that the ratio $\frac{S^2}{\bar{X}}$ is not useful for the analysis of his aggregated point patterns because of its systematic value of 1, and he is probably correct in making this statement. This kind of surprising result stresses the importance of the third type of heterogeneity, that due to autocorrelation, and the need to assess it for point patterns as well as for surface patterns. As a complement to the sample mean and variance of quadrat counts, the classical Pearson's r statistic calculated between the interior quadrat counts (e.g., on an 8×8 grid within a 10×10 grid) and the average of counts in the four neighboring quadrats in 2-D space provides a coarse but very useful measure of autocorrelation of quadrat counts. In fact, counts made in non-overlapping quadrats should not be correlated for a completely random point pattern (Section A3.1). By contrast, positive autocorrelation is expected for aggregated point patterns; empty quadrats should then be close to one another and quadrats covering the same cluster should all show high counts. Accordingly, the value of the statistic r introduced above is 0.086 and is not statistically different from 0 for the completely random point pattern in Fig. 3.1(a), but is 0.554 and is statistically different from 0 for the aggregated point pattern in Fig. 3.1(c). This statistic is used again in the next example.

Six 2-D spatial point patterns are displayed on 8×8 grids of square quadrats in Fig. 3.3. In each pattern, there are 32 points located in 32 different quadrats, leaving 32 empty quadrats. It follows that all six point patterns are characterized by a variance-to-mean ratio value of about 0.5 (exactly 32/63), whereas one pattern is completely random [Fig. 3.3(a)], another is regular [Fig. 3.3(b)], and the others are trendy [Fig. 3.3(c)] and aggregated [Fig. 3.3(d)–(f)]. This seems to contradict everything one reads in the literature about the variance-to-mean ratio and the related guidelines in terms of the characterization of point patterns, as well as the material presented so far in this chapter. Actually, the explanation is simple. Despite clear differences among the point patterns, the identical values of the variance-to-mean ratio result from the fact that each quadrat contains either zero or one point. In real-life situations, this constraint may apply to a regular point pattern, but not to a completely random point pattern in which some quadrats are expected to have

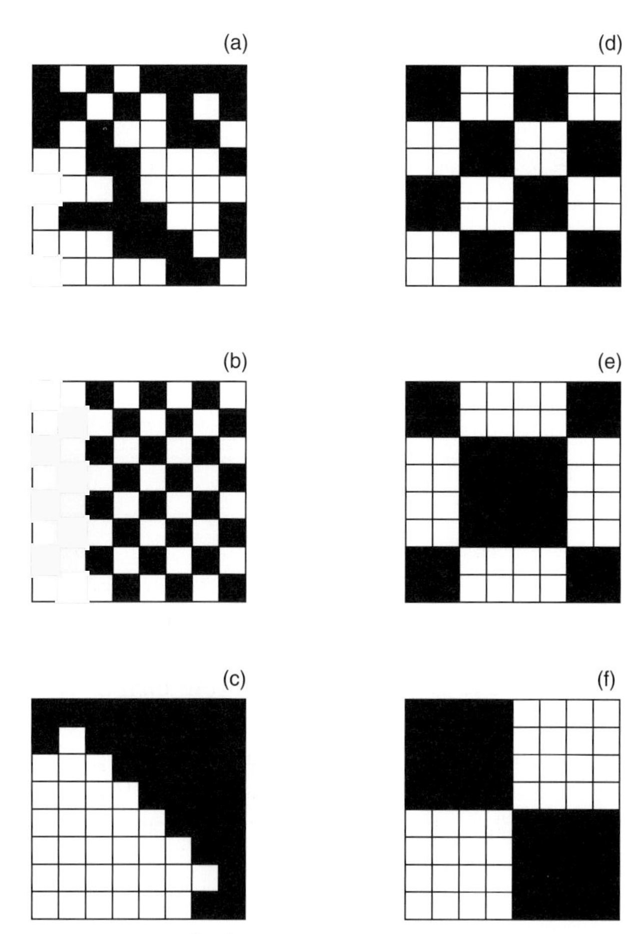

Fig. 3.3. Six artificial 2-D spatial point patterns on an 8 × 8 grid of unit quadrats: (a) completely random; (b) regular; (c) a bottom-left-to-top-right gradient; and (d)–(f) aggregated, with eight entire 2 × 2 clusters in (d), one entire 4 × 4 cluster and quarters of four 4 × 4 clusters in (e), and two entire 4 × 4 clusters in (f). There is one point in a filled unit quadrat, and zero in an empty unit quadrat.

more than one point; the constraint applies even less to trendy and aggregated point patterns in which some quadrats should show an even higher concentration of points (Subsection 3.2.1). Thus, a variance-to-mean ratio of 0.5 should be reserved for a regular point pattern in real-life situations. Very interestingly, the values of Pearson's r statistic introduced above differ among point patterns and thus provide a basis for a relevant classification: (a) 0.039 for the completely random pattern (no correlation with counts in neighboring quadrats); (b) −0.742 for the

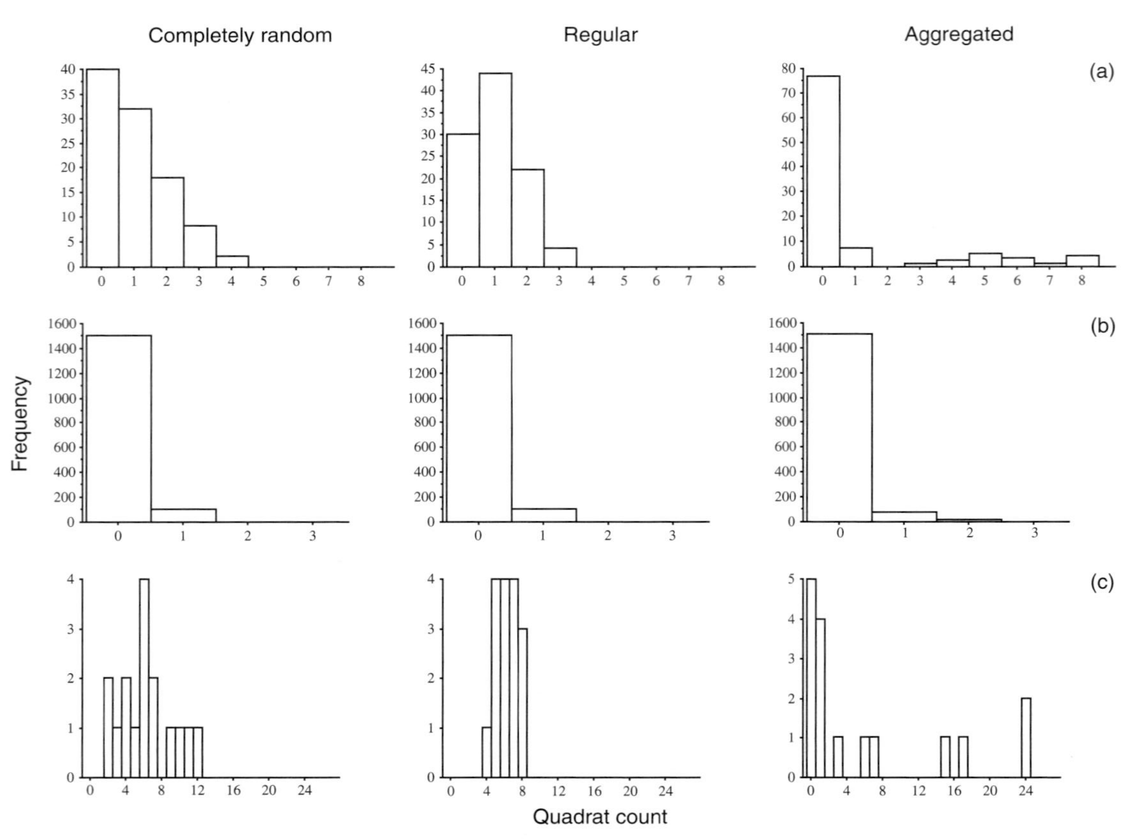

Fig. 3.4. Observed frequency distributions of quadrat counts for the three 2-D spatial point patterns mapped in Fig. 3.1, using quadrat sizes of (a) 1×1, (b) 0.25×0.25, and (c) 2.5×2.5.

regular pattern (negative correlation); and (c)–(f) 0.814, 0.468, 0.775, and 0.884 for the trendy and aggregated patterns (positive correlation) in Fig. 3.3. Alternatively, the use of larger quadrats (e.g., 2 × 2) allows the ratio $\frac{S^2}{\bar{X}}$ to distinguish the point patterns, with values of 0.8 for completely random, 0 for regular, and 3.467 and 4.067 for trendy and aggregated.

Summary: By definition (Section A3.1), quadrat counts are directly related to the first-order and second-order intensity functions. For the reasons given above, the variance-to-mean ratio, which is based on quadrat counts, must be used with due care, and should preferably be evaluated for different quadrat sizes in a multi-scale analysis within the same extent. The sign and value of the correlation between quadrat counts in close neighborhood provide a relevant classification of point patterns.

Guidelines for the characterization of spatial point patterns based on the variance-to-mean ratio are given below. These guidelines are applied to the simulated data mapped in Fig. 3.1. Effects of measurement scale and quadrat size are discussed on the same three 2-D spatial point patterns.

Normally, a completely random point pattern [Fig. 3.1(a)] is expected to provide a value close to 1 for the ratio $\frac{S^2}{\bar{X}}$; a regular point pattern [Fig. 3.1(b)], a value smaller than 1; and an aggregated point pattern [Fig. 3.1(c)], a value greater than 1. Contrary to $\mathrm{Var}[N(A)]/\mathrm{E}[N(A)]$ which is a ratio of two population parameters, $\frac{S^2}{\bar{X}}$ is the ratio of two sample statistics and, hence, a random variable. That is why "close to 1" is used for the completely random point pattern. Similarly, it is difficult to be more specific than "smaller than" and "greater than" for the two other point patterns, the importance of the departure from 1 being dependent on the scale at which heterogeneity is analyzed and the number of points observed.

The use of the 1 × 1 quadrat size yields three grids of 10 × 10 unit quadrats in Fig. 3.1. Given the number of simulated points (100), such a quadrat size is reasonable because it corresponds to a mean number of one point per quadrat. The frequency distribution of counts then decreases linearly from 0 to 4 points per quadrat for the completely random point pattern, presents a rough bell-shape over {0, 1, 2, 3} for the regular point pattern, and shows about 75% of empty quadrats and low but strictly positive frequencies at 1 and 3 to 8 points per quadrat for the aggregated point pattern [Fig. 3.4(a)]. At the 0.25 × 0.25 quadrat size, differences in quadrat counts are much smaller among point patterns,

Table 3.1. *Variance-to-mean ratio calculated from quadrat counts, in relation to quadrat size, for the three 2-D spatial point patterns of Fig. 3.1*

Quadrat size	Completely random	Regular	Aggregated
0.25 × 0.25	0.998	0.938	1.298
0.5 × 0.5	0.992	0.812	2.296
1 × 1	1.091	0.687	4.909
2 × 2	1.625	0.437	9.729
2.5 × 2.5	1.461	0.245	12.149

with about 1500 empty quadrats in all three cases; the observed values of quadrat counts are {0, 1, 2} for completely random, {0, 1} for regular, and {0, 1, 2, 3} for aggregated [Fig. 3.4(b)]. Note that the following frequencies are too small to be visible on the graph, but there were three quadrats with 2 points for the completely random point pattern and two with 3 points for the aggregated point pattern, and the 100 simulated points were located in 100 different 0.25 × 0.25 quadrats for the regular point pattern. At the other extreme, the use of the 2.5 × 2.5 quadrat size produces very large differences in quadrat counts among point patterns; the range of observed values of quadrat counts is narrow for regular (from 4 to 8 points per quadrat), wider for completely random (from 2 to 12, excluding 8), and even wider but unevenly represented, with still a good number of empty quadrats, for aggregated [Fig. 3.4(c)]. In comparison with the 1 × 1 quadrat size, there is no linear decrease of the frequency distribution of quadrat counts for the completely random point pattern but the frequency distribution is roughly bell-shaped again for the regular point pattern at the 2.5 × 2.5 quadrat size.

The visual observations above (for three quadrat sizes) are confirmed by the numerical results in Table 3.1 (for five quadrat sizes). Values of $\frac{S^2}{\bar{X}}$ smaller than 1 are observed at all quadrat sizes for the regular point pattern. Results for the completely random point pattern are in agreement with the guidelines at the 0.25 × 0.25, 0.5 × 0.5 and 1 × 1 quadrat sizes, with observed values of $\frac{S^2}{\bar{X}}$ close to 1 (between 0.9 and 1.1). Things are different at the 2 × 2 and 2.5 × 2.5 quadrat sizes, for which the observed values of the variance-to-mean ratio are clearly greater than 1 (1.625 and 1.461) due to the presence of quadrats with 10 to 12 points. The guidelines work without exception for the aggregated point pattern, with observed values of the ratio unequivocally greater than 1. If the quadrat size was greater than the cluster size, the variance-to-mean ratio would drop dramatically, and clusters would run undetected at that quadrat size (5 × 5 here). To prevent this, a practical recommendation is to work with quadrats small enough to be pooled if necessary. It is too late and useless to split large quadrats when counts are already made . . .

In closing, note that tests for the distribution of quadrat counts or of counts in general exist. Such a test is presented for the beta-binomial distribution in Section 3.4.

3.2.4 The distance-based approach

When looking at two points, as in the visual inspection of Fig. 3.1, does one first think about the distance between them or whether they are in the same quadrat? The answer is likely to be the former because, at first approximation, our eyes naturally act as a ruler, and to calculate the exact distance between two points, all that one needs is their coordinates. To know whether two points are in the same quadrat, one needs to know the spatial boundaries of the quadrat in addition to the location of the points, whether exact or approximate. In that sense, working with distances is more immediate than working with quadrat counts, for which the links with the first-order and second-order intensity functions are more direct. We will see that the distance-based approach allows a multi-scale type of analysis different from that based on quadrat counts.

Distances can be calculated and analyzed in many ways. All the distances that we will work with, in space as well as in time, will be Euclidean. The notation introduced in Chapter 2 for the Euclidean distance between two locations identified by indices s and s' is $\|s' - s\|$. In 2-D space, it is calculated as $\sqrt{(x' - x)^2 + (y' - y)^2}$, where (x, y) and (x', y') are the coordinate vectors of the two indices. Distances between all pairs of points (i.e., plants in the *Arisaema* example, epicenters in the example of California earthquakes) can be calculated that way, assuming the points were previously mapped. One may choose to incorporate all the distances in the analysis (option 1) or to work only with the nearest-neighbor distances (option 2). Assuming that a given number of locations has been sampled in the plot and the nearest point has been identified for each of them, a third option is to analyze the distances between sampling location and nearest point. Therefore, a sampling design, including a number of sampling locations and their positioning, is required. Systematic sampling designs and simple random sampling designs with five numbers of sampling locations equal to the numbers of quadrats in Table 3.1 are considered below. Note that "sampling location" and "point" here are called "point" and "plant" in Upton and Fingleton (1985), because the points of interest of the authors were plants! In general terms, each observed point is assumed to have been generated by a point process and be part of one of its partial realizations, whereas the number and positioning of sampling locations are defined by the experimenter.

The simulated 2-D spatial point patterns of Fig. 3.1 are used again for comparison and discussion purposes. By working with all the distances between pairs of points [i.e., $\frac{n(n-1)}{2}$ distances when there are n points], the differences between a completely random point pattern and a regular point pattern are small: the frequencies of small and intermediate distances are very slightly higher for the completely random point pattern and, as a result, those of large distances are slightly higher for the regular point pattern [Fig. 3.5(a)]. The frequency distribution of all the distances between pairs of points for the aggregated point pattern is distinct from those for the two other point patterns, with one peak over the interval [0, 2], to which belong the distances between points of the same cluster, and another peak centered around 5–6, which corresponds to the average distance between clusters. Thus, this first option in the distance-based approach to the heterogeneity analysis of point patterns does not appear worth pursuing because of the inherent risk of confounding two of the three basic point patterns. (Strictly speaking, the frequency plots in Fig. 3.5 are histograms because of the continuous quantitative nature of distances, whereas the frequency plots in Fig. 3.4 are bar charts because of the discrete quantitative nature of quadrat counts.)

When working with the frequency distributions of nearest-neighbor distances, things are different because one is then investigating the left tail of the frequency plots in Fig. 3.5(a), and in so doing the chances are better to distinguish the completely random and regular point patterns. In fact, frequencies are evenly distributed over the first three classes of nearest-neighbor distances for the completely random point pattern, whereas the frequency distribution is bell-shaped, the minimum distance between points (0.3) is apparent and the range of nearest-neighbor distance values is narrower for the regular point pattern [Fig. 3.5(b)]. As for the aggregated point pattern, the frequency distribution of nearest-neighbor distances is dominated by the first class, which corresponds, with the second class, to the points within clusters, the background points (i.e., the few points outside of the clusters) providing the positive frequencies at the other classes. Definitively, the analysis of nearest-neighbor distances is worth investigating further.

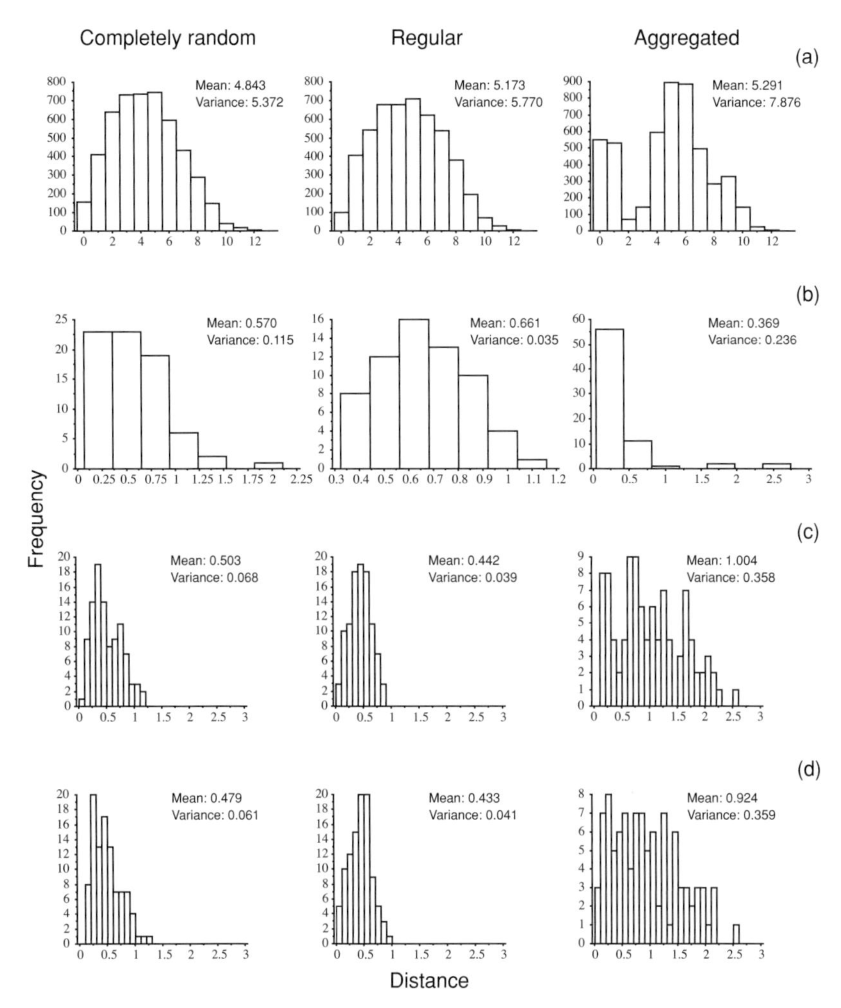

Fig. 3.5. Observed frequency distributions of (a) all the distances between pairs of points; (b) only the nearest-neighbor distances; and the distances between sampling location and nearest point in the case of (c) a systematic sampling with a 10 × 10 grid and (d) a simple random sampling with 100 locations, for the three 2-D spatial point patterns of Fig. 3.1. "Mean" and "Variance" refer to the sample mean and the sample variance, respectively.

The calculation and analysis of the distances between sampling location and nearest point were likely thought to overcome the mapping of all the points in a plot before the advent of spatial positioning technologies. Below, we discuss the question of systematic vs. simple random sampling design, leaving the question of the number of sampling locations until the testing of hypotheses is considered. Two different sets of 100 sampling locations were used to produce Fig. 3.5(c) and (d). The sampling locations were the centers of squares of unit side laid on a 10×10 regular grid for the systematic sampling in (c), whereas the distribution of sampling locations in the plot was completely random for (d). In both cases, the three basic point patterns can be distinguished from the frequency distribution of distances between sampling location and nearest point: it is skewed to the right and bimodal with a main peak and a secondary peak for completely random; it is bell-shaped over a narrow range of distance values for regular; and frequencies are lower and more evenly distributed and the range of distance values is wide for aggregated. Thus, this option in the heterogeneity analysis of point patterns is worth pursuing, and could be implemented in situations where it is not possible to map all the points in the plot. Overall, the systematic and simple random sampling designs provide similar frequency distributions of the distances between sampling location and nearest point, and their range, mean and variance differ more among patterns than between sampling designs [Fig. 3.5(c) and (d)]. However, the completely random distribution of sampling locations through the plot introduces one more random component in the analysis. This may explain why the two peaks for the completely random point pattern are more apparent in Fig. 3.5(c) (systematic sampling) than in Fig. 3.5(d) (simple random sampling), and why the bell-shaped frequency distribution for the regular point pattern is more symmetrical in the former figure. In practice, the simple random sampling design is also more difficult to implement than the systematic sampling design, even with modern techniques of spatial positioning.

Summary: Among the three types of distances considered above, two provide a good basis for heterogeneity analysis of point patterns because they allow one to distinguish the homogeneity of complete randomness from the heterogeneities of regularity and aggregation. These are the nearest-neighbor distances and the distances between sampling location and nearest point for a given sampling design.

Key note: A fourth type of distance indirectly related to the first-order intensity function will be considered in the discussion of point pattern modeling aspects (see Ripley's functions in Subsections 5.1.2 and 5.1.4).

In the analysis of point processes and outside of it (see the analysis of distance matrices in Chapters 7 and 8), tests based on distances can be classified in two categories: (1) the parametric tests, which are usually valid asymptotically (i.e., when the sample size is large) because of the application of the Central Limit Theorem, and (2) the randomization and permutational tests, which can be used with smaller sample sizes. Below, we outline a popular test that belongs to the first category. Thereafter the randomization testing procedure designed by Peter J. Diggle for the analysis of spatial point patterns is presented with nearest-neighbor distances first and then with distances between sampling location and nearest point; randomization arises from the simulation of independent partial realizations under the model of point process against which the observed point pattern is tested (Diggle, 2003). Variants and extensions apply to marked spatial point patterns (Section 3.3) and bivariate spatial point patterns (Section 3.5) and to temporal and spatio-temporal point patterns (see Chapter 4); SAS codes are available on the CD-ROM. The parametric test due to Clark and Evans (1954) is discussed in a preliminary step.

The Clark–Evans test is performed on nearest-neighbor distances. It is valid asymptotically (i.e., when the number of points is large) because it involves the Central Limit Theorem. Matérn (1972) and Donnelly (1978) modified it to account for edge effects (i.e., when the actual nearest neighbor is beyond the limits of the plot). In 2-D space, edge effects are more important when the mean number of points per area unit is smaller and there are more peripheral points, the perimeter and the area of the plot being equal.

Key note: The bias [resulting from edge effects] would be problematic if one was directly concerned with estimation (Diggle, 2003, p. 60; see Section 5.1 here).

Following Donnelly's (1978) modification of the Clark–Evans test, the mean value of nearest-neighbor distances can be regarded as an observation from a normal distribution with population mean

$$0.5\sqrt{\frac{PA}{n}} + 0.051\frac{PP}{n} + 0.042\frac{PP}{\sqrt{n^3}} \tag{3.2}$$

and population variance

$$0.070\frac{PA}{n^2} + 0.037\sqrt{\frac{PP^2 PA}{n^5}}, \tag{3.3}$$

if the point pattern is completely random and the number of points is sufficiently large (e.g., $n > 50$); PA above denotes the plot area, and

PP the plot perimeter. The standardized mean value of nearest-neighbor distances is 1.69 for the completely random point pattern, 4.81 for the regular point pattern, and –5.20 for the aggregated point pattern of Fig. 3.1. In absolute value, the second and third quantities are greater than the 0.975-quantile of the standard normal distribution Z (1.96). Thus, the regular and aggregated point patterns are declared significantly ($P < 0.05$) different from a completely random point pattern of 100 points on a plot with same area (100) and perimeter (40). This is good, of course, but a number of questions then arise. For example, where are the departures? Are they over the whole range of nearest-neighbor distance values or for some classes in particular? The first elements of a response are found by looking at the histograms of nearest-neighbor distances in Fig. 3.5(b). The regular point pattern is characterized by fewer small nearest-neighbor distances than the completely random point pattern, while the aggregated point pattern is characterized by more.

The results above motivate another question. Would it not be possible to make the differences more apparent in a graph that would cover the whole range of nearest-neighbor distance values? In the spatial framework, Peter J. Diggle answered this question by combining the cumulative relative frequency distribution of nearest-neighbor distances of the observed point pattern, with lower and upper envelopes obtained by simulating 99 independent partial realizations of a completely random process with same number of points on a plot with same area and perimeter.

Key note 1: Such a randomization procedure differs from a permutational procedure in which the same data are used at each permutation while the subscripts identifying them are randomly arranged.

Key note 2: The lower and upper envelopes mentioned above may be thought of as the lower and upper bounds of an approximate 99% acceptance region at a given cumulative relative frequency. An approximate 95% acceptance region may be obtained by simulating 999 independent partial realizations and calculating the empirical 2.5th and 97.5th percentiles (instead of the minimum and maximum in the original procedure) for each cumulative relative frequency.

To perform Diggle's basic randomization testing procedure with n points, one needs:

(i) to calculate the n nearest-neighbor distances of the observed point pattern and rank them in ascending order;

(ii) to plot the ith value of the cumulative relative frequency distribution $(\frac{i}{n})$ (vertical axis) against the nearest-neighbor distance ranked ith (horizontal axis) for $i = 1, \ldots, n$;

(iii) to calculate the corresponding values of the lower and upper envelopes from 999 independent partial realizations of a completely random point process (see *Key note 2* above);

(iv) to plot the ith value of the cumulative relative frequency distribution $(\frac{i}{n})$ (vertical axis) against the ith value of the lower and upper envelopes (horizontal axis) for $i = 1, \ldots, n$, and superimpose the result to the plot made in (ii).

When the cumulative relative frequency distribution of nearest-neighbor distances of an observed point pattern happens to be outside the envelopes, it indicates a departure from the model of point process for which 999 independent partial realizations were simulated. This departure can be located by reading the corresponding nearest-neighbor distances on the horizontal axis. For example, the cumulative relative frequency distributions of nearest-neighbor distances of the regular and aggregated point patterns of Fig. 3.1 are outside the envelopes over the first distances in Fig. 3.6(a). The former is below the envelopes, indicating a lack of small nearest-neighbor distances, while the latter is above them, indicating an excess of small nearest-neighbor distances, compared with completely random point patterns. There is no such departure for the completely random point pattern [see left panel in Fig. 3.6(a)].

Diggle's randomization testing procedure readily applies to distances between sampling location and nearest point for a given sampling design. The observed cumulative relative frequency distribution is then obtained by ranking the distances in ascending order and by associating the distance ranked ith with the value of the ith cumulative relative frequency, which this time is equal to i divided by the number of sampling locations. Using the same sampling design (with the same number of sampling locations, but not the same locations in the case of random sampling), the values of the lower and upper envelopes associated with the ith cumulative relative frequency are calculated from 999 independent partial realizations of the hypothetical model of point process, as the wanted empirical percentiles (e.g., the empirical 2.5th and 97.5th percentiles) of the distances between sampling location and nearest point ranked ith. This procedure was used to analyze the three 2-D spatial point patterns of Fig. 3.1, which were considered the observed patterns while complete randomness was the hypothetical model. Four regular grid sizes (i.e., 5×5, 10×10, 20×20, and 40×40) were tried in as many systematic sampling designs.

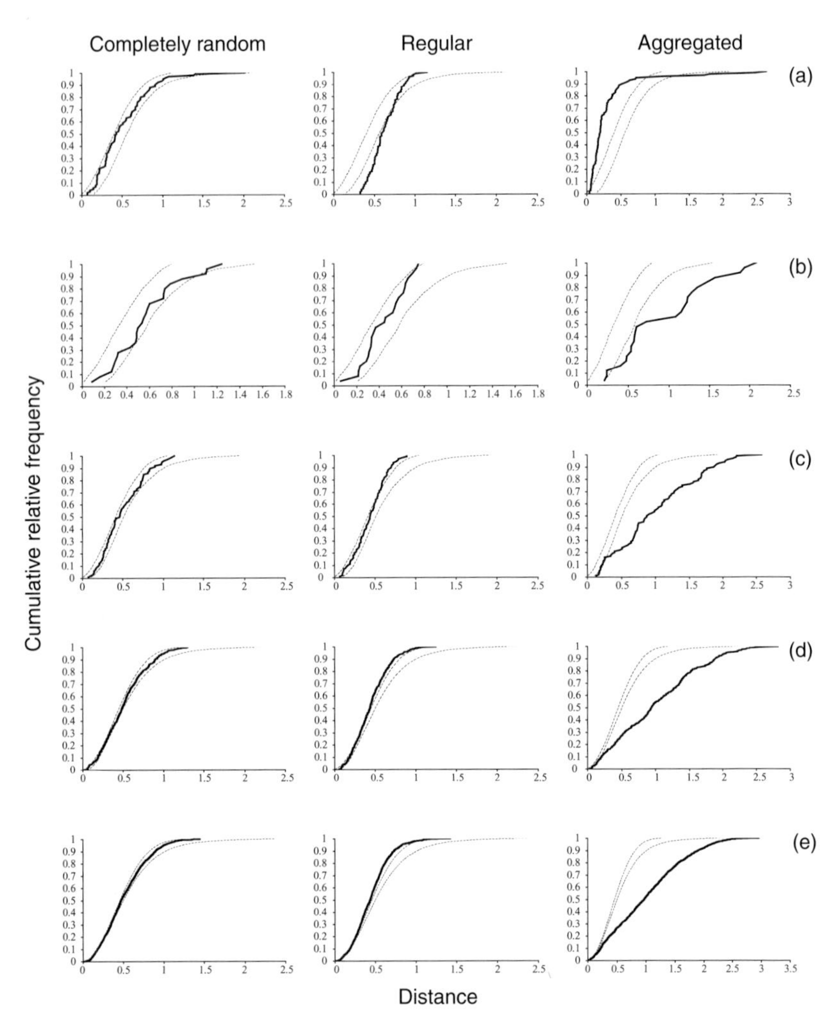

Fig. 3.6. Cumulative relative frequency distributions of (a) nearest-neighbor distances and (b)–(e) distances between sampling location and nearest point in the case of systematic sampling with (b) 5 × 5, (c) 10 × 10, (d) 20 × 20, and (e) 40 × 40 grids, for the three 2-D spatial point patterns of Fig. 3.1. The observed cumulative relative frequency distributions are shown as bold solid curves. Dashed curves represent the 2.5th and 97.5th percentile envelopes evaluated from 999 independent partial realizations of a completely random point process on the [0, 10] × [0, 10] sampling domain. The number n of points generated in each partial realization is the same as in the observed point patterns.

In view of the results in Fig. 3.6(b)–(e), the width of envelopes decreases when the number of sampling locations increases, and departures from complete randomness for the regular and aggregated point patterns, based on distances between sampling location and nearest point, are different from what they are with nearest-neighbor distances [Fig. 3.6(a)]. The regular point pattern is characterized by an excess of intermediate distances between sampling location and nearest point and a lack of large distances, and the aggregated point pattern by a lack of small and intermediate distances and an excess of large distances. Departure from complete randomness for the aggregated point pattern is very important at all grid sizes and covers the whole range of distance values for the 40×40 grid. The number of sampling locations strongly influences the position of the observed curve relative to the envelopes for the regular point pattern: the observed curve of the regular point pattern is within the envelopes for the 4×4 and 5×5 grids, begins to be above the envelopes at large distances for the 10×10 grid, and shows an increasing departure for the 20×20 and 40×40 grids. As expected, there is no noticeable departure for the completely random point pattern for any grid size.

In Diggle's (2003) landmark book on the statistical analysis of spatial point patterns (i.e., first edition in 1983), the completely random or simple Poisson process is used as the hypothetical model, or null hypothesis, against which the observed point pattern is tested in a preliminary stage. Below, we will see that Diggle's randomization testing procedure is general and flexible enough to allow the assessment of hypotheses other than complete randomness, which is standard but biologically uninformative. Note that no model is fitted in what follows. The example is inspired from the partial realizations of 2-D spatial Poisson cluster processes in Fig. 3.2(a) and (d). Both aggregated point patterns have the same number of clusters, numbers of points per cluster, and centers and areas of clusters. They differ in the distribution of points within clusters, which is concentric in Fig. 3.2(a) and uniform in Fig. 3.2(d). This difference motivates the question: how could one test either of the two observed point patterns against either of the two hypothetical models of point process? To be complete, such an assessment would require a total of four tests about the distribution of points within clusters: concentric (observed pattern) against concentric (hypothetical model), concentric against uniform, uniform against uniform, and uniform against concentric. Therefore, 999 independent partial realizations of a Poisson cluster process [like the one of Fig. 3.2(a), or (d)] were simulated under the

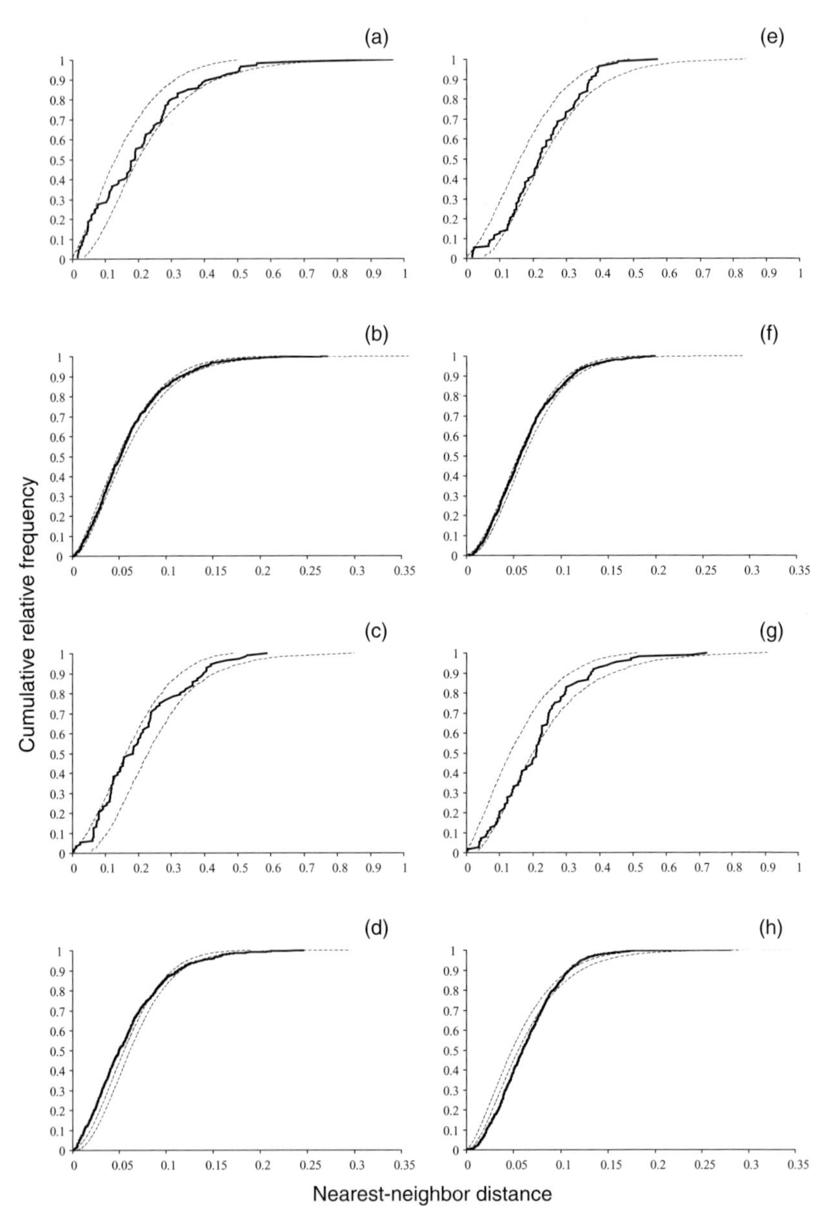

Fig. 3.7. Observed cumulative relative frequency distributions (bold solid curves) of nearest-neighbor distances, with the 2.5th and 97.5th percentile envelopes (dashed curves) evaluated from 999 independent partial realizations of a hypothetical model of point process. The observed 2-D spatial point patterns are aggregated with five clusters of the same area and same center as in Fig. 3.2(a) and (d). The spatial distribution of points within clusters (i.e., concentric or uniform) or the total

hypothetical model, and the aggregated point pattern of Fig. 3.2(a), or (d), was used as the observed point pattern. This was done by dividing the number of observed points by 10 ($n = 112$ instead of 1120) and by working with $n = 1120$, in order to discuss a possible effect on the results. In general and simple terms, the null hypothesis should be accepted in Fig. 3.7(a) and (e) because the observed point pattern is a partial realization of the point process under the hypothetical model. This is mostly the case, although the observed cumulative relative frequency distribution of nearest-neighbor distances approaches or even touches the 95% envelopes in a few places. One interpretation is that 112 points distributed into four clusters are not sufficient to represent the corresponding model faithfully. Because of the larger number of points, envelopes in Fig. 3.7(b) and (f) are narrower and the observed curve does not touch the 95% envelopes where it does it in Fig. 3.7(a) and (e). In general and simple terms again, the null hypothesis should be rejected in Fig. 3.7(c) and (g) because the observed point pattern is not a partial realization of the point process under the hypothetical model. This is not really the case, since the crossing of the envelopes by the observed cumulative relative frequency distributions is just slightly more pronounced than in Fig. 3.7(a) and (e). This time, the number of points (112) is not large enough to make a clear distinction between the observed pattern and the tested model. Finally, an excess of small nearest-neighbor distances in the concentric distribution of points within clusters for the observed pattern, compared with the uniform distribution for the tested model, is clearly observed in Fig. 3.7(d), and the excess of small distances is replaced by a lack in Fig. 3.7(h).

Summary: Diggle's randomization testing procedure allows the assessment of homogeneity and heterogeneity hypotheses on point patterns and can be used with nearest-neighbor distances as well as distances between sampling location and nearest point. The number of observed points on the one hand and the number of sampling locations on the other hand define the number of distances to work with in Diggle's procedure, and therefore have a direct effect on our ability to accept or reject hypotheses on point patterns, especially when these are refined and involve modeling aspects.

Fig. 3.7. (*cont.*) number of points (i.e., $n = 112$ or 1120), or both, change from panel to panel: (a) concentric (observed pattern) against concentric (hypothetical model), $n = 112$; (b) concentric against concentric, $n = 1120$; (c) concentric against uniform, $n = 112$; (d) concentric against uniform, $n = 1120$; (e) uniform against uniform, $n = 112$; (f) uniform against uniform, $n = 1120$; (g) uniform against concentric, $n = 112$; and (h) uniform against concentric, $n = 1120$.

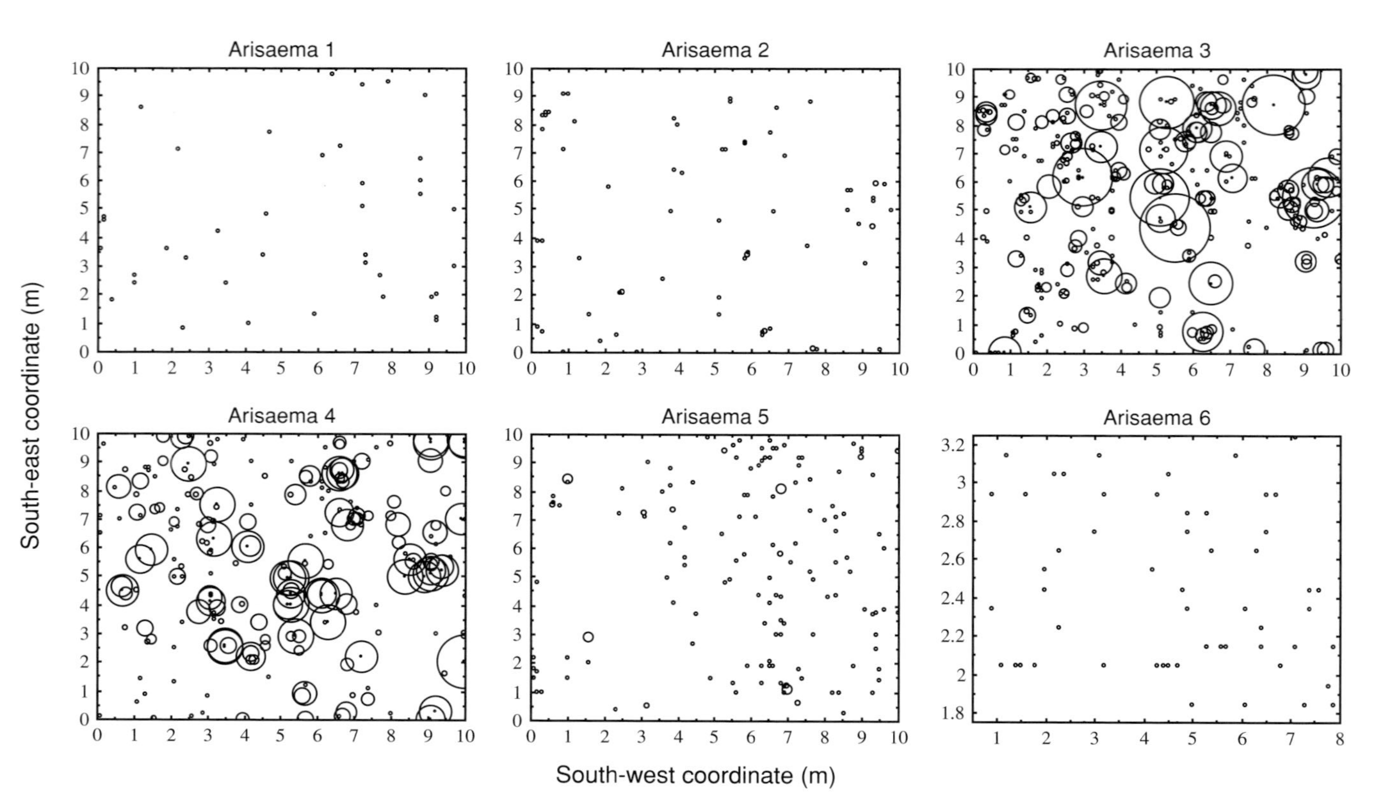

Fig. 3.8. Arisaema example in space: observed 2-D spatial point patterns for six datasets named Arisaema 1 to 6. Note that the size of a circle (i.e., the symbol used to represent a "point" in a map) is directly proportional to the number of plants found in the same dm^2; see text for further explanation.

3.2.5 Examples with real data

Below, the count-based and distance-based approaches are illustrated and compared with real data in two examples. For each dataset, a type of spatial point pattern is identified among the types that were introduced and discussed with simulated data. Results of the data analyses are interpreted in terms of each example.

The Arisaema *example*

Arisaema triphyllum (L.) Schott, commonly known as "Jack-in-the-pulpit," is a perennial forest herb that belongs to the family Araceae (Fig. 1.2). Hereafter, the plant will be called simply "*Arisaema*," and the datasets will be given the same name but unitalicized and numbered. The data used in this section and in Chapters 4 and 5 were collected in the understory of an old-growth deciduous forest at the Gault Nature Reserve on Mont-Saint-Hilaire, 32 km southeast of Montréal (Québec, Canada). In the spring and summer of 1988, 1989, 1990, and 1991, a census of *Arisaema* was conducted in 11 10 m × 10 m plots. The spatial location of each plant identified as an *Arisaema* was recorded at the 1 dm × 1 dm scale. When several individuals were found in the same 1-dm^2 square, they were all counted, but their spatial location was considered the same, namely the center of the 1-dm^2 square. A number of characteristics (e.g., stage of development: juvenile or adult; gender: male, female, or undetermined; health status: healthy or infected by a rust fungus; mortality: dead or alive) were determined for each individual.

The codes Arisaema 1–6 used for the datasets here correspond to the following censuses, with n, the number of points or identified *Arisaema* plants: Plot 4, summer 1988 (Arisaema 1, n = 39); Plot 3, spring 1991, plants alive other than juveniles (Arisaema 2, n = 70); Plot 3, spring 1991, juveniles (Arisaema 3, n = 924); Plot 3, spring 1988, juveniles (Arisaema 4, n = 863); Plot 9, summer 1989 (Arisaema 5, n = 155); and a part of Plot 5, summer 1988 (Arisaema 6, n = 56). In Fig. 3.8, our eyes do not recognize any particular 2-D spatial point pattern for Arisaema 1, and might find some trace of regularity in Arisaema 6. By comparison, the spatial distribution of juveniles in Arisaema 3 and Arisaema 4 is clearly aggregated, and aggregation is much less important for Arisaema 2, where the mapped population is composed of adults. The 2-D spatial point pattern of Arisaema 5 is closer to partial realizations of a heterogeneous Poisson process [Fig. 2.2(a)–(c)], with a non-monotonic trend in point density from left to right in this case.

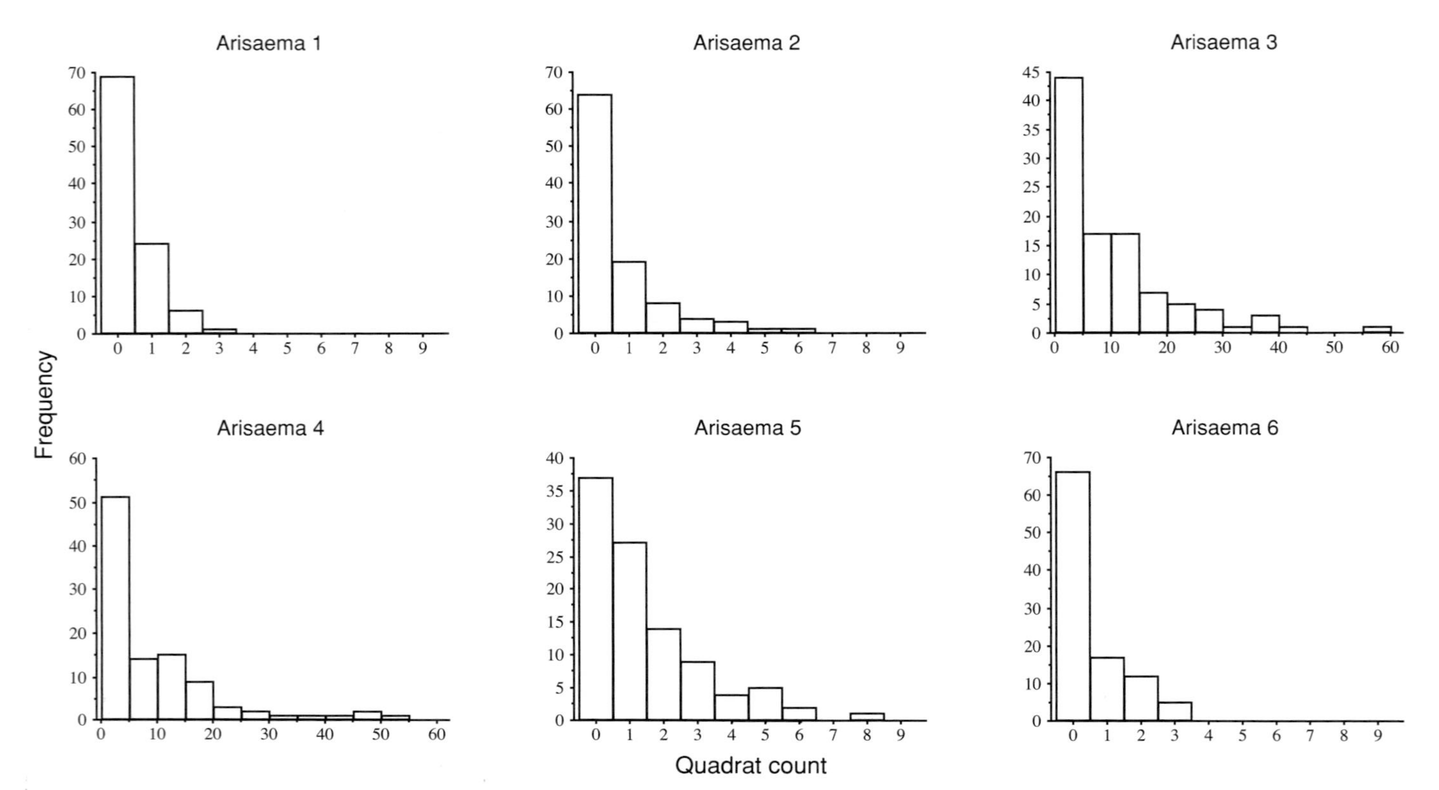

Fig. 3.9. Arisaema example in space: observed frequency distributions of quadrat counts for the six 2-D spatial point patterns of Fig. 3.8. Plants were counted in quadrats of 1 m × 1 m.

Quadrats of 1 m × 1 m were used to produce the frequency distributions of quadrat counts presented in Fig. 3.9, so that a frequency of 1 represents 1% of the quadrats (except for Arisaema 6) because the complete plots are 10 m × 10 m, and count values were pooled into classes of 5 for Arisaema 3 and Arisaema 4. At first sight, all the frequency distributions look the same. This impression comes from the very large numbers of quadrats containing five individuals or less in Arisaema 3 and Arisaema 4 and the numerous empty quadrats in the other plots. By looking at the graphs more closely, the range of count data in Arisaema 3 and Arisaema 4 (from 0 to 55 or 60) is different from what it is in the other datasets (from 0 to 3, 6, or 8). This confirms the strong aggregation that our eyes perceived for Arisaema 3 and Arisaema 4 in Fig. 3.8. In fact, about 10% of the quadrats in Arisaema 3 and 5% of the quadrats in Arisaema 4 contain 25 plants or more, which means a high concentration of individuals in some parts of the two plots. Arisaema 1 and Arisaema 6 have the same range of count values (from 0 to 3), and show small differences in the frequency of each count value. The range of count values is slightly wider for Arisaema 2, with a few quadrats with more than three individuals. As for Arisaema 5, the maximum number of plants found in a quadrat is 8 and frequencies decrease more slowly with increasing count values than for the other datasets. From the comparison with the frequency distributions of quadrat counts for the completely random and aggregated point patterns in Fig. 3.4(a), it follows that the 2-D spatial point patterns of Arisaema 3 and Arisaema 4 can correctly be identified as aggregated, whereas the identification of a completely random point pattern for Arisaema 5 would not be completely justified because there are several quadrats with more than 4 points in the case of Arisaema 5.

To assess possible effects of the quadrat size on the identification of a type of spatial point pattern for the Arisaema 1–6 datasets, the traditional variance-to-mean ratio was evaluated by using different quadrat sizes (Table 3.2). The quadrat sizes chosen for 10 m × 10 m plots correspond to those used with simulated data in 10 × 10 plots (Table 3.1), which facilitates the comparison between the two sets of results. Overall, results are consistent. At all quadrat sizes, variance-to-mean ratio values greater than 1 point to the aggregated spatial point pattern for Arisaema 2–5. For Arisaema 1, the variance-to-mean ratios calculated for four of the five quadrat sizes are close to 1, suggesting that the spatial point pattern is completely random. The most puzzling result was obtained for Arisaema 6, for which only two of the quadrat sizes could

Table 3.2. *Variance-to-mean ratio calculated from quadrat counts, in relation to quadrat size (m^2), for the six 2-D spatial point patterns of Fig. 3.8 in the* Arisaema *example in space*

Quadrat size	Arisaema 1	Arisaema 2	Arisaema 3	Arisaema 4	Arisaema 5	Arisaema 6
0.25 × 0.25	0.976	1.700	7.880	8.464	1.485	1.0529
0.5 × 0.5	1.059	1.858	9.065	11.371	1.843	0.736
1 × 1	1.082	2.121	12.712	14.045	2.559	N/A
2 × 2	0.645	2.827	16.458	10.574	2.473	N/A
2.5 × 2.5	1.092	1.794	16.532	18.786	3.932	N/A

N/A = not applicable because of the size of the plot (i.e., 1.5 × 7.5).

be tried: the variance-to-mean ratio indicates complete randomness for one (0.25 × 25) and regularity for the other (0.5 × 0.5). The Pearson's r statistic used as an autocorrelation measure helps clarify things, with negative (non-significant) values suggesting regularity at both quadrat sizes (0.25 × 0.25: −0.089 and 0.5 × 0.5: –0.106). While the 2-D spatial point patterns of Arisaema 2 and Arisaema 5 are very different, their variance-to-mean ratios are similar at most scales. Again, Pearson's r helps distinguish between the two types of pattern, with smaller positive values indicating weaker aggregation for Arisaema 2 (e.g., 0.5 × 0.5: 0.059) and greater positive values indicating stronger aggregation for Arisaema 5 (e.g., 0.5 × 0.5: 0.146).

We have seen that the regular point pattern is characterized by a lack of small nearest-neighbor distances, and the aggregated point pattern by an excess of such distances, compared to the completely random point pattern [Fig. 3.6(a)], and when distances between sampling location and nearest point are calculated following a systematic sampling, the regular point pattern tends to be characterized by an upward departure from complete randomness at intermediate distances (excess) and large distances (lack), and the aggregated point pattern by a downward departure over most distances [Fig. 3.6(b)–(f)]. In view of Figs. 3.10 and 3.11, where the observed cumulative relative frequency distributions are plotted with envelopes, (i) Arisaema 1 does not show any evidence of departure from complete randomness; (ii) Arisaema 6 presents a slight departure from complete randomness at small distances – this departure might be attributed to competitive interactions that resulted in individuals avoiding growing too close together; (iii) Arisaema 3 and Arisaema 4 are clearly

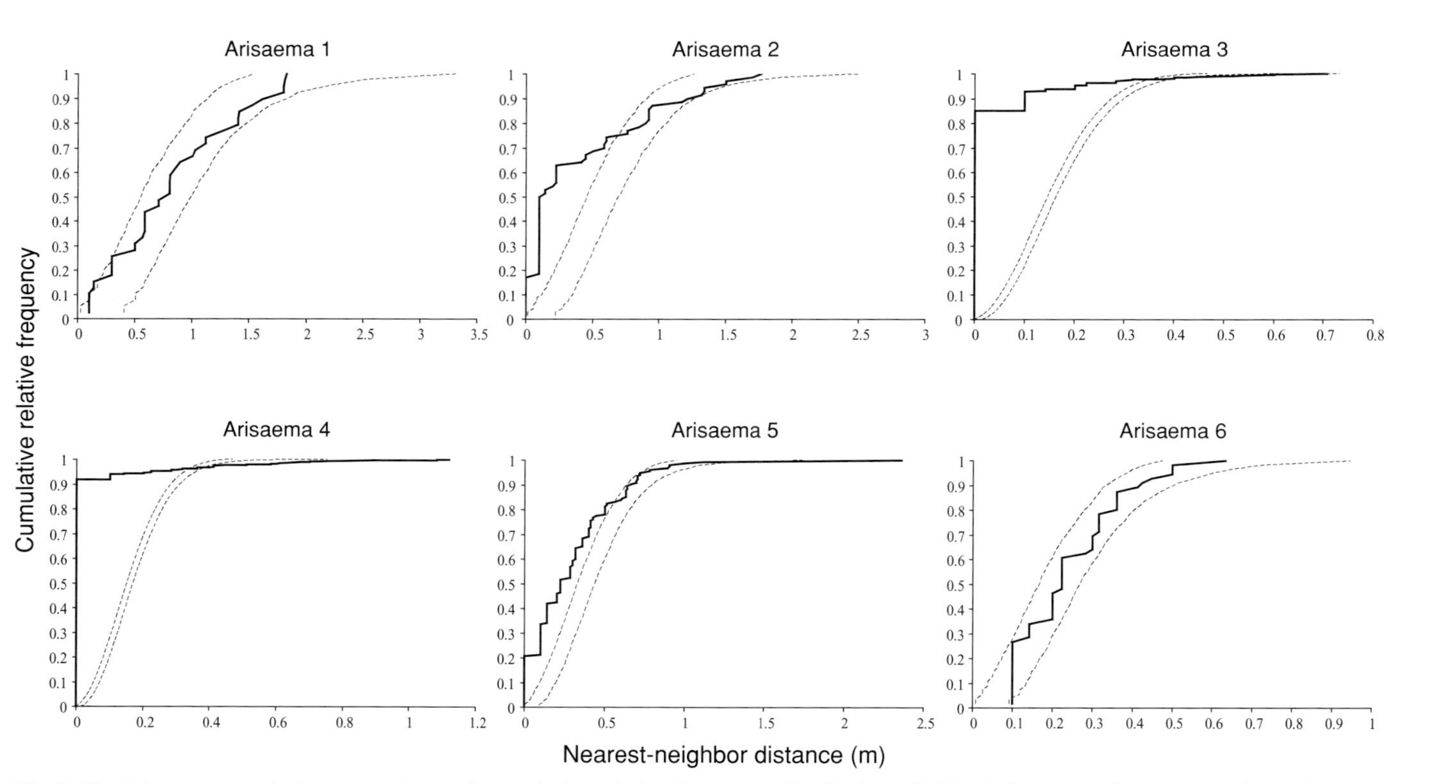

Fig. 3.10. Arisaema example in space: observed cumulative relative frequency distributions (bold solid curves) of nearest-neighbor distances, with the 2.5th and 97.5th percentile envelopes (dashed curves) evaluated from 999 independent partial realizations of a completely random point process, for the six 2-D spatial point patterns of Fig. 3.8.

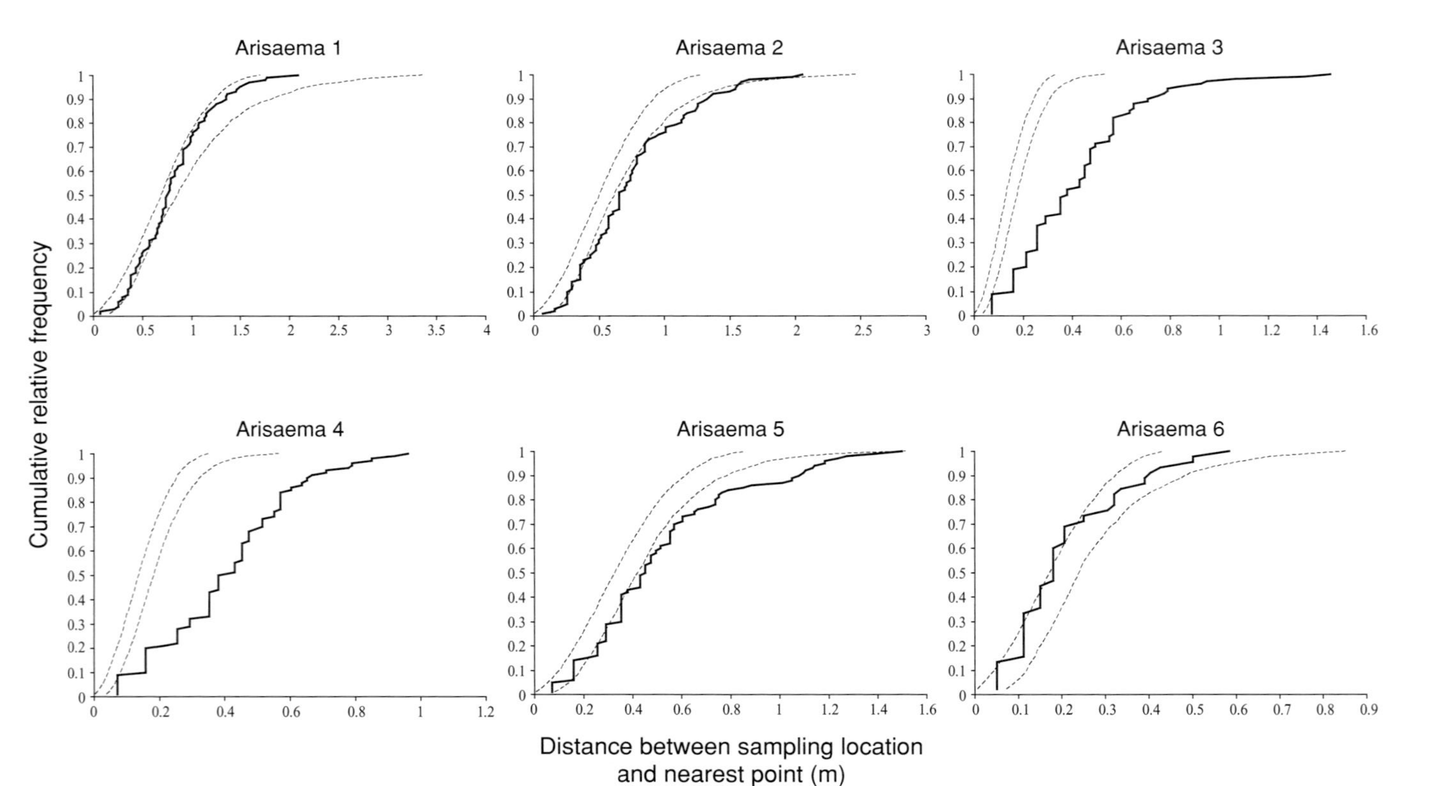

Fig. 3.11. Arisaema example in space: observed cumulative relative frequency distributions (bold solid curves) of distances between sampling location and nearest point, with the 2.5th and 97.5th percentile envelopes (dashed curves) evaluated from 999 independent partial realizations of a completely random point process, for the six 2-D spatial point patterns of Fig. 3.8. The grid used for systematic sampling was 10 × 10 for Arisaema 1–5 and 3 × 15 with adjacent nodes separated by 0.5 m for Arisaema 6.

aggregated, which suggests a patchy distribution of offspring clusters in favorable environments – in particular, 85% of the plants in Arisaema 3 had their nearest neighbor in the same 1-dm^2 square and the proportion of zero nearest-neighbor distances even exceeds 90% for Arisaema 4 (this information is not available in the distances between sampling location and nearest point); (iv) departures from complete randomness for Arisaema 2 and Arisaema 5 are more apparent on nearest-neighbor distances than on distances between sampling location and nearest point – by contrast with variance-to-mean ratios, departure is not the same for the two spatial point patterns, with a jump at the 0.1-m nearest-neighbor distance for Arisaema 2 and a quasi-linear increase of cumulative relative frequencies above the 2.5th percentile envelope from 0.1–0.7 m for Arisaema 5.

Summary: Besides moderate effects of quadrat size in the former approach, the count-based and distance-based approaches mostly agree on the spatial point pattern types identified for the Arisaema 1–6 datasets; these are mainly aggregated (especially for the spatial distribution of juveniles) and a minority are completely random or regular. The Pearson's r *statistic used as an autocorrelation measure is a very helpful complement to the traditional variance-to-mean ratio in the count-based approach.*

The example of California earthquakes

The earthquakes of magnitude 5 and more, which occurred between 114° and 124° longitude west and 32° and 42° latitude north from January 1, 1940, to December 31, 1999, are used for this example (source of the data: Northern California Earthquake Data Center). They are called "California earthquakes," even though the territory (i.e., the plot) includes parts of the states of Nevada, Arizona and Mexico. A seismic event is neither concentrated on one point in space nor instantaneous in time, so the epicenter was used as the reference point in space and the maximum magnitude as the reference point in time.

Clearly, the spatial distribution of the 404 California earthquakes that occurred over the period 1940–1999 is not completely random, but the seismic events tend to be located in specific zones [Fig. 3.12(a)]. These zones correspond to the intersection of the North American and Pacific tectonic plates. The aggregated point pattern of the California earthquakes in space is confirmed by the frequency distribution of quadrat counts: there has been no earthquake of magnitude 5 or more in almost half of the 100 quadrats of 1 degree of longitude by 1 degree of latitude

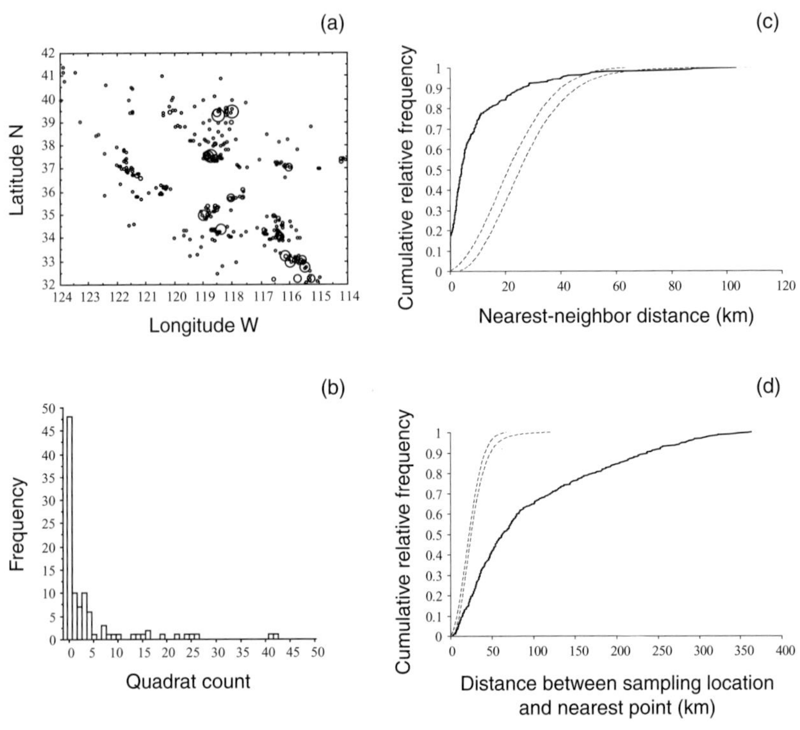

Fig. 3.12. Example of California earthquakes in space: (a) 2-D spatial point pattern observed over the period 1940–1999; (b) observed frequency distribution of quadrat counts; and observed cumulative relative frequency distributions (bold solid curves) of (c) nearest-neighbor distances and (d) distances between sampling location and nearest point, with the 2.5th and 97.5th percentile envelopes (dashed curves) evaluated from 999 independent partial realizations of a completely random point process. Earthquakes of magnitude 5 or more were counted in quadrats of 1 degree of longitude by 1 degree of latitude in (b), whereas quadrat centers were used as sampling locations in (d).

into which the territory can be divided, while 13% of them experienced 10 seismic events or more [Fig. 3.12(b)]. Spatial aggregation is further confirmed by the nearest-neighbor distances and the distances between sampling location and nearest point [Fig. 3.12(c) and (d)].

Key note: The longitude and latitude coordinates of epicenters (in degrees, minutes, seconds) were transformed to Universal Transverse Mercator (UTM) coordinates (in km) prior to calculating distances.

The proportion of small nearest-neighbor distances for California earthquakes in space is much greater than for a completely random point pattern. Indeed, there are more than 15% of zero nearest-neighbor distances because it happened that several seismic events occurred at the same spatial location [see the bigger symbols in Fig. 3.12(a)].

The variance-to-mean ratio was evaluated for different quadrat sizes (in degrees of longitude by degrees of latitude). The following values were obtained: 3.90 (quadrat size of 0.25 × 0.25), 6.98 (0.5 × 0.5), 15.70 (1 × 1), 22.08 (2 × 2), and 24.50 (2.5 × 2.5). These are consistent with the results reported above and the aggregation of the point pattern of California earthquakes in space, as well as with the Pearson's r statistic values of 0.571 (0.25 × 0.25), 0.636 (0.5 × 0.5), 0.265 (1 × 1), and 0.158 (2.5 × 2.5) (there are only four interior points at the 2.5 × 2.5 quadrat size). Finally, the increase of variance-to-mean ratio values with increasing quadrat sizes was investigated through fractals and the box-counting procedure of fractal dimension estimation (Foroutan-pour *et al.*, 1999). If the 2-D spatial point pattern of Fig. 3.12(a) is seen as a shape, the fractal dimension will be closer to 1 if the shape is closer to a straight line, or 2 if the shape is closer to the whole plane. For a given quadrat size (0.25 here), the number N of boxes intersected by the shape (i.e., quadrats in which at least one earthquake occurred) was counted, and this counting was repeated for increasing quadrat sizes (by powers of 2), so the regression $\log(N) = a + b\log(1/\text{size})$ could be fitted. The fractal dimension of the shape is given by the slope b. In this case, the slope estimate is 1.10 ($R^2 = 0.98$), indicating that California earthquakes are almost aligned in space, as suggested by our preliminary inspection of Fig. 3.12(a).

3.3 Marked spatial point patterns

Qualitative or quantitative marks are often recorded for the points found in the plot, in addition to their spatial location. The combination of a point process and a mark is called "marked point process," and one of its partial realizations is a "marked point pattern." Marked point patterns are observed in space as well as in time (see Chapter 4). It is important not to dissociate the mark from the companion point process. Indeed, there can be no mark if there is no point! It follows restrictions and constraints on the mapping and interpolation of quantitative marks.

In the *Arisaema* examples of Subsection 3.3.2, the healthy plants and the infected plants do not represent the components of a bivariate point process. In fact, the infected plants have been healthy at an earlier stage, and later have become infected by the rust fungus. The argument applies *a fortiori* to dead plants and plants alive. In both cases, one population of plants established in the plot is classified in two categories. Accordingly, two separate analyses of quadrat counts, one for infected plants and one for healthy plants (or one for dead plants and one for plants alive), are not as appropriate as a single statistical analysis like the one that is presented below, for testing an ecological hypothesis of interest about the spatial distribution of infected (or dead) plants. Leaving the discussion of quantitative marks for temporal point patterns (Section 4.1), here we concentrate on qualitative marks and the question: Are the points of the category of interest randomly distributed among all the points in the plot?

3.3.1 The distance-based approach

A very appropriate statistical method for the kind of question addressed above is inspired from the use of the observed cumulative relative frequency distribution of nearest-neighbor distances with lower and upper envelopes in Subsection 3.2.4. Thus, it belongs to Diggle's (2003) approach and assumes that all the points in the plot have been previously mapped. There are two important differences with the statistical procedure of Subsection 3.2.4, though. These differences, which follow from the question addressed, are: (1) distances are calculated between the n_1 points of the category of interest (e.g., infected plants) and the nearest points among the n_0 other points (e.g., healthy plants), with $n_1 + n_0 = n$; (2) the lower and upper envelopes are defined by permutations – in one permutation, the observed values of the qualitative mark (e.g., n_1 "infected" values and n_0 "healthy" values) are randomly distributed among the n points, while keeping their spatial location unchanged; this is repeated 999 times, independently from one permutation to the other.

3.3.2 Other examples with real data

The Arisaema *example – infected vs. healthy plants*

Four *Arisaema* datasets are used in this example; one follows from the combination of Arisaema 2 and Arisaema 3 in Subsection 3.2.5, and the other three are Arisaema 4–6 from the same subsection. Note that the plot for Arisaema 6 is 10 m × 10 m here. If n_0 and n_1 denote, respectively, the observed number of healthy plants in a plot and the observed number of plants infected by the rust fungus in the same plot, then $n_0 = 834$ and $n_1 = 164$ in Arisaema 2 and Arisaema 3 combined; $n_0 = 751$ and $n_1 = 186$ in Arisaema 4; $n_0 = 102$ and $n_1 = 53$ in Arisaema 5; and $n_0 = 130$ and $n_1 = 24$ in Arisaema 6. The number of infected plants is relatively high in the four plots, which is good statistically speaking because this will provide relatively smooth envelopes in the cumulative relative frequency distribution graphs. Differences in the total number of points (n) between some of these datasets and those in Subsection 3.2.5 follow from the non-inclusion of dead plants in the latter, whereas dead plants for which the infection status and spatial coordinates were available are included here. It was not possible to map healthy plants and infected plants together without missing the location of infected plants, because of the highly aggregated type of spatial point patterns (especially for Arisaema 2 and Arisaema 3 combined and Arisaema 4). This is why healthy plants and infected plants are mapped separately in Fig. 3.13.

In view of Fig. 3.13(a)–(d), infected plants do not seem to be especially distant from healthy plants and there is apparently no plot in which there are infected plants with no healthy plant nearby. To draw sound conclusions about the randomness of the distribution of infected plants among the plants mapped, however, the statistical analysis of distances between infected plant and nearest healthy plant is needed (Fig. 3.14). The most interesting result comes from Arisaema 5, for which only 5% of the infected plants are found in 1-dm^2 squares with at least one healthy plant [Fig. 3.14(c)]. This positions the observed curve below the lower envelope obtained from 999 random distributions of n_1 infected plants among the n plants present. Although the departure extends to 0.25 m, one would hesitate to speak of isolation of infected plants from healthy plants or of aggregation of infected plants. Results for Arisaema 2 and Arisaema 3 combined are similar to those for Arisaema 4 [Fig. 3.14(a) and (b)]. Because of the strong aggregation of plants in both plots, the proportion of infected plants found in 1-dm^2 squares with at least one healthy plant is very high (>75%), but slightly lower than expected for a random distribution of infected plants among all the plants of the plot. The departure extends to 0.25 m for Arisaema 2 and Arisaema 3 combined and to 0.2 m for Arisaema 4. No evidence of departure from randomness is observed for Arisaema 6 [Fig. 3.14(d)].

The Arisaema *example – dead plants vs. plants alive*

Four other *Arisaema* datasets are used in this example. These originate from the 1990 censuses of Plot 6 in the spring (Arisaema 7) and summer (Arisaema 8), of Plot 3 in the summer (Arisaema 9), and of Plot 8 in the summer (Arisaema 10). If n_0 and n_1 denote, respectively, the observed number of plants alive in a plot and the observed number of dead plants in the same plot, then $n_0 = 269$ and $n_1 = 14$ in Arisaema 7; $n_0 = 63$ and $n_1 = 7$ in Arisaema 8; $n_0 = 421$ and $n_1 = 27$ in Arisaema 9; and

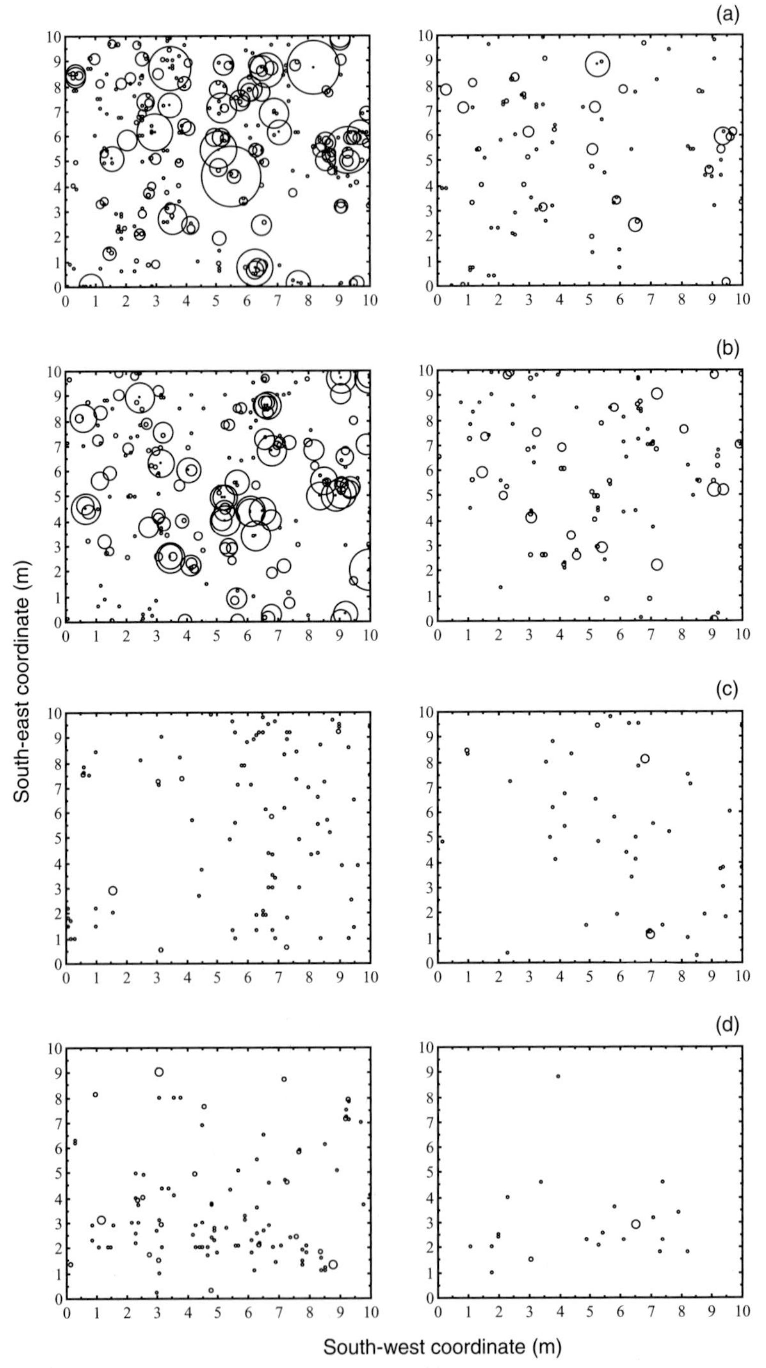

Fig. 3.13. Arisaema example of marked spatial point pattern (qualitative mark): observed 2-D spatial point patterns of healthy plants (left panels) and plants infected by a rust fungus (right panels) for (a) datasets Arisaema 2 and 3 combined; (b) Arisaema 4; (c) Arisaema 5; and (d) Arisaema 6, using data for the whole 10 m × 10 m plot.

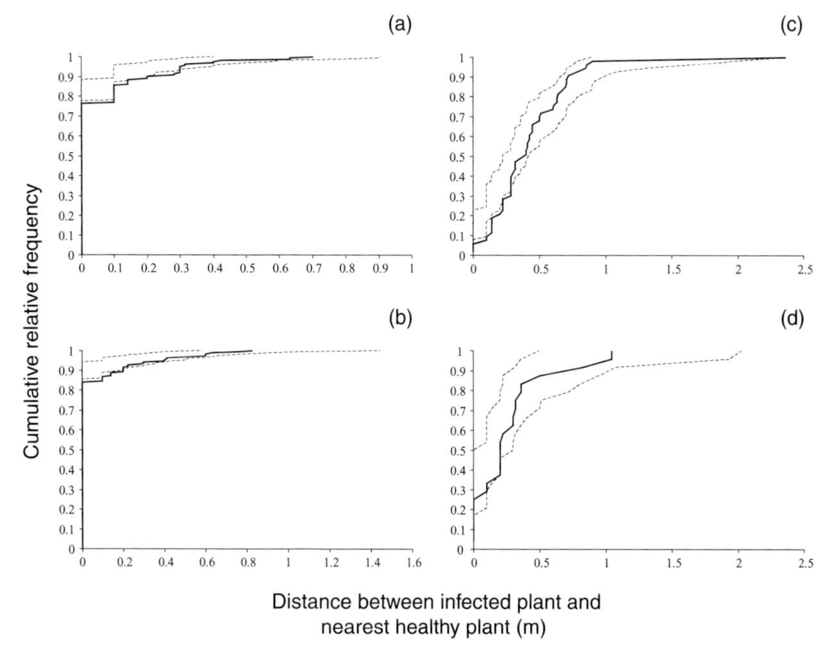

Fig. 3.14. Arisaema example of marked spatial point pattern (qualitative mark): observed cumulative relative frequency distributions (bold solid curves) of distances between infected plant and nearest healthy plant, with the 2.5th and 97.5th percentile envelopes (dashed curves) evaluated from 999 random permutations, for the four 2-D spatial marked point patterns of Fig. 3.13. In one permutation, infected plants were randomly distributed among all plants surveyed. This was repeated 999 times, independently from one permutation to the other.

$n_0 = 357$ and $n_1 = 53$ in Arisaema 10. Plants alive and dead plants in a given plot could have been mapped in the same panel, but the separate mapping in Fig. 3.15 is consistent with what was done in Fig. 3.13.

Clearly, the dead plants in Arisaema 10 are located in specific (central) parts of the plot, where only a few plants alive are found [Fig. 3.15(d)]. *Arisaema* has been submitted to particularly dry conditions in that environment in the summer of 1990, which explains the high mortality. Visually, there seems to be one dead plant in a 1-dm^2 square with one plant alive in Arisaema 8, and the two categories of plants do not appear to be really distant from each other in this plot [Fig. 3.15(b)]. More dead plants and plants alive are found in the same 1-dm^2 square in Arisaema 7 and Arisaema 9 [Fig. 3.15(a) and (c)].

The information obtained by visual inspection of the point patterns is refined by the results of the statistical analysis of distances (Fig. 3.16).

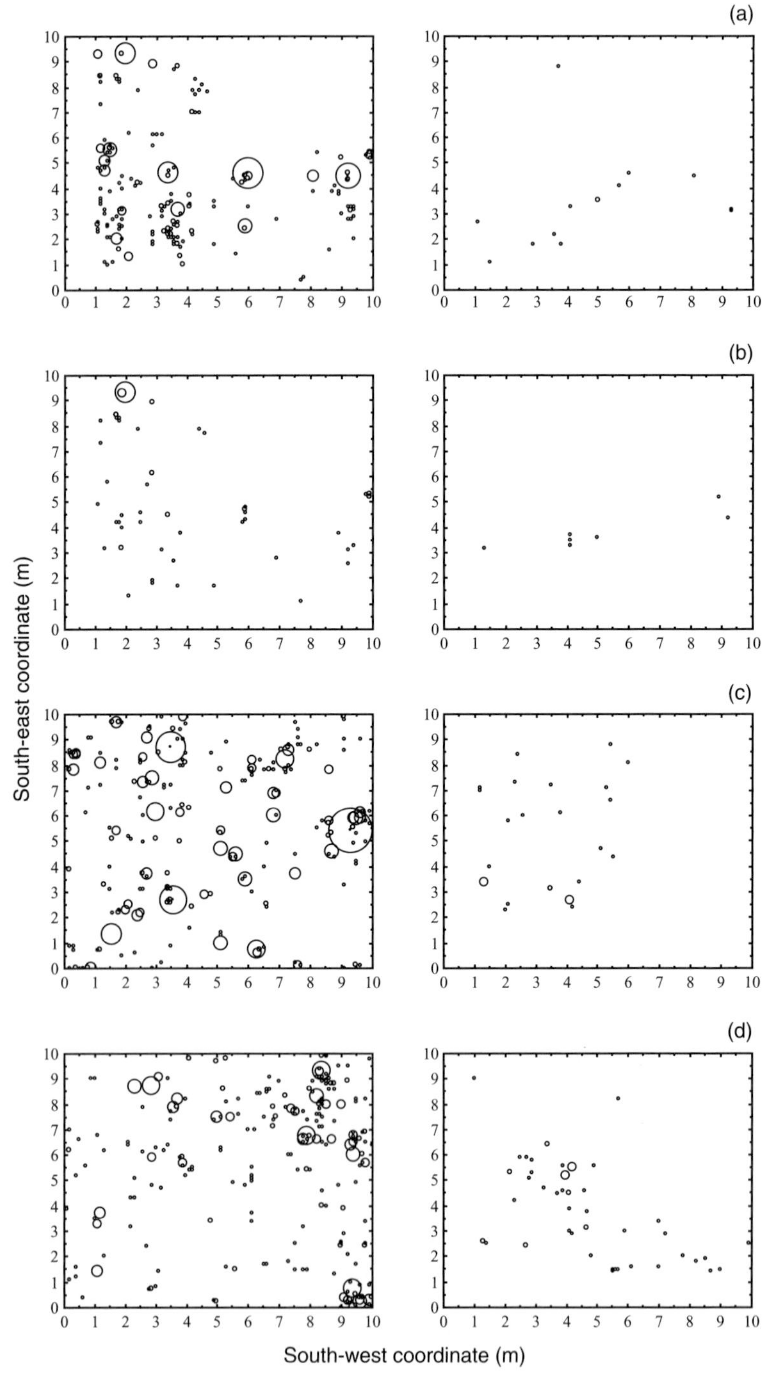

Fig. 3.15. Other *Arisaema* example of marked spatial point pattern (qualitative mark): observed 2-D spatial point patterns of plants alive (left panels) and dead plants (right panels) for (a)–(d) the four datasets named Arisaema 7 to 10.

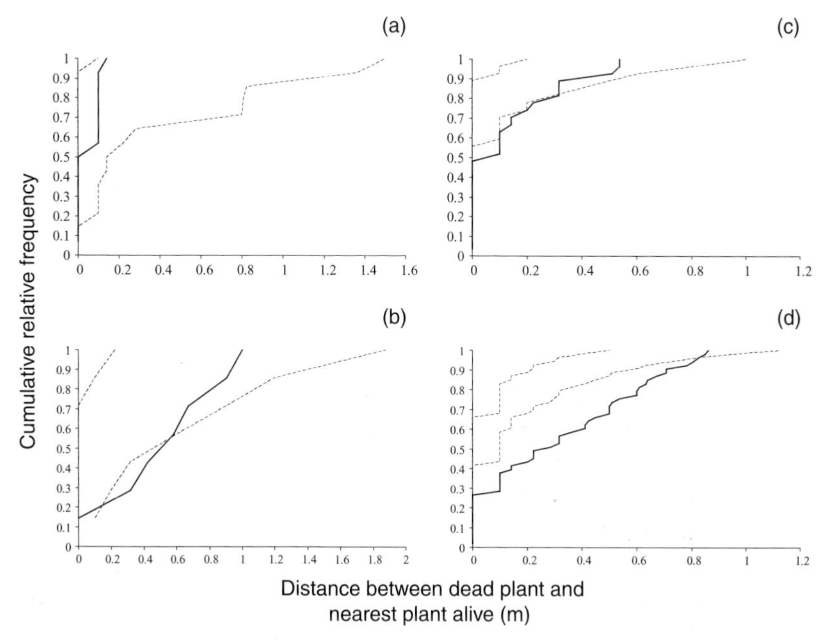

Fig. 3.16. Other *Arisaema* example of marked spatial point pattern (qualitative mark): observed cumulative relative frequency distributions (bold solid curves) of distances between dead plant and nearest plant alive, with the 2.5th and 97.5th percentile envelopes (dashed curves) evaluated from 999 random permutations, for the four 2-D spatial marked point patterns of Fig. 3.15. In one permutation, dead plants were randomly distributed among all plants surveyed. This was repeated 999 times, independently from one permutation to the other.

Dead plants are not randomly distributed among the plants surveyed in Arisaema 10, as their separation from plants alive is characterized by fewer small distances between dead plant and nearest plant alive than expected for a random distribution [Fig. 3.16(d)]. Despite wide envelopes due to the small value of n_1, the observed curve for Arisaema 8 departs downwards from 0.2 to 0.5 m [Fig. 3.16(b)]. There is no evidence of a departure from randomness in Arisaema 7, for which there are only three different values of the distance between dead plant and nearest plant alive [Fig. 3.16(a)]. Those values (0, 0.1 and 0.1414 m) indicate that a dead plant and the nearest plant alive were either in the same 1-dm^2 square or in neighboring or adjacent squares. Like Arisaema 10, Arisaema 9 lacks small distances, but unlike Arisaema 10, this lack is compensated by a larger number of intermediate distances and the absence of large distances [Fig. 3.16(c)].

Key note: The smoothness and thickness of the envelopes in Figs. 3.14 and 3.16 depend on the observed number of points of the category of interest (n_1). Increasing the number of random permutations (999) will not completely remove the jumps in the envelopes, and changing the category of interest for the category with the greater observed number of points does not appear natural because of the sequence of events in time: plants were healthy before being infected and all the plants were alive at first . . .

3.4 The beta-binomial distribution

The statistical distribution presented in this subsection is very useful in the analysis of point patterns because one of its parameters is a measure of aggregation. It is also very general because, as we will see, it can be applied to spatial, temporal and spatio-temporal point processes, whether these are marked or not. The reasons for its name "beta-binomial" are: its probability function may be written in terms of the "beta function" (Smith, 1983), and the binomial distribution which is taught in introductory statistics courses is a particular case of the beta-binomial distribution, for which the value of the parameter of aggregation is zero (complete randomness).

The beta-binomial distribution may take a finite number of numerical values, so it is discrete quantitative. It applies to counts of "successes" among n "trials," with a probability of success associated with each trial; the number of successes, x, is limited downwards by 0 and upwards by the number of trials ($x = 0, 1, \ldots, n$). In the general case, the probability of success may change from trial to trial, following the beta function. In the particular case of the binomial distribution, the probability of success is equal to p and is thus constant over the n trials. As a discrete quantitative distribution, the beta-binomial is characterized by a probability function, $P(X = x)$ for $x = 0, 1, \ldots, n$. This probability function is written below in the form that allows a direct link with the binomial distribution through the parameter of aggregation, although this form does not use the beta function explicitly:

$$P(X = x) = \frac{n!}{x!(n-x)!} \frac{\prod_{i=1}^{x} (p + (i-1)\theta) \prod_{i=1}^{n-x} (1 - p + (i-1)\theta)}{\prod_{i=1}^{n} (1 + (i-1)\theta)}$$

$$\text{for } x = 0, 1, \ldots, n, \tag{3.5}$$

where Π is the product operator and $\theta \geq 0$ is the parameter of aggregation. Concerning interpretation rules, the value of θ increases when the aggregation increases (Madden *et al.*, 1995). (Setting θ at 0 in equation (3.5) provides the classical formula for the probability function of the binomial distribution: $P(X = x) = \frac{n!}{x!(n-x)!} p^x (1 - p)^{n-x}$ for $x = 0, 1, \ldots, n$.)

The beta-binomial distribution has been introduced in plant pathology by Gareth Hughes and Laurence Madden, to describe the aggregated pattern of disease incidence (Hughes and Madden, 1993). To my knowledge, it has been used mainly in the 2-D spatial framework, and in Chapter 4, we will see how to apply it to temporal and spatio-temporal point processes. In Hughes and Madden's work, the sampling unit is a quadrat; a diseased plant found within a quadrat is a "success," so the diseased plants are counted among the n plants ("trials") surveyed in a quadrat; and n may vary from quadrat to quadrat. It follows that the BBD analysis (from the name of the software of Madden and Hughes, 1994) is readily applicable to the *Arisaema* data on infected vs. healthy plants (Fig. 3.13) and on dead plants vs. plants alive (Fig. 3.15). Since count data are used in a BBD analysis, results are dependent on the quadrat size. Here, the BBD analyses of *Arisaema* datasets have been restricted to one quadrat size because of moderate numbers of infected plants and dead plants. Accordingly, each plot was divided into 25 quadrats of 2 m × 2 m. Results are reported and discussed below. BBD analyses were also performed for the example of California earthquakes in space [Fig. 3.12(a)]. In this case, a quadrat is a part of the California territory; each quadrat is subdivided into n cells ("trials"); and a cell that experienced at least one earthquake is a "success." The beta-binomial distribution can be used to assess the spatial aggregation of earthquakes. In addition, the large number of seismic events in the case of California allows a multi-scale analysis. Accordingly, the plot in Fig. 3.12(a) was divided into 25 quadrats of 2 degrees of longitude by 2 degrees of latitude, and each quadrat into 4 2 × 2 cells (scale 1) or 16 4 × 4 cells (scale 2). At either scale, the cells that experienced at least one earthquake were counted in each quadrat ($x = 0, 1, \ldots, 4$ or $x = 0, 1, \ldots, 16$), and the frequencies of these counts were used as input for a BBD analysis. Figure 3.17 shows how each spatial dataset (*Arisaema*, California earthquakes) was respectively prepared. Version 1.2 of the BBD software (Madden and Hughes, 1994) was used to fit a beta-binomial distribution to the count frequencies; an equivalent SAS code written by Laurence Madden is distributed on the CD-ROM accompanying this book.

From the BBD analyses, it appears that spatial aggregation is moderate in the infection data of the *Arisaema* example. With the standard errors in parentheses, the estimates of θ are: 0.057 (0.025) for Arisaema 2

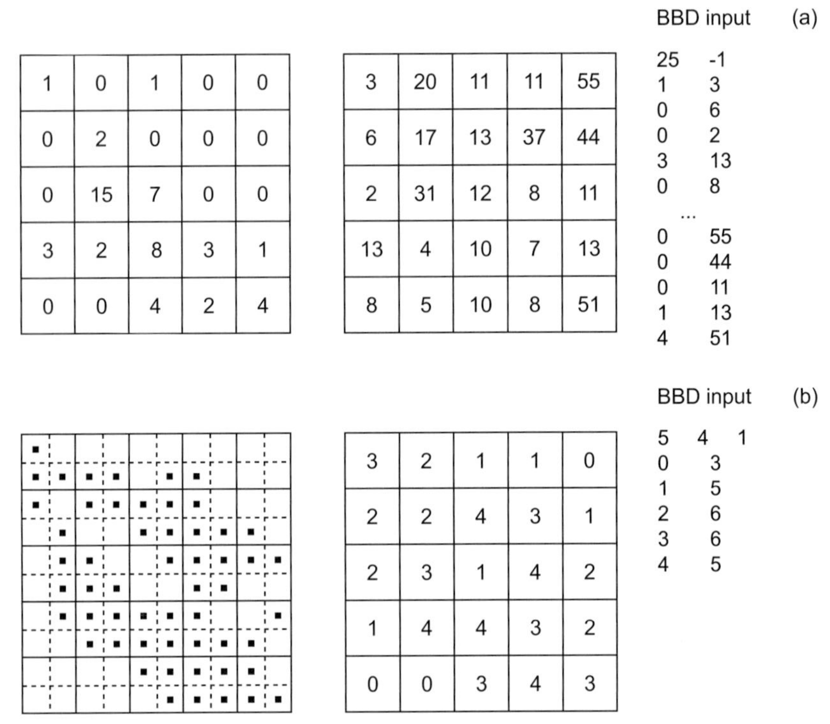

Fig. 3.17. Illustration of the preparation of the input data table prior to fitting a beta-binomial distribution to 2-D spatial binary data, from (a) the number of points with a given characteristic per quadrat (left square) among a total number of points per quadrat (right square) or (b) the number of cells with at least one point with a given characteristic (i.e., these cells are indicated by a dot) among a given number of cells per quadrat (e.g., four). The data used for illustration in (a) and (b) come from the *Arisaema* example and that of California earthquakes, respectively.

and Arisaema 3 combined; 0.026 (0.021) for Arisaema 4; 0.055 (0.081) for Arisaema 5; and 0.025 (0.039) for Arisaema 6. With the exception of Arisaema 2 and Arisaema 3 combined, the estimate of θ is never greater than two times the standard error. In the mortality data, evidence for spatial aggregation is the highest when mortality is the highest, with the following estimates and standard errors: 0.003 (0.019) for Arisaema 7; 1.325 (1.230) for Arisaema 8; 0.125 (0.065) for Arisaema 9; and 0.508 (0.196) for Arisaema 10. All these results are in agreement with those obtained with the distances between infected plant and nearest healthy plant (Fig. 3.14) and the distances between dead plant and nearest plant alive (Fig. 3.16). Note that (i) zero-zero data (i.e., no infected plant in a

Table 3.3. *Summary of BBD outputs for the example of California earthquakes in space, using 25 quadrats of 2 degrees of longitude by 2 degrees of latitude [see Fig. 3.12(a)] and dividing each of them into (a) 2 × 2 cells and (b) 4 × 4 cells*

(a) $\hat{p} = 0.550$ SE($\hat{p}$) = 0.065

$\hat{\theta} = 0.303$ SE($\hat{\theta}$) = 0.182

x	Observed frequency	BBD frequency	Binomial frequency
0	3	3.04	1.02
1	5	4.92	5.01
2	6	5.97	9.19
3	6	6.11	7.49
4	5	4.96	2.29

BBD goodness-of-fit: $\chi^2_{obs} = 0.001$

Binomial goodness-of-fit: $\chi^2_{obs} = 1.901$

(b) $\hat{p} = 0.287$ SE($\hat{p}$) = 0.042

$\hat{\theta} = 0.220$ SE($\hat{\theta}$) = 0.082

x	Observed frequency	BBD frequency	Binomial frequency
0	3	2.57	0.11
1	3	2.94	0.72
2	2	2.95	2.18
3	2	2.80	4.08
4	2	2.57	5.33
5	3	2.29	5.14
6	3	2.00	3.79
7	4	1.70	2.18
8	0	1.42	0.98
9	1	1.14	0.35
10	1	0.89	0.10
11	0	0.67	0.02
12	0	0.47	0.00
13	0	0.31	0.00
14	1	0.18	0.00
15	0	0.09	0.00
16	0	0.03	0.00

BBD goodness-of-fit: $\chi^2_{obs} = 3.965$

Binomial goodness-of-fit: $\chi^2_{obs} = 4.688$

quadrat without plant) represent undesirable indetermination and were therefore discarded prior to performing the BBD analyses; (ii) expected frequencies for the beta-binomial and binomial distributions could not be computed because of different values of n among quadrats.

An almost perfect fit ($\chi^2_{obs} = 0.001$) combined with a high estimate of θ (Table 3.3) clearly shows that the beta-binomial is appropriate for the spatial distribution of California earthquakes at scale 1 (a). At scale 2 (b), the ratio of the estimate of θ to the standard error remains high, but the estimate of θ is lower and the goodness-of-fit for the beta-binomial is not so good ($\chi^2_{obs} = 3.965$), though better than for the binomial.

The results of the BBD analyses above give us the opportunity to discuss further the concept of autocorrelation in the context of spatial point patterns: What does spatial autocorrelation mean for infected or dead plants and for earthquakes? Assume infected plants are highly aggregated among the plants present in a plot. If the beta-binomial distribution is used to count infected plants, it will have a greater variance than that of the binomial distribution with same mean because of higher frequencies of the extreme counts (i.e., 0 and n); due to their aggregation in specific parts of the plot, infected plants tend to be with infected plants, so that chances are good to find another infected plant near an infected plant, and the same for healthy plants. In other words, the qualitative mark with two values, infected or healthy, is positively autocorrelated in space. Similarly, in the presence of positive autocorrelation, dead plants should be grouped in a limited number of quadrats. There should be quadrats with many plants alive and no dead plant and quadrats in which all the plants are dead. In such a case, the probability of "success" (i.e., mortality) is likely to change from quadrat to quadrat, and the beta-binomial distribution applies. There is no mark in the example of California earthquakes in space, but if earthquakes are grouped specifically in some quadrats, the numbers of quadrats with zero cell having experienced earthquakes and quadrats with all the cells having had this experience will be relatively large. In the absence of positive autocorrelation, the joint observation of such quadrats is unlikely.

Key note: The beta-binomial distribution provides unique insight on the positive autocorrelation of qualitative marks in point pattern analysis. It is also useful to study aggregation in non-marked point patterns, but unlike the Pearson's r *statistic introduced in Subsection 3.2.1, the beta-binomial distribution does not cover regularity and negative autocorrelation because its parameter θ is restricted to be non-negative.*

3.5 The multivariate case

A bivariate point process is the first step in the extension of a univariate to a general multivariate point process. It is different from a marked point process in several aspects. First of all, one would deal with a marked point process if all the earthquakes that had occurred on some territory over a given period of time were classified according to their magnitude: less than 5 vs. 5 and more. One deals with a bivariate point process when two tree species (i.e., species 1 and species 2) are surveyed in the same plot at a given time or when the same tree species is surveyed in a given plot at two times (i.e., time 1 < time 2). The first example is purely spatial and the second involves space and time, but the same statistical method of data analysis can be used in both examples. This method is based on the calculation of spatial "inter-distances" between the points of one type (i.e., species 1 in the first example, time 1 in the second) and the nearest points of the other type (i.e., species 2, time 2). However, the construction of envelopes in the relevant extension of Diggle's randomization testing procedure needs to be modified because of a new type of null hypothesis.

In the statistical analysis of inter-distances for the species 1-species 2 bivariate point pattern, the species 1 point pattern could be kept fixed while simulating 999 independent partial realizations of a hypothetical model of point process for species 2, but this procedure would mainly concern the species 2 univariate point pattern while using the points of species 1 as sampling locations. Instead, the species 1 point pattern is kept fixed and the other is submitted to 999 independent "toroidal shifts," which takes into account the initial distribution of species 2 points while relocating them in a well-defined way. This procedure allows the testing of the null hypothesis: the points of the species 1 univariate point pattern are not more or less distant from those of the species 2 univariate point pattern than expected if the species 2 point pattern as a whole had been observed elsewhere in space. Following the 999 independent toroidal shifts, the computation of the 2.5th and 97.5th percentile envelopes is ad hoc. A full description of the procedure of toroidal shift falls beyond the scope of this subsection, and I refer to Upton and Fingleton (1985, pp. 253–254) for it. Here, its description is limited to the following simple example. Imagine a 10 × 10 square. Under the toroidal model, its four corners coincide. Assume the square is submitted to a translation defined by the vector (5, 5). It follows that the bottom left quarter is entirely moved to the top right quarter. Under the toroidal model, the three other quarters are redistributed inside the square as follows: a point that would initially be at (9, 9), near the top right corner, is moved to (4, 4), in the bottom left quarter of the square near the center (i.e., where the four coinciding initial corners are translated); a point initially at (1, 9), near the top left corner, is moved to (6, 4), in the bottom right quarter of the square; and a point initially at (9, 1), near the bottom right corner, is moved to (4, 6), in the top left quarter of the square. The SAS code for this extension of Diggle's randomization testing procedure is also on the CD-ROM.

Three datasets are used to illustrate the analysis of multivariate point patterns in 2-D space and explain how to interpret the results of Diggle's randomization testing procedure in this case. The data were collected in the frame of forest ecology studies conducted at the Arboretum of McGill

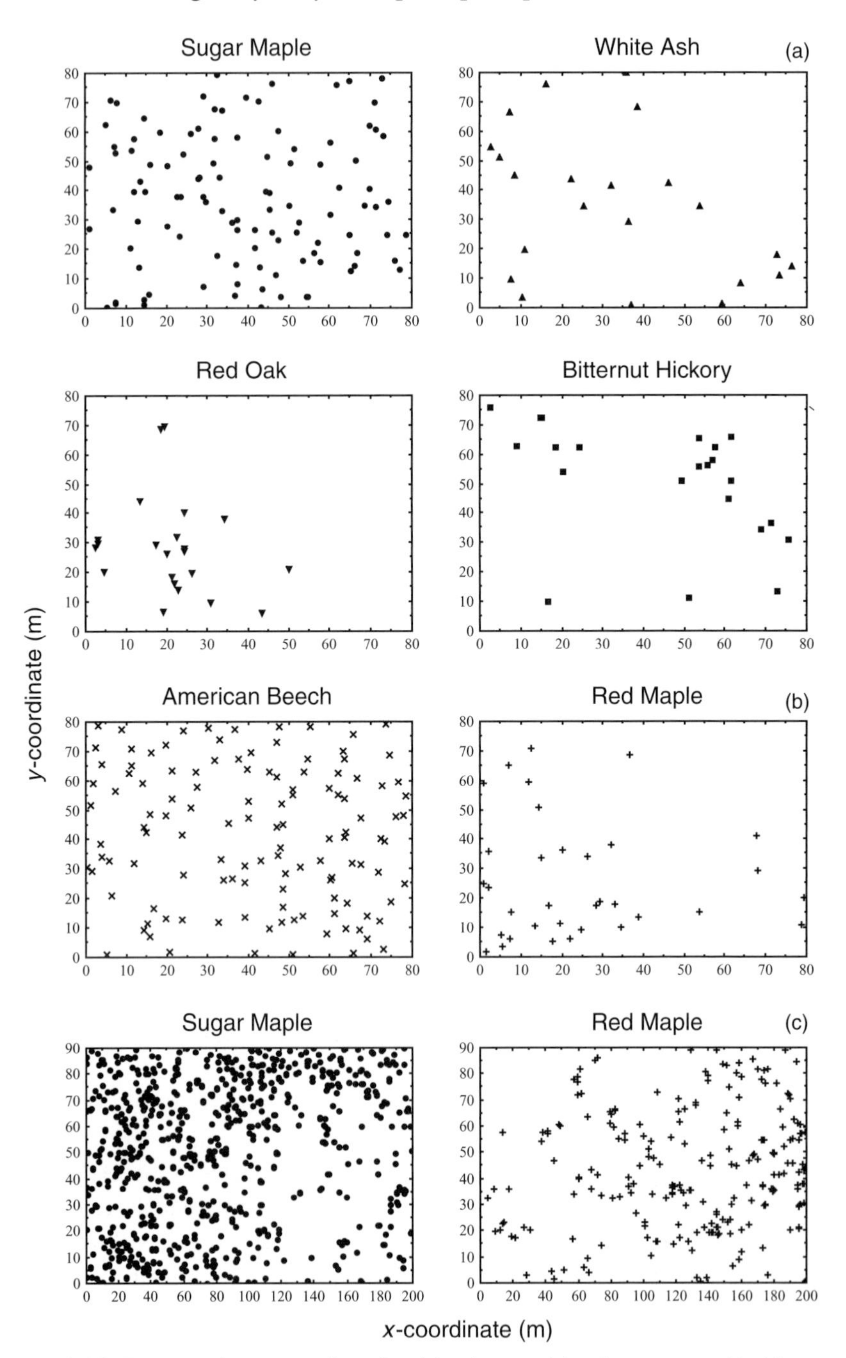

Fig. 3.18. Forest ecology examples of multivariate spatial point patterns: (a)–(c) observed 2-D spatial point patterns of various tree species in datasets 1 to 3. Note that point patterns in (b) and (c) are bivariate.

University, Ste-Anne-de-Bellevue (datasets 1 and 2), and at Station de Biologie des Laurentides, Saint-Hippolyte (dataset 3), both in Québec, Canada. The plots are 80 m × 80 m for datasets 1 and 2, and 200 m × 90 m for dataset 3. Only trees alive with a diameter at breast height >15 cm were included in the analyses, which provided the following observed numbers of points for the most abundant and second most abundant species in the three datasets: $n = 106$ for Sugar Maple and 22 for White Ash, Red Oak and Bitternut Hickory (i.e., there is a triple tie) in dataset 1; $n = 122$ for American Beech and 34 for Red Maple in dataset 2; and $n = 823$ for Sugar Maple and 279 for Red Maple in dataset 3. The null hypothesis reads as follows: Given the 2-D spatial point pattern observed for the most abundant tree species (species 1), its proximity from the one observed for the second most abundant species (species 2) is not different from what it would be if the latter was relocated within the limits of the plot. Of course, the forest ecologist would like to reject this hypothesis, which would mean that the actual proximity is particular, and be able to explain it by environmental factors at the site. Note that, for dataset 1, spatial distances will also be computed and analyzed between two point patterns even if the point process is 4-variate.

From the univariate spatial point patterns in Fig. 3.18(a), it appears that Sugar Maple (species 1) is present in a consistent manner throughout the plot in dataset 1, following a pattern between complete randomness and regularity; the spatial distribution of White Ash is completely random; Red Oak is concentrated into one cluster in about one half of the plot, within which the points seem to be randomly distributed; and the spatial distribution of Bitternut Hickory falls between the latter two. With one exception, there is no particular spatial association between Sugar Maple and the other species [Fig. 3.19(a)]. The exception is provided by an excess of distances around 5 m for White Ash, which is difficult to explain. In dataset 2 [Fig. 3.18(b)], American Beech (species 1) follows a spatial distribution similar to Sugar Maple in dataset 1, while Red Maple (species 2) resembles the mirror image of Bitternut Hickory in dataset 1, and no particular spatial association between species 1 and species 2 is to be reported [Fig. 3.19(b)]. In Fig. 3.18(b), Red Maple tends to be present in parts of the plot where American Beech is less abundant. To a greater degree, this is the case for Sugar Maple vs. Red Maple in dataset 3 [Fig. 3.18(c)]. Accordingly, the cumulative relative frequency distribution of inter-distances highlights the competition between the two maple species or their tendency to not co-exist in this plot, with

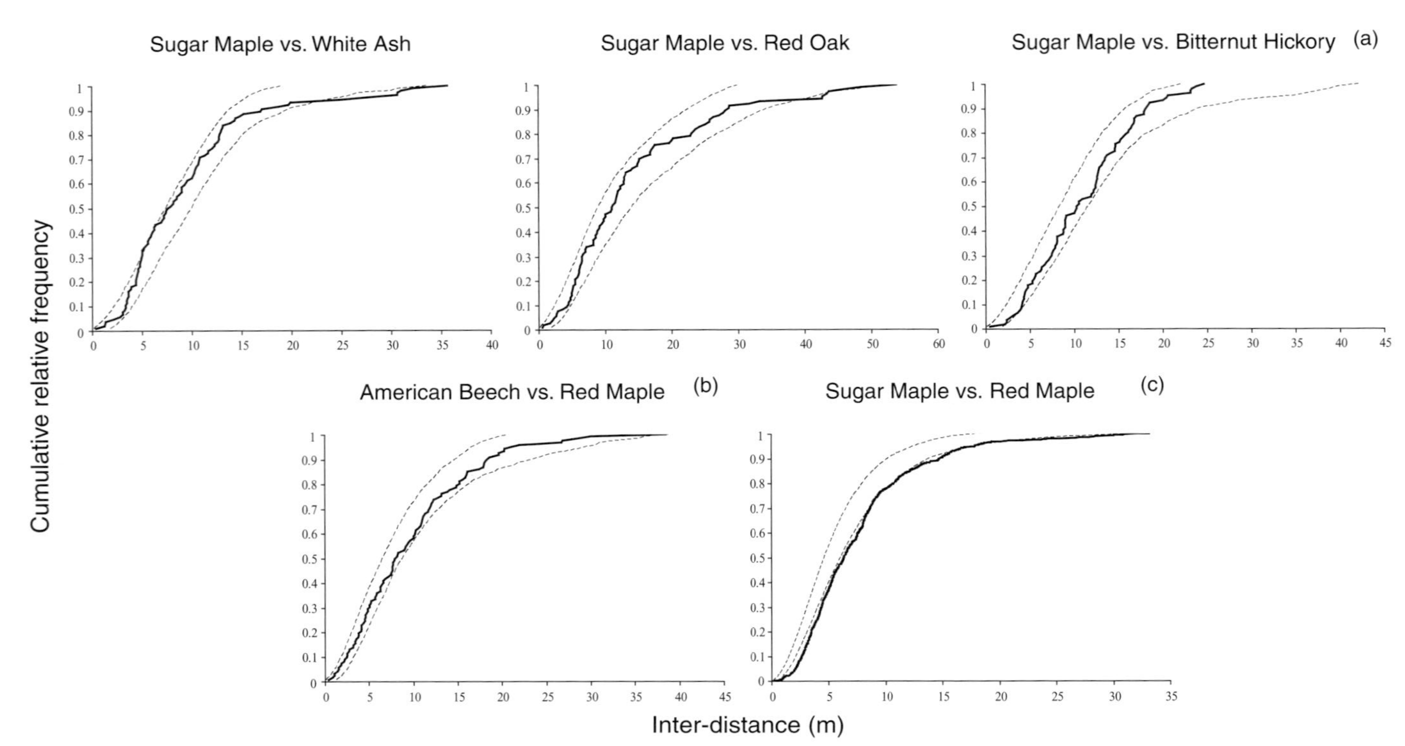

Fig. 3.19. Forest ecology examples of multivariate spatial point patterns: observed cumulative relative frequency distributions (bold solid curves) of distances from trees of the most abundant species to the nearest trees of the other species, with the 2.5th and 97.5th percentile envelopes (dashed curves) evaluated from 999 independent toroidal shifts of the observed 2-D spatial point patterns of the other species. Panels (a)–(c) correspond to datasets 1 to 3 in Fig. 3.18.

a lack of very small distances and a few excessively large distances [Fig. 3.19(c)].

Summary: In the distance-based approach, the analysis of multivariate spatial point patterns differs from the analysis of univariate spatial point patterns, by the use of spatial inter-distances to assess the spatial proximity (i.e., the cross-correlation) between point patterns.

Appendix A3: Intensity functions and simulation procedures

A number of mathematical and computational questions are discussed here because they would have disrupted the flow of the discourse if they had been included in the body of the chapter.

A3.1 Calculation of population parameters for quadrat counts

In continuous space or time, any observed point pattern represents a finite number of points in an infinitely large number of possible locations, even if the interval, the area or the volume within which points are observed is of finite size. The counting of points in quadrats (i.e., portions of the interval, area or volume) provides the experimenter with quadrat counts. Below, it is shown how the first two moments of the probability distribution of quadrat counts (i.e., the population mean and the population variance and covariance) can be calculated theoretically, using the first-order and second-order intensity functions.

Classically in statistics, the values of a probability density function are not probabilities. This is why the probability density function $f(u)$ is integrated over some interval of values of u to obtain the probability of interest when the random variable U is continuous quantitative. Similarly, the values of the first-order intensity function and the second-order intensity function are not population means, variances or covariances. To obtain population parameters in this case, the intensity functions need to be integrated over subsets of S, which assumes that the expression of each function to integrate is known and the set of possible values for the index s is continuous.

As an intensity, the first-order intensity function is used to express the expected number of points in a quadrat, $\mathrm{E}[N(A)]$, relative to the size of the quadrat, $|A|$. Thus, its definition is based on the ratio of $\mathrm{E}[N(A)]$ to $|A|$, and $\mathrm{E}[N(A)]$ could be seen as the product of $|A|$ times

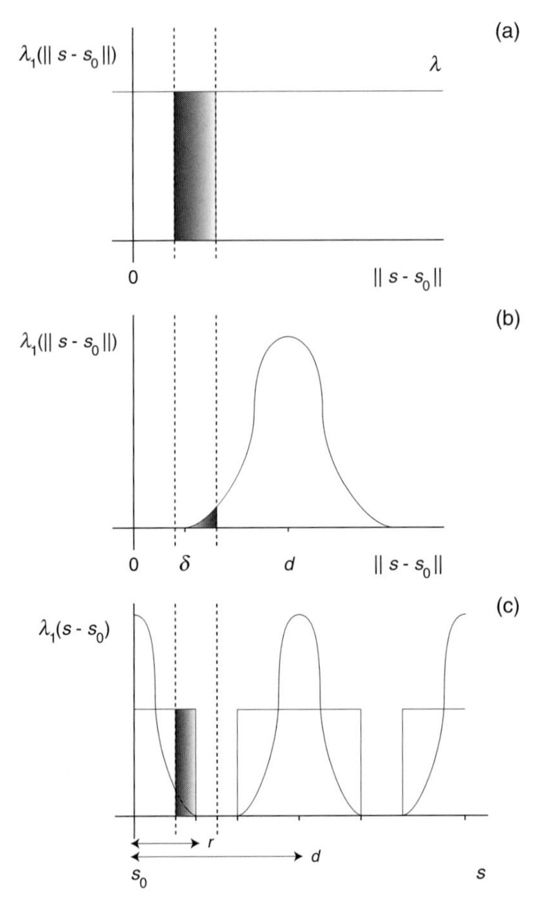

Fig. A3.1. First-order intensity functions for (a) a simple Poisson process; (b) a simple inhibition process with minimum permissible distance δ; and (c) a Poisson cluster process with a uniform distribution (rectangles) or a concentric distribution (bell-shaped curves) of points within clusters. In (a), s_0 and s denote a location of reference and any location in the sampling domain, respectively. In (b), d denotes the mean distance between nearest neighbors, while s_0 and s denote an existing point in the sampling domain and its nearest neighbor, respectively. In (c), the horizontal axis represented by s relates the centers of three assumed aligned clusters and s_0 is the center of one cluster, while d and $2r$ denote the mean distance between the centers of neighboring clusters and the mean width of a cluster, respectively. Depending on the dimensionality of the sampling domain, the shaded area under a curve is equal or proportional to the expected number of points in a quadrat which has the two dashed lines as limits.

the first-order intensity function. Strictly speaking, however, $E[N(A)]$ is equal to this product only if the first-order intensity function is a constant; otherwise, the product must be replaced by an integral. As for the second-order intensity function, it is involved in the calculation of the expected values of the square, $E[\{N(A)\}^2]$, in $\mathrm{Var}[N(A)] = E[\{N(A)\}^2] - \{E[N(A)]\}^2$, and the cross-product, $E[N(A)N(B)]$, in $\mathrm{Cov}[N(A), N(B)] = E[N(A)N(B)] - E[N(A)]E[N(B)]$, so the simple integral is replaced by a double integral in the general case.

Formally, the expected number of points in quadrat A, namely, the population mean of quadrat count $N(A)$, is given by

$$E[N(A)] = \int_A \lambda_1(s)ds, \tag{A3.1}$$

where $\lambda_1(s)$ denotes the value of the first-order intensity function at point s and $\int_A \cdot ds$ represents the simple integration operator over A for any nonrandom function of s; the population variance of quadrat count $N(A)$ and the population covariance between quadrat counts $N(A)$ and $N(B)$ are expressed in terms of the second-order intensity function $\lambda_2(s, s')$ as follows:

$$\mathrm{Var}[N(A)] = \int_A \int_A \lambda_2(s, s')ds\, ds' + E[N(A)]\{1 - E[N(A)]\} \tag{A3.2}$$

and

$$\mathrm{Cov}[N(A), N(B)] = \int_A \int_B \lambda_2(s, s')ds\, ds' + E[N(A \cap B)] - E[N(A)]E[N(B)], \tag{A3.3}$$

where $\int_A \int_B \cdot ds\, ds'$ represents the double integration operator over A and B for any nonrandom function of s and s', and $A \cap B$ denotes the intersection of quadrats A and B. Note that equation (A3.2) is a particular case of (A3.3) with $A = B$.

Key note: When an intensity function is random (see the Cox process in Section 2.4, Case 1), realizations of it over quadrats, which are nonrandom by definition, should be used in the integrals above.

The first-order intensity functions of a simple Poisson process, a simple inhibition process and a Poisson cluster process are represented graphically in Fig. A3.1(a)–(c). For the simple Poisson process (completely random point patterns), the expected number of points in a quadrat depends only on the size of the quadrat [see the shaded area under the curve in Fig. A3.1(a)], that is, $E[N(A)] = E[N(B)]$ for any quadrats A and B centered at $s \neq s'$ provided $|A| = |B|$. For the simple inhibition process (regular point patterns), the expected number of points in a quadrat of moderate size is close to zero when the quadrat is near an existing point, increases after the minimum permissible distance δ from the existing point is reached [see the shaded area under the curve in Fig. A3.1(b)]), and will be maximum around the average distance d between nearest neighbors. For the Poisson cluster process (aggregated point patterns), the first-order intensity function takes high values within clusters and low values outside of them, since the values of $\lambda_1(s)$ outside of clusters are exactly zero if there is no background point. Accordingly, the expected number of points in a quadrat is large within clusters, small outside of them, and intermediate when the quadrat partially overlaps a cluster [see the shaded area under the curve in Fig. A3.1(c)]. Clearly, the type of distribution of points within clusters (e.g., concentric vs. uniform) also has an effect on the shape of $\lambda_1(s)$ (e.g., bell-shaped vs. rectangular) and the value of $E[N(A)]$.

Summary: A quadrat of a given size at a given location may be expected to contain very different numbers of points depending on the type of point process.

A point process is stationary up to order 2 if (i) $\lambda_1(s) = \lambda_1$ for all s [i.e., the rate of occurrence of points per space or time unit is constant, so the expected number of points in any quadrat A is $\lambda_1|A|$] and (ii) $\lambda_2(s, s') = \lambda_2(s' - s)$ for any (s, s') [i.e., the value of the second-order intensity function depends on the difference $s' - s$ instead of the pair of indices]. Even though $\lambda_1(s)$ and $\lambda_2(s, s')$ are not a population mean and a population variance or covariance, this definition is consistent with the general definition of weak stationarity for a stochastic process in Chapter 2. In fact, $\lambda_1(s) = \lambda_1$ implies that the expected numbers of points in quadrats of same size are equal and $\lambda_2(s, s') = \lambda_2(s' - s)$ implies that the population variances of counts in quadrats of same size are equal and the population covariance between counts in two quadrats depends only on the vector linking their centers, $s' - s$. One could show similarly that the definition of isotropic point process, that is, $\lambda_2(s, s') = \lambda_2(\|s' - s\|)$, is consistent with the general definition given in Chapter 2. For weakly stationary and isotropic point processes, counts in two disjoint quadrats are uncorrelated if $\lambda_2(s, s') = \lambda_1^2$.

The simple Poisson process is the least heterogeneous of the three point processes discussed above. It is actually a model of homogeneity, which is stationary up to order 2 and isotropic. In fact, $\lambda_1(s) = \lambda_1$ for all s [Fig. A3.1(a)] and $\lambda_2(s, s') = \lambda_1^2$, so $\mathrm{Var}[N(A)] = \mathrm{E}[N(A)] = \lambda_1|A|$ for any quadrat A [from equation (A3.2)] and $\mathrm{Cov}[N(A), N(B)] = 0$ for any two disjoint quadrats A and B [from equation (A3.3)]. The equality $\mathrm{Var}[N(A)] = \mathrm{E}[N(A)]$ is the theoretical basis for a traditional index of heterogeneity (i.e., the variance-to-mean ratio).

A3.2 Simulation of partial realizations of point processes

In Section 3.2 (see also Chapters 4 and 5), a few simulated partial realizations of spatial and temporal point processes are represented in maps and plots, and much larger numbers are used in the calculations of inferential procedures to assess heterogeneity. In this section, it is explained how to write SAS codes to simulate such partial realizations. These include the completely random, regular and aggregated point patterns, as partial realizations of the simple Poisson, simple inhibition and Poisson cluster processes. They also include the true and false trends in the distribution of points, generated, respectively, by a heterogeneous Poisson process and a doubly stochastic Cox process (see Section 2.4, Case 1).

It is easy to simulate a completely random point pattern, with a given number of points inside an interval, an area or a volume. It suffices to have a function that generates pseudo-random numbers from a uniform distribution over the interval [0, 1], such as RANUNI in SAS (SAS Institute Inc., 2009). Assume one wants to simulate a partial realization from a simple Poisson process on the rectangle $[a, b] \times [c, d]$ in 2-D space. In SAS, the simulation of such a completely random pattern of n points is written as a DO-END loop with three commands executed n times:

```
SET abcd;
DO i = 1 TO n;
x = a + (b – a)*RANUNI(seed);
y = c + (d – c)*RANUNI(seed);
OUTPUT;
END;
```

where "abcd" is the name of the DATA step that contains the values of lower and upper bounds of the intervals $[a, b]$ and $[c, d]$ defining the rectangle, and "seed" (i.e., the argument of the function RANUNI) is an integer number, positive or negative. Note that, despite the use of the same seed, the x- and y-coordinates are simulated independently, for a given point and from point to point. In the case of a non-rectangular plot in 2-D space, one proceeds similarly by working with the smallest rectangle $[a, b] \times [c, d]$ that includes the plot in question until n points have fallen inside the plot; the DO-END loop is then replaced by a DO-UNTIL or DO-WHILE.

Key note: Because the first-order intensity function of the simple Poisson process is constant over the whole plot (Section A3.1), the numbers of points in two quadrats of same size are expected to be the same. It does not mean that the numbers of points observed in quadrats of same size will be the same, since the number of points in quadrat A *is random and follows a Poisson distribution with parameter* $\mathrm{m} = \lambda_1|\mathrm{A}|$ *under the model of the simple Poisson process.*

It might seem that regularity would be simpler to simulate than complete randomness. In reality, this is not the case, except if one wants to reproduce a tree plantation on a grid, for example, but this represents a deterministic instead of random distribution of points. There are different approaches to the simulation of a regular point pattern. The approach based on simple inhibition processes (Diggle, 2003, pp. 72–74) is one of them. The following biological example illustrates it well. Imagine a colony of marine birds on an island. The first arriving pair of birds has installed its nest at the most appropriate place. The second pair to arrive does not fight for or claim the nesting place of the first pair, and chooses to install its nest in the second best place, and every new pair installs its nest where the other pairs already in place allow it. Eventually, each nest is located at a minimum permissible distance δ from its nearest neighbor. The successive arrivals of pairs of birds and the 2-D spatial point pattern resulting from their choice of a territory for nesting (i.e., 1 point = 1 nest) can be simulated with the function RANUNI of SAS, as we did for the completely random point pattern. Here, however, a newly simulated point will be discarded if it falls within a predetermined distance δ from one of the previously simulated points. It follows that the values of δ and n cannot be too large when the interval, area or volume is small, because if this was the case, the simulation of n points would be impossible to complete. Although it reproduces the marine bird colony example well, this simulation procedure can be time-consuming because it requires the calculation of nearest-neighbor distances. For not too large values of δ

and n, the simulation of the last points will always take more time (i.e., more trials) than that of the first. Alternatively, a regular point pattern can be simulated by moving the nodes of a regular grid (not necessarily rectangular in 2-D or right parallelepipedic in 3-D) by small distances randomly chosen. Such a simulation procedure does not have the requirement of the former in terms of distance calculation, since small random distances can be drawn directly from a uniform or normal distribution with given parameter values. It does not guarantee the absence of points within a given distance, however, and can be written in a few lines in a SAS code, as shown below.

```
SET n;
DO i = 1 TO n;
y = INT((i – 1)/10) + 0.25 + RANUNI(1);
x = i – 1 – INT((i – 1)/10)*10 + RANUNI(1) – 0.25;
IF (x < 0) THEN x = 0.25*RANUNI(1);
IF (y > 10) THEN y = 10 – 0.25*RANUNI(1);
OUTPUT;
END;
```

The latter procedure was used to simulate the 2-D spatial regular point pattern in Fig. 3.1(b), where the smallest distance between points is 0.33. If this seems too close for two bird nests (although the distance unit is not specified), nests located at such a distance from one another in 2-D space (i.e., horizontally) might be at different heights, or might be separated by a thin rocky structure. Recently, the point process model found appropriate for the spatial distribution of polygon trough intersection points on Earth in the Arctic and at the surface of Mars was assessed with a similar simulation procedure, using the RANNOR function of SAS to generate small random distances from a normal distribution (see Subsection 5.1.3).

As for aggregated point patterns, the simulation of a partial realization of a Poisson cluster process requires a number of clusters, the spatial boundaries of each cluster, and the type of spatial distribution of points within each cluster. Each of the two aggregated point patterns in Fig. 3.2(a) and (d) can be seen as, and actually is, a partial realization of a 2-D spatial Poisson cluster process because (1) the number of clusters in the plot follows a Poisson distribution, (2) the spatial distribution of cluster centers is completely random, (3) the number of points per cluster is proportional to the size of the cluster, and (4) the distribution of points within clusters is the same or similar over clusters. In the *Arisaema* example, several point patterns can be thought of as partial realizations of a 2-D spatial Poisson cluster process, as juvenile plants tend to be grouped in areas of the plot where adult plants established in previous years. The distribution of points within clusters may differ between Poisson cluster processes. For example, it is uniform when the density of points is constant [Fig. 3.2(e)], or concentric if the density of points decreases from the center of the cluster [Fig. 3.2(b)]. In SAS, these two distributions of points may be simulated within a circular cluster with center (a, b) and radius r as follows:

Uniform distribution	Concentric distribution
SET abr;	SET abr;
DO i = 1 TO n;	DO i = 1 TO n;
uni1 = RANUNI(seed1);	uni1 = RANUNI(seed1);
uni2 = RANUNI(seed2);	uni2 = RANUNI(seed2);
x = a + r*uni1*COS(8*ATAN(1)*uni2);	x = a + r*2*(uni1 – 0.5);
y = b + r*uni1*SIN(8*ATAN(1)*uni2);	factor = SQRT(1 – (x – a)*(x – a));
OUTPUT;	y = b + r*2*(uni2 – 0.5)*factor;
END;	OUTPUT;
	END;

Here "seed1" and "seed2" are two different seeds and $8 * \text{ATAN}(1) = 2\pi$.

True vs. false trends in 2-D spatial point patterns were discussed in view of Fig. 2.2, in which two types of point processes were illustrated in 2-D space: inhomogeneous Poisson processes (true trends) and Cox processes (false trends). The expected number of points in a given area is defined by a deterministic first-order intensity function for the former, whereas it is defined by a partial realization of a stochastic process for the latter because Cox processes are characterized by a first-order intensity function denoted $\Lambda_1(s)$ that itself is a stochastic process. For a given size $|A|$ of quadrats (i.e., squares of unit side in Fig. 2.2), $E[N(A)]$ was calculated directly by evaluating the second-order polynomial function $\lambda_1(s)$ at the center of each quadrat for the inhomogeneous Poisson process in Fig. 2.2(a)–(c); this calculation resulted in one set of values of $E[N(A)]$ for the three panels because of the deterministic nature of $\lambda_1(s)$. For the Cox process, there were three different sets of values of $E[N(A)]$ in Fig. 2.2(d)–(f), since the three partial realizations of the 2-D spatial AR(1) process $\Lambda_1(s)$ (evaluated at the center of each quadrat) were different.

Thereafter, the RANPOI function of SAS was used for each quadrat in each panel, to generate one observation from a Poisson distribution Po(m) with $m = \mathrm{E}[N(A)]$, and the resulting number of points were randomly distributed within the quadrat.

Many other point processes can be simulated with more sophisticated procedures.

4 · *Heterogeneity analysis of temporal and spatio-temporal point patterns*

In this chapter, the main procedures presented for heterogeneity analysis of spatial point patterns in Chapter 3 are transposed to the temporal framework (Section 4.1) and the space-time case (Section 4.2). This means that counts in well-defined quadrats and cells and various types of distances (e.g., distance to the next event in time, inter-distance in space-time) will continue to play a key role in the heterogeneity analysis of point patterns. Connected statistical methods to be seen as extensions or variants (i.e., the analysis of time-dependent transition probabilities, Subsection 4.1.3; point movement analysis with the beta-binomial distribution, Subsection 4.2.2) will also be presented. Introductory examples with simulated data and illustrative examples with real data will be discussed.

4.1 Temporal point patterns

In Subsection 4.1.1, we will first learn how to visualize points on the time axis, instead of looking for them in space. In this "eye training," simulated data will be used, with illustrative examples in the background to help relate to potential applications. Then, in Subsection 4.1.2, real data will be analyzed. More specifically, the California earthquakes will be studied in time, without and with a mark as well as in an application of the beta-binomial distribution in the temporal framework.

The statistical procedures presented in the spatial framework (Section 3.2 and following sections in Chapter 3) apply readily to the heterogeneity analysis of temporal point patterns, so the presentation of analytical procedures will not be detailed again in this section. One new statistical method, however, will be presented; it concerns state transitions (Subsection 4.1.3). A state transition requires the temporal axis to occur, with or without a spatial component, as will be illustrated by the insect outbreak, sheep, and mosquito examples.

4.1.1 The univariate case

The three temporal point patterns displayed in Fig. 4.1 could represent: (a) rainy days in a desert; (b) daily meals of a crab on seashore; and (c) attempts of a predatory fish to catch prey in a lake at sunset and sunrise on four days. These interpretations are likely if the time unit on the horizontal axis is the day in (a) and (b) and the hour in (c), and one assumes that a vertical bar represents one rainy day in (a), one meal of the crab in (b), and one attempt of the predatory fish in (c). The three temporal point patterns are distinctly different, and so are the three biological examples associated with them. Periods without rain in a desert can be long, but it can also rain on several occasions within a limited period of time. This happens without any consistent pattern, that is, completely randomly [Fig. 4.1(a)]. By contrast, a crab on seashore may expect to have zero meal on bad days, one on most days and two when food is abundant. More importantly, tides regulate the activity of crabs, so the temporal point pattern of their feeding activity is regular or "periodic" [Fig. 4.1(b)]. In freshwater lakes, predatory fish are usually more active at sunset and sunrise, and the capture of a prey may require more than one attempt. It follows that the attempts of a predatory fish tend to be aggregated at those two times of day [Fig. 4.1(c)].

The characteristics of the completely random, regular and aggregated point patterns and the procedures presented for their analysis in the space case apply to the time case. First of all, how about the variance-to-mean ratio and the statistic of autocorrelation based on Pearson's r for the three temporal point patterns of Fig. 4.1? Using the 100 simulated points and the frequencies plotted in Fig. 4.2(a), the observed values of $\frac{S^2}{\bar{X}}$ and Pearson's r are 1.051 and 0.141 for the completely random pattern, 0.343 and –0.582 for the periodic pattern, and 2.384 and 0.622 for the aggregated pattern, which confirms the guidelines given in Subsection 3.2.3 for the characterization of the corresponding spatial point patterns.

Consider next the frequency distributions of all the distances between pairs of events, only the nearest-neighbor distances (i.e., the distances between one event and the nearest on the time axis) and the distances from one event to the next. There are two reasons for considering the third type of distances: the use of sampling locations is less justified in time where the observation of points is generally easier than in space; and given the ordered nature of time, there might be differences in the results obtained with distances to the nearest point, which may precede

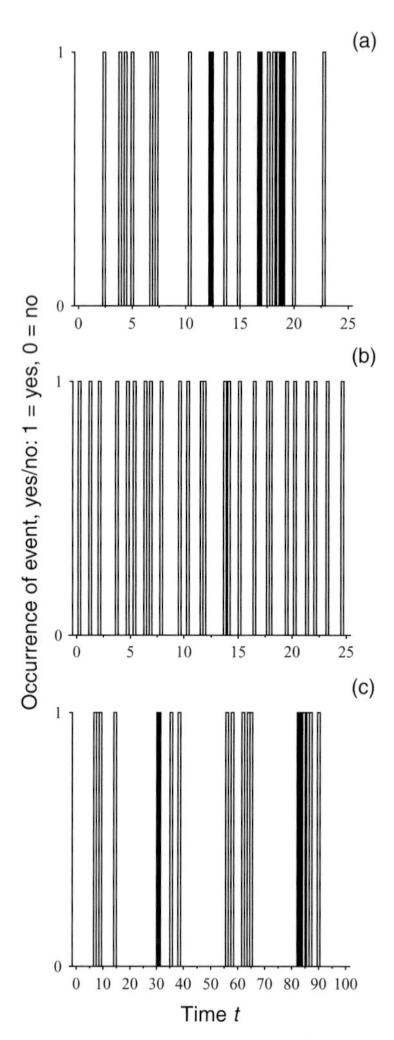

Fig. 4.1. Partial realizations of three temporal point processes: (a) a completely random or simple Poisson process; (b) a simple inhibition process showing regularity, called periodicity in the temporal framework; and (c) a Poisson cluster process showing aggregation. One hundred points were simulated in the continuous interval [0, 100]. For clarity reasons, only the points belonging to the interval [0, 25] are shown in (a) and (b), and 24 points (i.e., six per cluster) are displayed in (c).

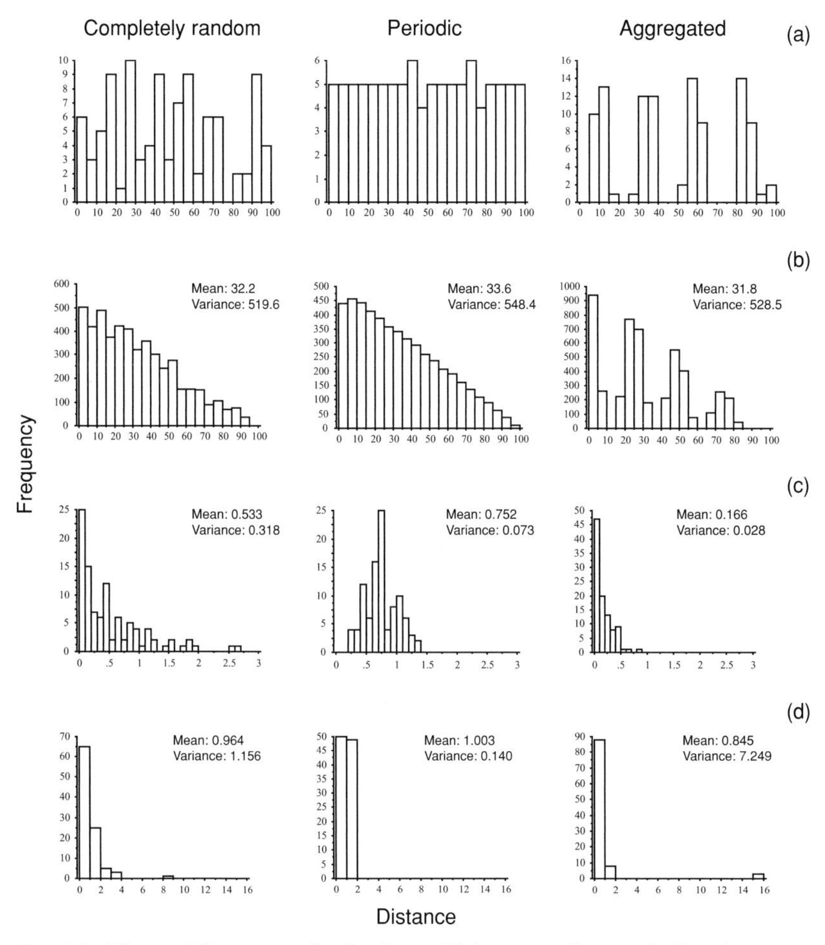

Fig. 4.2. Observed frequency distributions of (a) counts of events in 5-unit intervals starting from time 0; (b) all distances between pairs of events; (c) nearest-neighbor distances; and (d) distances from one event to the next, for the three temporal point patterns of Fig. 4.1 (without "background points" for the aggregated pattern). The 100 simulated points were used in the calculations.

or follow the point considered, and distances to the next point, which necessarily follows the current point.

The lack of similarity between Fig. 3.5(a) and Fig. 4.2(b), concerning the frequency distributions of all distances between pairs of points for the completely random, regular and aggregated point patterns in space and in time, is to be attributed to the difference in dimension between the frameworks (i.e., 2-D vs. 1-D). In the temporal aggregated pattern,

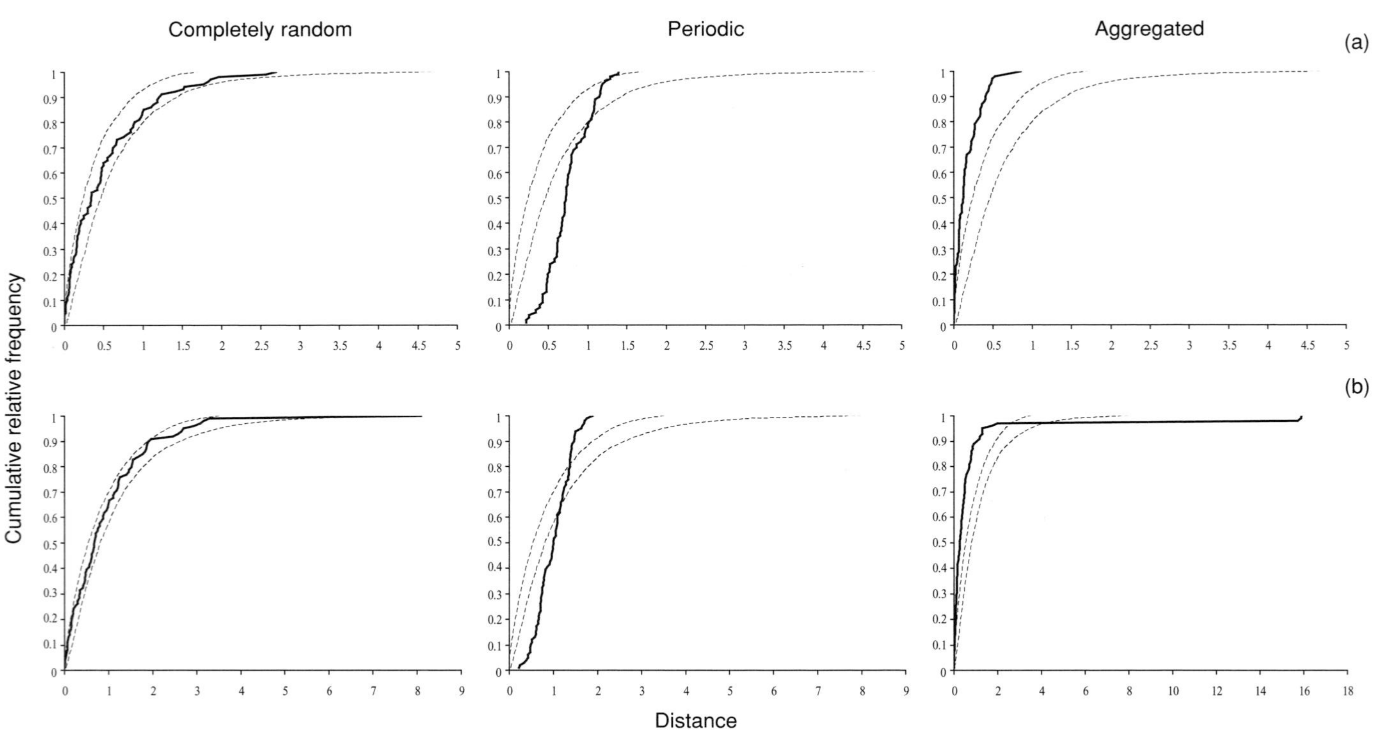

Fig. 4.3. Observed cumulative relative frequency distributions (bold solid curves) of (a) nearest-neighbor distances and (b) distances from one event to the next, for the three temporal point patterns of Fig. 4.1 (without "background points" for the aggregated pattern). The 2.5th and 97.5th percentile envelopes (dashed curves) were evaluated from 999 independent partial realizations of a completely random point process over the sampling interval [0, 100]. The number n of points in each partial realization is the same as in the entire observed point patterns (i.e., $n = 100$).

distances between events of the same cluster are small and those between events of different clusters are intermediate to large or very large, depending on the distance between cluster centers. This explains why four peaks are observed in Fig. 4.2(b). In the 2-D spatial aggregated pattern, distances between points of the same cluster are small, but those between points of different clusters cannot be classified as intermediate, large and very large. So, there are only two peaks, a secondary and a main, in the frequency distribution [Fig. 3.5(a)]. A similar reasoning can be followed for the two other point patterns. As far as the distances between all pairs of points are concerned, the resemblance between 1-D space and time is greater than between 2-D space and time and between 3-D space and time in this order. To be complete, the temporal periodic pattern is characterized by a linear decrease of frequency with increasing distance between events, and to a good degree, this behavior is shared by the temporal completely random pattern [Fig. 4.2(b)].

There are several similarities between Figs. 3.5(b) and 3.6(a) and Figs. 4.2(c) and 4.3(a), suggesting that nearest-neighbor distances are "dimension-free," contrary to distances between all pairs of points. Distances to the next event, through their frequency distribution (cumulative or not), appear to help distinguish the three temporal point patterns [Figs. 4.2(d) and 4.3(b)]. Similar departures from complete randomness are observed on nearest-neighbor distances and on distances to the next event [Fig. 4.3(a) and (b)]. In all cases, the periodic point pattern is characterized by a lack of small and large distances, and the aggregated one, by an excess of small distances [Fig. 4.2(c) and (d)]. The observed cumulative relative frequency distribution of distances to the next event for the periodic point pattern even crosses the 2.5th percentile envelope eventually, and that for the aggregated point pattern finally rejoins and passes the 97.5th percentile envelope. Thus, the use of the cumulative relative frequency distribution of nearest-neighbor distances with lower and upper envelopes to assess the appropriateness of a point process model for an observed point pattern in space extends to time, with distances to the next event as a possible substitute to nearest-neighbor distances in time.

4.1.2 Applications to real data

Below, the 70 California earthquakes that occurred between January 1, 1990, and December 31, 1999, are used for illustration. In view of Fig. 4.4(a), a large number of these earthquakes (24) occurred in 1992;

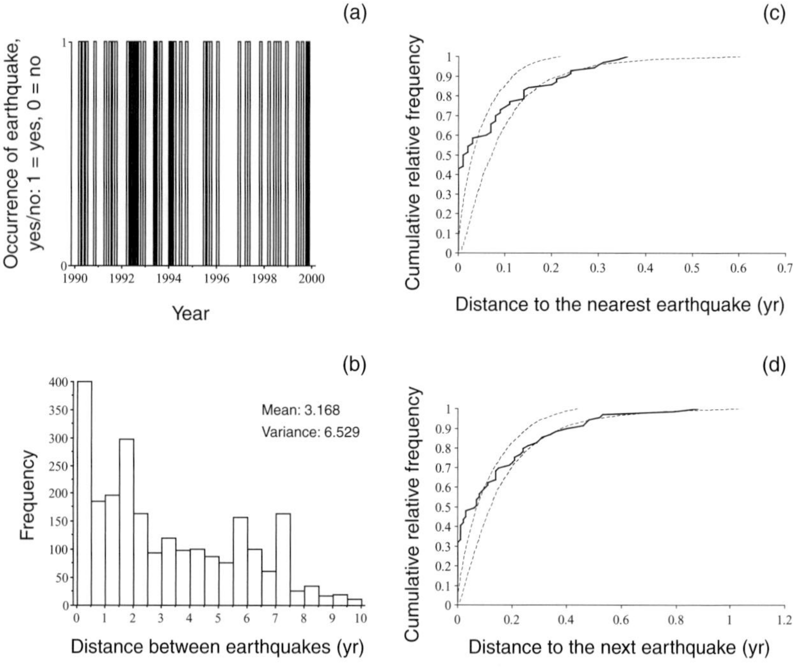

Fig. 4.4. Example of California earthquakes in time: (a) observed temporal point pattern during the period 1990–1999 ($n = 70$); (b) observed frequency distribution of all distances between earthquakes; and observed cumulative relative frequency distributions (bold solid curves) of (c) nearest-neighbor distances and (d) distances from one earthquake to the next, with the 2.5th and 97.5th percentile envelopes (dashed curves) evaluated from 999 independent partial realizations of a completely random point process.

the year with the second largest number (11) is 1994; and some years, successive (1995–1997) or not (1991, 1993), experienced four earthquakes or fewer. Aggregation or some form of randomness may thus be expected from the quantitative analyses. There is no real preliminary evidence for periodicity.

Variance-to-mean ratios evaluated from interval counts tend to support a temporal aggregated pattern, with values of 3.906 at the scale of the month, 4.593 at 3 months, 4.376 at 6 months, and 6.127 at the scale of the year. The statistic of autocorrelation based on Pearson's r moderately supports aggregation with positive values of 0.170, 0.220, and 0.190 at the scales of 1, 3, and 6 months; its negative value of −0.293 at the scale of the year originates from the low–high–low and high–low–high sequences in the numbers of earthquakes in 1991–1993 and 1992–1995.

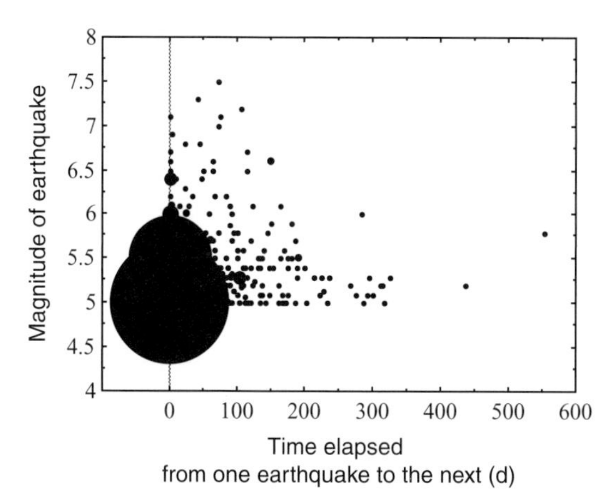

Fig. 4.5. Scatter plot of the magnitude of California earthquakes of magnitude 5 or more during the period 1940–1999 against the time elapsed until the next earthquake. The size of a dot in the plot is directly proportional to the number of earthquakes that have same magnitude and same temporal proximity (in days) with the next earthquake.

Actually, the observed values of $\frac{S^2}{\bar{X}}$ are quasi-constant from scale to scale, which contrasts with the sharp increase of ratio values with increasing quadrat sizes in Tables 3.1 and 3.2. The explanation is in the occurrence of 12 earthquakes in only one month (June 1992), which created a temporal pseudo-aggregated pattern with one cluster. The observed frequency distribution of all distances between pairs of events reflects this intermediate situation, with peaks at 1.5–2, 5.5–6, and 7–7.5 years combined with an overall decrease in frequency with increasing distance [Fig. 4.4(b)]. The cumulative relative frequency distributions of nearest-neighbor distances and distances to the next event help resolve the question, with a clear excess of near-zero distances (less than a month) [Fig. 4.4(c) and (d)].

The concept of mark is essentially the same, whether it is associated with a point process in space or in time, and the magnitude provides a quantitative mark for California earthquakes in time as well as in space. In the following application, the correlation between the time elapsed between two successive earthquakes and the magnitude of the second earthquake of the pair is assessed. A threshold of 5 is applied to the mark values and the period covered is January 1, 1940, through December 31, 1999, so $n = 404$; the events are the same as in Subsection 3.2.5, except that they are analyzed in time instead of space. In view of Fig. 4.5, the two variables are not correlated. In fact, the observed values of Pearson's r (calculated classically) and Spearman's rank-based correlation coefficients are –0.006 and –0.021, and the probabilities of significance are well above 0.05. These results do not support the perception that a longer period without earthquake presages a more damaging earthquake to come, but one cannot draw general conclusions from them. Note that the biggest point in Fig. 4.5 represents 20 earthquakes of magnitude 5 that followed an earthquake of magnitude 5 or more on the same day.

In the following application of the beta-binomial distribution, temporal binary data provided by the example of California earthquakes in time are used; recall that the original data cover the period 1990–1999 and are characterized by an excess of near-zero distances to the nearest earthquake and to the next earthquake, compared with a temporal completely random point pattern [Fig. 4.4(c) and (d)]. To apply the beta-binomial distribution here, the period of 10 years was divided into 10 intervals of 1 year and each 1-year interval was subdivided into 12 monthly cells. The binary response (1/0) then followed from the occurrence (yes/no) of at least one earthquake of magnitude 5 or more in a cell. The results of the BBD analysis are summarized in Table 4.1. Though not very high, the estimate of θ is close to two times the standard error, and the goodness-of-fit of the beta-binomial is slightly better than that of the binomial. To some degree, these results obtained at the scale of the month confirm those supporting a temporal aggregation of California earthquakes at small scale [Fig. 4.4(c) and (d)]. Such aggregation may originate from "aftershocks," that is, short sequences (over a few days) of seismic shakes, which occur in the same place and are, actually, part of the same earthquake.

4.1.3 Time-dependent transition probabilities

A transition probability is the probability associated with a particular type of event consisting of a state transition. Hereafter, we first illustrate with an example that state transitions are related to marked temporal point patterns. Then, the basic elements for the quantitative analysis of state transitions are introduced, and a statistical method designed to take into account the temporal heterogeneity of transition probabilities, which makes them time-dependent, is presented. Three examples, in forest ecology (the insect outbreak example), animal science (the sheep example), and entomology (the mosquito example), are discussed.

Table 4.1. *Summary of the BBD output for the example of California earthquakes in time, using 10 yearly intervals [see Fig. 4.4(a)] and dividing each of them into 12 monthly cells*

$\hat{p} = 0.325$	$SE(\hat{p}) = 0.050$		
$\hat{\theta} = 0.030$	$SE(\hat{\theta}) = 0.056$		
x	Observed frequency	BBD frequency	Binomial frequency
0	0	0.19	0.09
1	0	0.73	0.51
2	2	1.46	1.36
3	3	1.99	2.19
4	3	2.04	2.38
5	1	1.64	1.84
6	0	1.06	1.04
7	0	0.56	0.43
8	0	0.23	0.13
9	1	0.08	0.03
10	0	0.02	0.00
11	0	0.00	0.00
12	0	0.00	0.00

BBD goodness-of-fit: $\chi^2_{obs} = 0.072$

Binomial goodness-of-fit: $\chi^2_{obs} = 0.122$

Imagine a forest in which trees experience defoliation occasionally due to the outbreak of some insect population; note that an insect population outbreak may last more than a year. In the years of insect population outbreak, trees suffer from defoliation. In the years preceding one outbreak, they did not suffer from defoliation; in the years following it, they will typically rebuild a new canopy. In this context, a point is the first year of an insect population outbreak; a mark is the number of years covered by an outbreak; and the two possible states for the forest are "defoliated" in the years of insect population outbreak and "not defoliated" in the other years. Accordingly, there are two state transitions in this example. These are defined by the passage from a year without outbreak to a year with outbreak, and vice versa. Data on larch sawfly outbreaks in subarctic Québec (Filion and Cournoyer, 1995) are presented in Fig. 4.6. As indicated by the main and secondary peaks in the frequency plot, insect population outbreaks appear to be often separated by 5 and 32 years in this case.

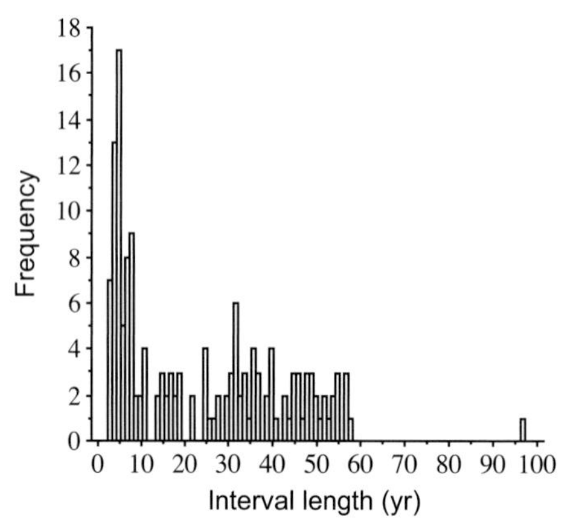

Fig. 4.6. Observed frequency distribution of the number of years separating two successive sawfly outbreaks in 24 eastern larch stands in subarctic Québec from 1743 to 1993.

Classically, an estimated transition probability is obtained by dividing the number of occurrences of the state transition of interest by the total number of state transitions observed, and an estimate of the probability of being in a given state is calculated by adding up all the estimated transition probabilities having that state as initial state of the transition. This procedure is based on the basic statistical principle of estimating the probability of an event by the observed relative frequency. When events are observed in time, as is the case with transition probabilities, the procedure is valid under two assumptions: first, the occurrence of the state transition is not dependent on the time of observation (i.e., a stationarity assumption; Section 2.3); second, the current state is dependent only on the previous state (i.e., the Markovian assumption). These assumptions are not very realistic in applications, and alternative procedures relaxing them to various degrees were proposed in the literature (e.g., Oden, 1977; Rook and Penning, 1991; Brownie *et al.*, 1993). Below, focus is on the method of time-dependent transition probabilities (Dutilleul *et al.*, 2000a), which is presented in the frame of the example that motivated its development (i.e., the chewing behavior of sheep). It can be applied to study many other animal behaviors, provided animals are followed individually; the breeding behavior of birds in captivity and the aggressive behavior of fish in aquarium are two other possible examples.

The method of time-dependent transition probabilities is based on the collection of replicated behavioral sequences in continuous time

[Fig. 4.7(a1) and (a2)]. In the sheep example, this means the chewing behavior of each individual sheep was recorded continuously during a number of days (1 day = 1 replicate). Recall that the repertoire of chewing behavior of the sheep is made of three exhaustive and mutually exclusive states: eating (state 1), ruminating (state 2), and idling (state 3). From the recorded behavioral sequences, elementary binary responses are defined for three types of time-dependent transition probabilities: the probabilities of being in a state (PBS), the probabilities of staying in a state (PSS), and the probabilities of changing of state (PCS). Note that being in a state is not a transition per se. For a given time-dependent transition probability, the elementary binary response (1/0) is evaluated in discrete time at multiples of a sampling time interval defined by the experimenter; it is equal to 1 if the event of interest is observed at the sampling time considered, and 0 otherwise [Fig. 4.7(b1) and (b2), (c1) and (c2), and (d1) and (d2)]. Histograms of the time spent in each state (i.e., the mark associated with the temporal point pattern) are very helpful in the definition of the sampling time interval. One minute was found to be an appropriate sampling time interval in the sheep example (Dutilleul *et al.*, 2000a). Finally, the probabilities PBS, PSS, and PCS are calculated by averaging the elementary binary responses over the replicates at each multiple of the sampling time interval [Fig. 4.7(d1)–(d3)].

(The calculations described with words above can be formalized in equations by using notations from Section 2.5. If $U_{frk}(t)$ denotes the elementary binary response corresponding to the state transition of interest in experimental period f for sheep r on day k at time t, then

$$U_{frk}(t) = \mu_f(t) + \varepsilon_{frk}(t),$$

by direct application of equation (2.14) and considering the sheep is a random factor. After averaging both sides of this equation over replicates (i.e., the days), it follows that

$$\bar{U}_{fr}(t) = \mu_f(t) + \bar{\varepsilon}_{fr}(t),$$

which provides the probability estimator $\bar{U}_{fr}(t)$, as an average of ones and zeroes; its expected value is $\mu_f(t)$, which contains the main effects of the time of day (through t) and the experimental period (through f), and their interaction; and its variance and temporal autocorrelation are that of the mean of random errors, $\bar{\varepsilon}_{fr}(t)$. The larger the number of replicates, the lower the variance and the higher the precision of the probability estimator at each time t.)

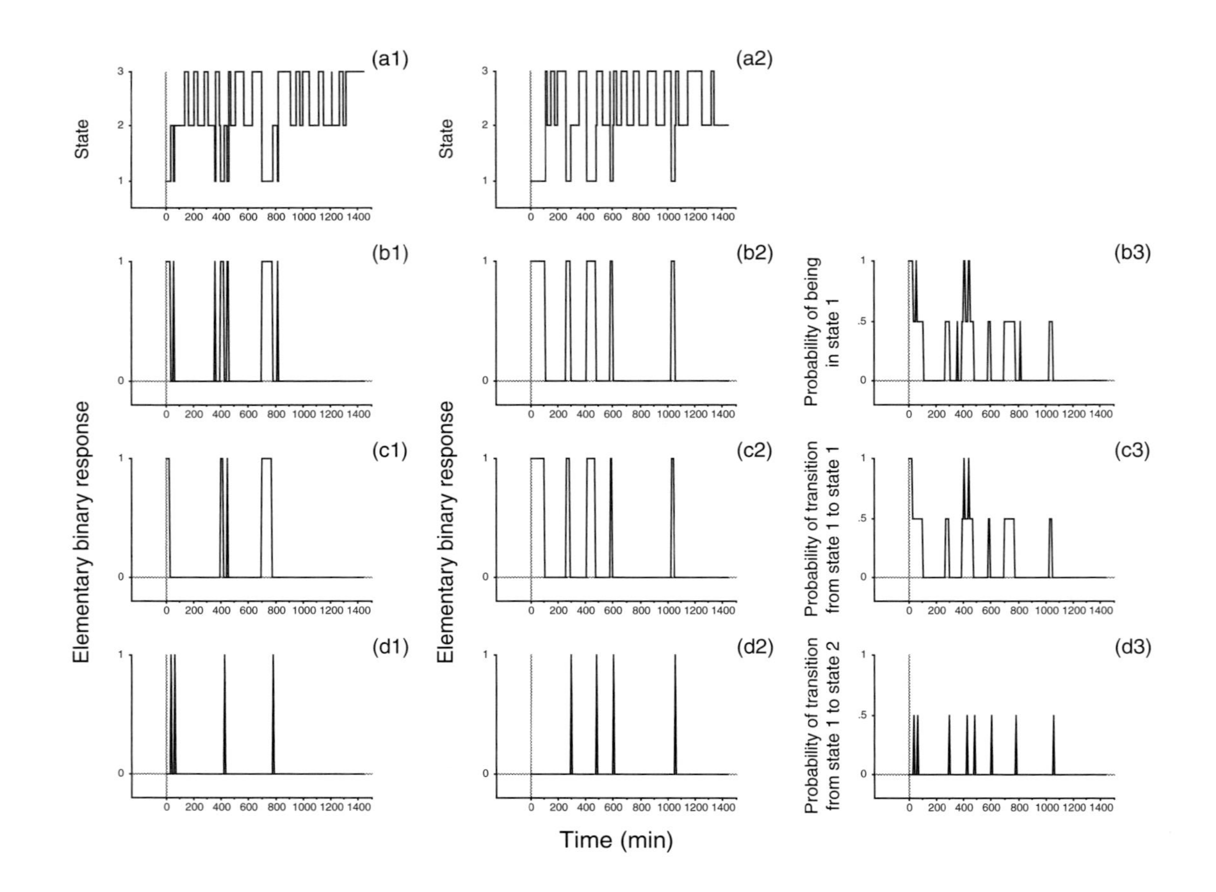
(a1)
(a2)
(b1)
(b2)
(b3)
(c1)
(c2)
(c3)
(d1)
(d2)
(d3)
State
Elementary binary response
Probability of being in state 1
Probability of transition from state 1 to state 1
Probability of transition from state 1 to state 2
Time (min)

The time-dependent transition probabilities PBS, PSS, and PCS, which were calculated for one of the two experimental periods in Dutilleul *et al.* (2000a), are plotted in Figs. 4.8 and 4.9. The slight differences between the probabilities PBS and the corresponding PSS follow from the use of the minute as sampling time interval. The temporal heterogeneity of the mean is very clear for most of the time-dependent transition probabilities. Peaks in eating correspond to troughs in ruminating and idling (Fig. 4.8). Also, the eating–idling and idling–eating transitions are much more frequent during the 10 hours following the first feeding than during the rest of the day (see PCS13 and PCS31 in Fig. 4.9). By comparison, the classical transition probabilities calculated under the stationarity and Markovian assumptions (Table 4.2) are much less informative.

Because they are calculated at multiples of a sampling time interval fixed by the data analyst, time-dependent transition probabilities are time series instead of temporal point patterns. As time series, they can be submitted to temporal autocorrelation analysis, which will be covered in Chapter 6; among other things, this method allows the identification of a range of autocorrelation different from the range of one time unit that is used under the Markovian assumption. As replicated time series, time-dependent transition probabilities can also be analyzed by the method of repeated measures ANOVA (Chapter 9).

There are at least four differences between the original method of time-dependent transition probabilities designed for the sheep example and the variant used for the mosquito example and illustrated in Fig. 4.10.

←

Fig. 4.7. Illustrative example of the calculation of time-dependent transition probabilities from replicated qualitative behavioral sequences. (a1) and (a2) Behavioral sequences recorded for the same individual during two days; 1 day = 1 replicate. At any time, the individual is in a given state from a repertoire of three exhaustive and mutually exclusive states. In the following panels, behavior is quantified at multiples of a time interval of fixed length, Δt. (b1) and (b2) Elementary binary response, which is equal to 1 at time t if the individual is found to be in state 1 at time t, and 0 otherwise. (c1) and (c2) Elementary binary response, which is equal to 1 at time t if the individual is found to be in state 1 at times t and $t + \Delta t$, and 0 otherwise. (d1) and (d2) Elementary binary response, which is equal to 1 at time t if the transition from state 1 to state 2 is observed at times t and $t + \Delta t$, and 0 otherwise. (b3), (c3), and (d3) Time-dependent probabilities of being in state 1 (i.e., PBS1), staying in state 1 (PSS1), and changing from state 1 to state 2 (PCS12), obtained by averaging the corresponding elementary binary responses over the two replicates. From Dutilleul *et al.* (2000a).

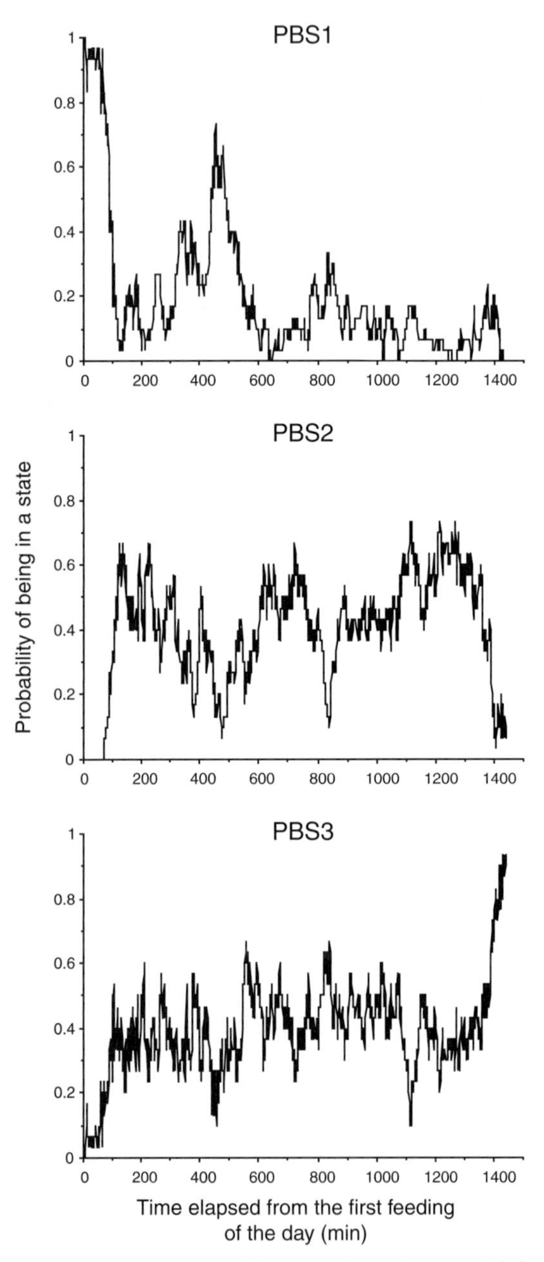

Fig. 4.8. Mean series of the time-dependent probabilities of being in the three chewing behavioral states of the sheep example: state 1 = eating; state 2 = ruminating; and state 3 = idling. Time 0 on the horizontal axis corresponds to the first feeding of the day at 09:00 h, and time 420 to the second feeding at 16:00 h. Modified from Dutilleul *et al.* (2000a).

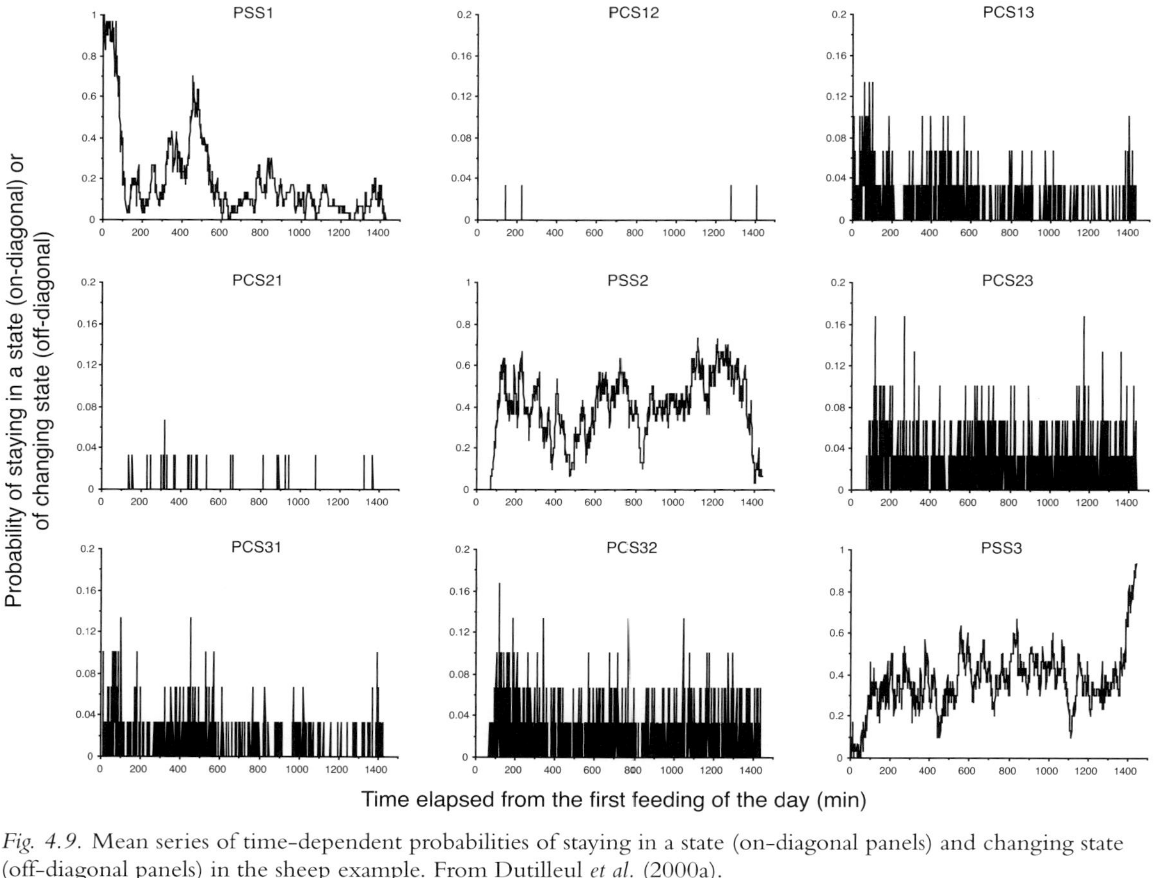

Fig. 4.9. Mean series of time-dependent probabilities of staying in a state (on-diagonal panels) and changing state (off-diagonal panels) in the sheep example. From Dutilleul *et al.* (2000a).

Table 4.2. *Overall means of the probabilities of being in a state (PBS), staying in a state (PSS), and changing of state (PCS), with the corresponding standard errors (in parentheses), for Period 1 in the sheep example (modified from Dutilleul* et al. *2000a)*

	PBS		
State	Eating	Ruminating	Idling
	0.204 (0.006)	0.404 (0.005)	0.393 (0.004)
	PSS (on-diagonal) and PCS (off-diagonal)		
Current state:	Eating	Ruminating	Idling
Previous state:			
Eating	0.193 (0.005)	0.001 (< 0.001)	0.010 (0.001)
Ruminating	0.002 (< 0.001)	0.385 (0.004)	0.017 (0.001)
Idling	0.009 (< 0.001)	0.017 (0.001)	0.366 (0.004)

The standard errors reported here are to be considered as rough measures of the variability of probabilities over time. In the presence of positive autocorrelation, this measure is known to suffer from a downward bias (Diggle, 1990, pp. 87–93). Differences between overall means of PBS and row or column totals of PSS and PCS, on the one hand, and between row totals of PSS and PCS and the corresponding column totals on the other hand, are mainly due to rounding to three digits.

First, mosquitoes, contrary to sheep, were not followed individually in Nguyen *et al.* (2002). Therefore, the time-dependent transition probabilities for mosquitoes were calculated from daily counts of first, second, third, and fourth instar larvae, pupae, and adults. Second, the scale of the day was somehow imposed to the experimenter in the mosquito example, whereas the sampling time interval used to calculate time-dependent transition probabilities within a day had to be thoughtfully determined in the sheep example. Third, a sheep may eat, ruminate, and eat again, but a first instar larva that has transformed to the second instar stage will never back-transform to first instar. It follows that there is an important reduction of the number of state transitions and transition probabilities to consider. Fourth, the time-dependent transition probabilities of changing state are interpreted differently in the two examples. Whether they are high or low, the probabilities PCS in Fig. 4.9 indicate the times of day at which state transitions in the chewing behavior of sheep are the most or the least likely to occur. The probabilities PSS1 and PCS12 in Fig. 4.10 evaluate the chances for first instar mosquito larvae to remain first instars and to transform to second instars on a daily basis. Both probabilities are heterogeneous in time and their values even differ whether the state transitions are considered within 24 or 48 hours. The probability

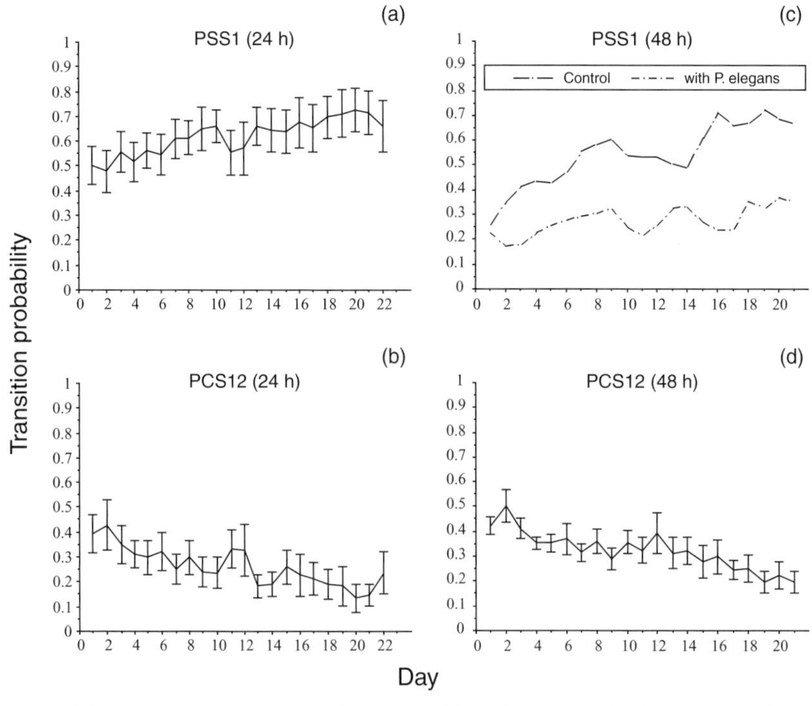

Fig. 4.10. Time-dependent transition probabilities that mosquito (*Aedes aegypti*) first instars (a) remain first instars within 24 h; (b) become second instars within 24 h; (c) remain first instars within 48 h, without vs. with exposure to the parasite *Plagiorchis elegans*; and (d) become second instars within 48 h. Modified from Nguyen *et al.* (2002).

that first instars remain first instars after 24 hours is greater than after 48 hours, both for the control group and for the group infected by the parasite *Plagiorchis elegans* [see PSS1 (24 h) vs. PSS1 (48 h)]; conversely, the probability that first instars transform to second instars within 24 hours is smaller than within 48 hours [see PCS12 (24 h) vs. PCS12 (48 h)].

Summary: The fundamental concepts related to the heterogeneity of point patterns and the main procedures designed to analyze it in the spatial framework (Section 3.2 and following sections in Chapter 3) are applicable in the temporal framework (this section). However, some of the questions that are of interest in the latter framework are specific to time because it is ordered. Such questions are addressed in the analysis of time-dependent transition probabilities, which can be seen as an extension of or a follow-up to the analysis of marked temporal point patterns when the mark is qualitative.

4.2 Spatio-temporal point patterns

The space-time case is not as difficult as it may seem, provided the methods and techniques applied separately in the purely spatial and purely temporal frameworks have been well understood. In fact, the cumulative relative frequency distribution of some type of distance and the beta-binomial distribution, with well-defined quadrats and cells, are readily applicable in space-time when the plot in which spatial point patterns are observed is entirely surveyed on a number of occasions. Excluding fractional surveys (i.e., when different parts of the plot are surveyed at different times), we will see that spatio-temporal point patterns can be analyzed via distances (Subsection 4.2.1), as if they were multivariate spatial point patterns. In Subsections 4.2.1 and 4.2.2, the beta-binomial distribution will be applied to spatio-temporal point patterns, in particular to analyze the movement of points (Subsection 4.2.2). Two examples with real data (*Arisaema* and California earthquakes) and one example with realistic scenarios (wolves) will be discussed.

4.2.1 Distances and counts

The question of a general distance between points in space-time remains open (Sklar, 1984, p. 58; see also Section 8.1 here), but such a distance is not needed in what follows. Consider two spatial point patterns observed in the same plot at two different times (i.e., time 1 $<$ time 2), as the components of a bivariate spatial point pattern. Then, the nature of the distance between a point at time 1 and the nearest point at time 2 is spatial instead of spatio-temporal, because the two spatial point patterns are observed in the same plot.

Below, the questions addressed are: How are the points of the spatial pattern at time 2 distributed relative to those of the spatial pattern at time 1? More specifically, are the points observed at time 2 closer to, or farther from, the points at time 1 than they would be if the spatial point pattern at time 2 was relocated in the plot by keeping the same distances between points of this pattern under the toroidal model (i.e., by using a toroidal shift)? In the statistical analysis, these two questions become: Where does the cumulative relative frequency distribution of the distances from each point at time 1 to the nearest point at time 2 (i.e., the inter-distances) position itself compared to the lower and upper envelopes evaluated by using 999 independent toroidal shifts of the spatial point pattern at time 2? Note that a departure of the observed curve from the envelopes in this case does not imply that the spatial point pattern at time 2 is regular or aggregated. To see this, imagine two spatial completely random point patterns such that for each point of one pattern, there is a point of the other pattern very close. Compared with the envelopes, the observed cumulative relative frequency distribution will show an excess of small inter-distances and a lack of large inter-distances. Both spatial point patterns are, however, completely random. Simply, they are not independent of each other, or the two distributions of points are positively correlated in space. Correlation, when it exists between spatial point patterns at two different times, is not restricted to be positive. For example, consider two spatial aggregated point patterns with clusters located in different parts of the plot from time 1 to time 2. In this case, the observed cumulative relative frequency distribution will show a lack of small inter-distances and an excess of large inter-distances. Thus, the correlation between spatial point patterns is negative, and means a complementary occupancy of space from time 1 to time 2, as reflected by the increase of inter-distance values. Formal tests of independence for spatial point patterns exist, but require that the underlying point processes be near complete randomness (Upton and Fingleton, 1985, p. 243). In the following two examples, spatio-temporal point patterns are analyzed with the multivariate version of Diggle's randomization testing procedure using the inter-distances (see the multivariate case in Section 3.5).

The *Arisaema* example in space-time is based on the four 2-D spatial point patterns that arose from the survey of healthy juvenile plants in Plot 8 in the summer of 1988 ($n = 371$), 1989 ($n = 349$), 1990 ($n = 337$), and 1991 ($n = 399$). The aggregation of 2-D spatial point patterns is clear (Fig. 4.11). The ecological hypothesis to be tested concerns the population dynamics of juvenile plants over the years in response to mortality and maturation: juveniles in 1991 (the last year) would colonize different sites than juveniles in 1988 (the first year), so the closeness in space to be expected for juveniles of 1988 and 1989 would be replaced by a different spatial point pattern in 1991. The data analyses support the hypothesis in that the observed cumulative relative frequency distribution of inter-distances between 1988 and 1989 crosses the 2.5th percentile envelope often, and is close to it throughout [Fig. 4.12(a)]. They do not support the hypothesis in that the three observed curves behave similarly with respect to their respective 2.5th percentile envelopes [Fig. 4.12(b) and (c)], despite a relatively important mortality in 1990 [see Arisaema 10 in Subsection 3.3.2]. There are merely

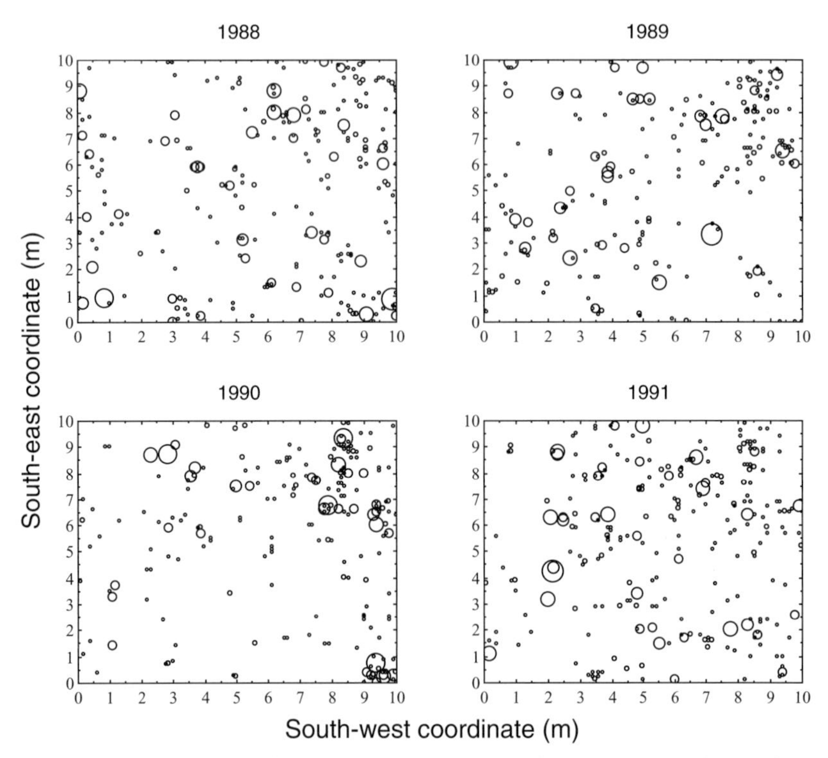

Fig. 4.11. Arisaema example in space-time: 2-D spatial point patterns observed in a given plot in the summer of years 1988–1991.

fewer zero inter-distances between 1988 and 1991 (2.5%) than between 1988 and 1990 (7.0%). Such constancy in space occupancy by juvenile plants over the years is likely to follow from environmental conditions favorable for seed germination and early development of *Arisaema* that are only met in specific parts of the plot.

The example of California earthquakes in space-time results from the splitting of the original dataset into three 20-year periods, so the superimposing of the three 2-D spatial point patterns mapped in Fig. 4.13 provides the 2-D spatial point pattern of Fig. 3.12(a). Some consistency in the spatial location of earthquakes is expected from one 20-year period to another because 60 years should not be long enough to observe a change in the positioning of tectonic plates. Clearly, fewer earthquakes of magnitude 5 or more occurred in 1960–1979 ($n = 85$) than in 1940–1959 ($n = 166$) and 1980–1999 ($n = 153$), and the earthquakes of each period are aggregated in space [Fig. 4.13(a)–(c)]. Though mainly located

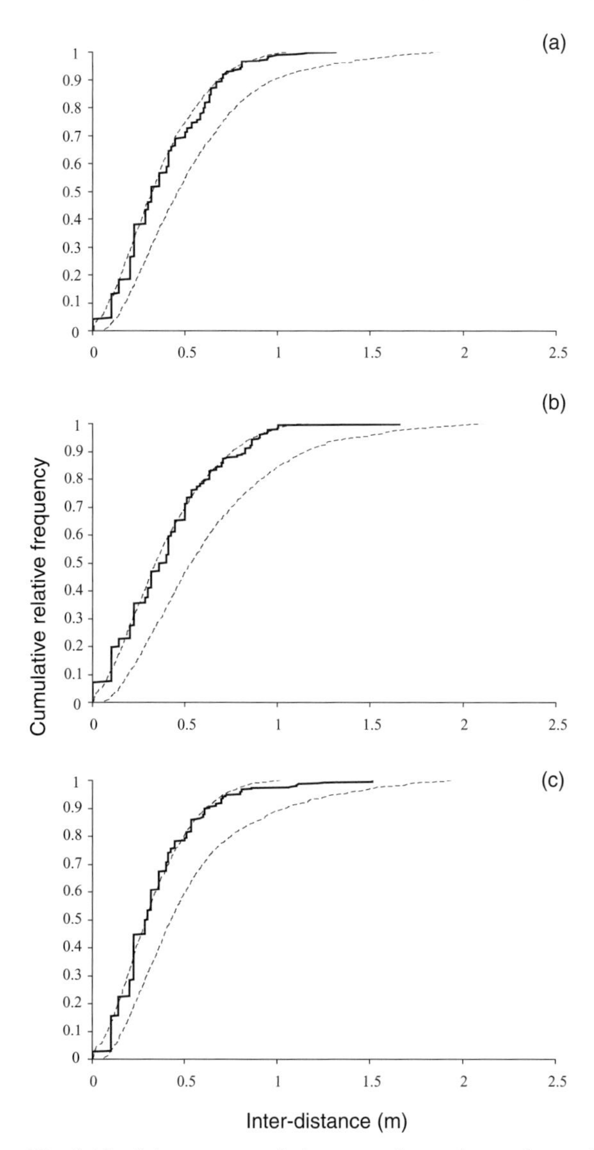

Fig. 4.12. Arisaema example in space-time: observed cumulative relative frequency distributions (bold solid curves) of spatial distances from points in 1988 to the nearest points in (a) 1989; (b) 1990; and (c) 1991, with the 2.5th and 97.5th percentile envelopes (dashed curves) evaluated from 999 independent toroidal shifts of the 2-D spatial point patterns observed in 1989–1991 (Fig. 4.11).

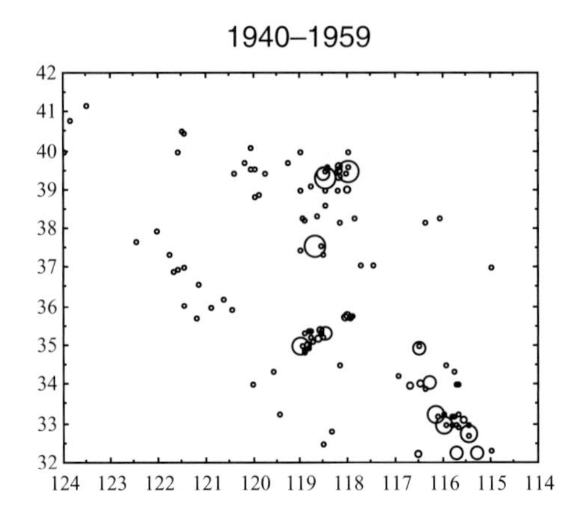

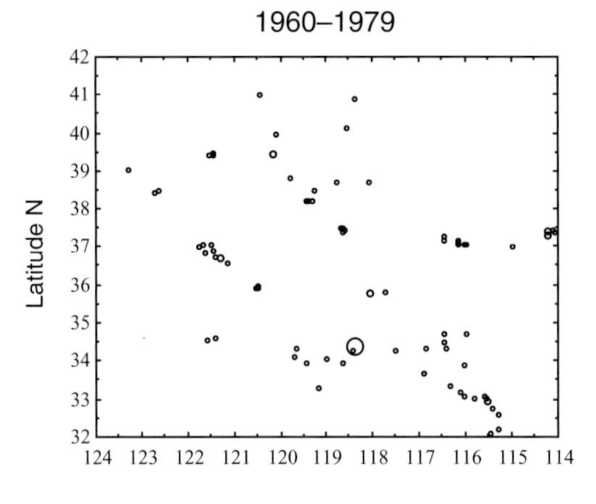

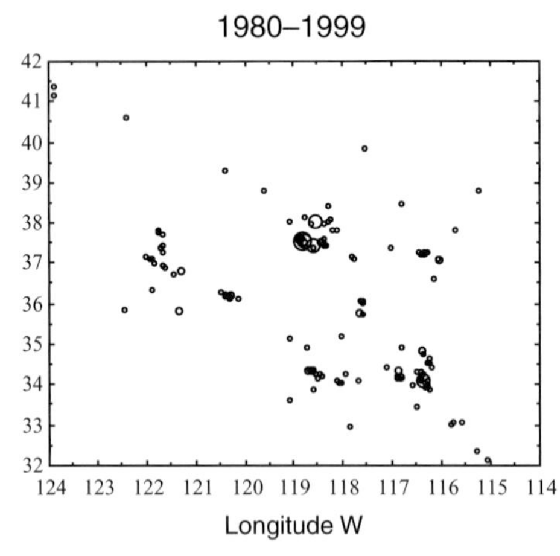

Fig. 4.13. Example of California earthquakes in space-time: observed 2-D spatial point patterns of earthquakes of magnitude 5 or more in successive 20-year periods from January 1, 1940, to December 31, 1999.

in the same zones, some reduction in size or some displacement of the clusters is observed from one period to another. The data analyses support the expectation in that the three comparisons between periods produce similar results: the observed cumulative relative frequency distribution of inter-distances is above the 2.5th percentile envelope over most of the inter-distances from 0 to 50, 100, or 150 km depending on the comparison (Fig. 4.14). It means that the spatial point pattern observed in the latter of the two periods compared is closer to the one in the former period, than the great majority of its toroidal shifts. The smaller number of earthquakes in 1960–1979 may explain the fluctuations of the observed curve, from 50–100 km or 50–150 km, in the two comparisons where that period is involved [Fig. 4.13(a) and (c)].

In the count-based approach, the beta-binomial distribution can be applied as follows to study aggregation in spatio-temporal point patterns. This application requires an appropriate definition of quadrats and of spatio-temporal cells into which quadrats are divided. The data of the example of California earthquakes in space-time are used for illustration. Recall that these data cover a territory of 10 degrees of longitude $\times$ 10 degrees of latitude in space and a period of 60 years in time. For our purpose, the territory is split into 10 $\times$ 10 2-D spatial cells of 1 degree of longitude $\times$ 1 degree of latitude and the period of 60 years is divided into six 10-year intervals [Fig. 4.15(a)]. One spatio-temporal cell follows from the crossing of one spatial cell of 1 degree of longitude $\times$ 1 degree of latitude with one temporal interval of 10 years. The quadrat size is fixed to eight (i.e., 2 $\times$ 2 spatial cells crossed with two successive 10-year intervals), which is the minimum required for a spatio-temporal BBD analysis like that performed here. One can then determine whether at least one earthquake of magnitude 5 or more has occurred in each spatio-temporal cell, count the number of these cells in each quadrat, and report the counts in three 5 $\times$ 5 squares representing the 75 quadrats [Fig. 4.15(b)]. In view of the counts listed in the BBD input, one could expect the beta-binomial distribution to perform better than the binomial distribution because the observed frequencies of more extreme counts (i.e., zero and 4–7) are too high for the binomial to fit well. The BBD output (i.e., the estimate $\hat{\theta}$ and its standard error, the BBD frequencies) confirms the better fit of the beta-binomial (Table 4.3). Whether the aggregation of earthquakes in space-time is more spatial than temporal, or the reverse, is a good question. Since temporal aggregation was found only at the scale of the month over the period 1990–1999 (Subsection 4.1.2), while the spatial analyses showed aggregation at several scales

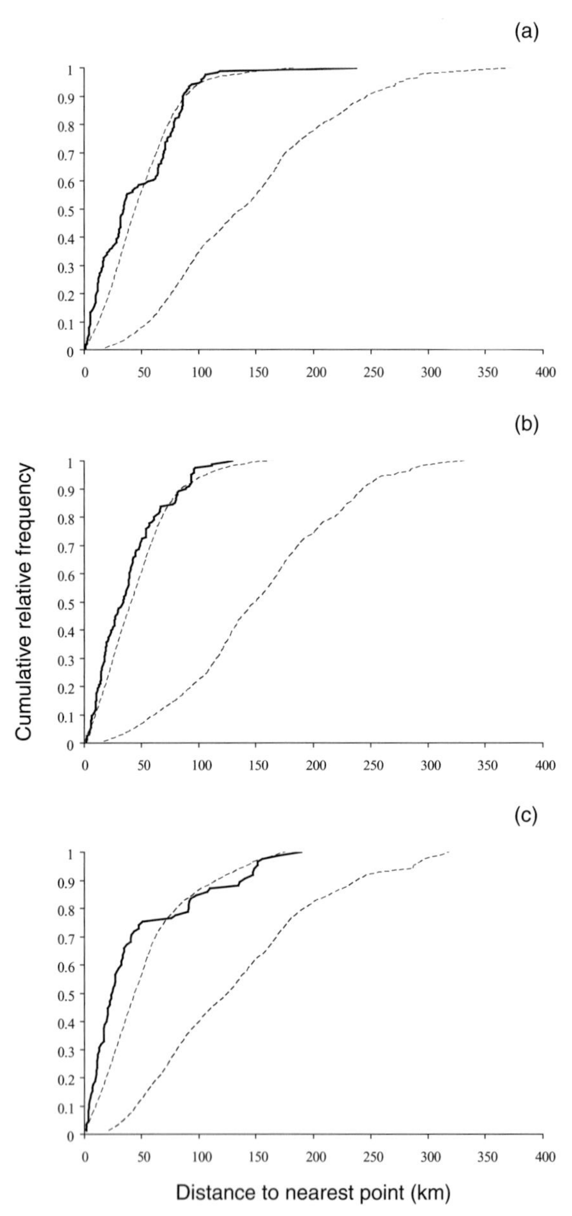

Fig. 4.14. Example of California earthquakes in space-time: observed cumulative relative frequency distributions (bold solid curves) of spatial distances (a) from points in 1940–1959 to nearest points in 1960–1979; (b) from points in 1940–1959 to nearest points in 1980–1999; and (c) from points in 1960–1979 to nearest points in 1980–1999, with the 2.5th and 97.5th percentile envelopes (dashed curves) evaluated from 19 999 independent toroidal shifts of the 2-D spatial point pattern observed in the latter 20-year period (Fig. 4.13).

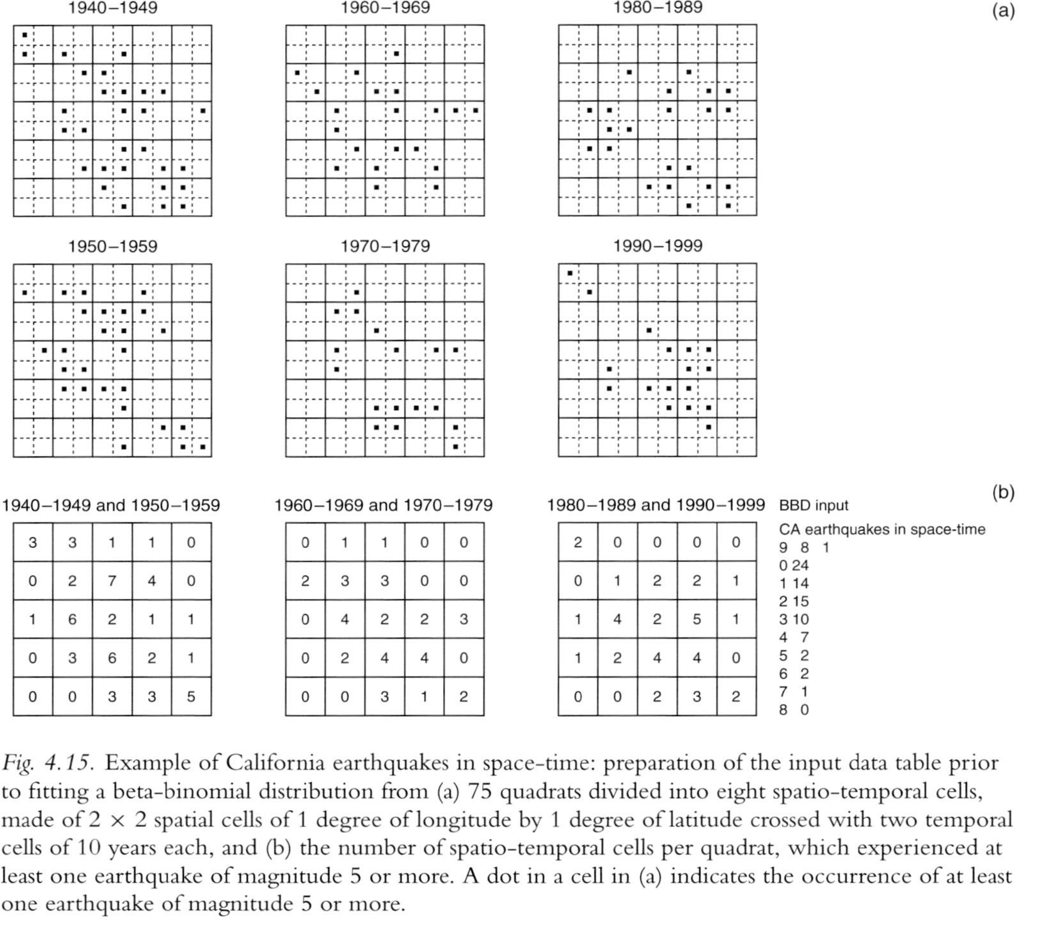

Fig. 4.15. Example of California earthquakes in space-time: preparation of the input data table prior to fitting a beta-binomial distribution from (a) 75 quadrats divided into eight spatio-temporal cells, made of 2×2 spatial cells of 1 degree of longitude by 1 degree of latitude crossed with two temporal cells of 10 years each, and (b) the number of spatio-temporal cells per quadrat, which experienced at least one earthquake of magnitude 5 or more. A dot in a cell in (a) indicates the occurrence of at least one earthquake of magnitude 5 or more.

Table 4.3. *Summary of the BBD output for the example of California earthquakes in space-time, using 75 quadrats of 2 degrees of longitude × 2 degrees of latitude × 20 years (Fig. 4.13) and dividing each of them into four spatial cells of 1 degree of longitude × 1 degree of latitude crossed with two temporal intervals of 10 years*

$\hat{p} = 0.217$ SE($\hat{p}$) $= 0.024$

$\hat{\theta} = 0.205$ SE($\hat{\theta}$) $= 0.062$

x	Observed frequency	BBD frequency	Binomial frequency
0	24	22.74	10.55
1	14	17.85	23.45
2	15	13.11	22.80
3	10	9.10	12.67
4	7	5.91	4.40
5	2	3.51	0.98
6	2	1.83	0.14
7	1	0.76	0.01
8	0	0.20	0.00

BBD goodness-of-fit: $\chi^2_{obs} = 1.729$

Binomial goodness-of-fit: $\chi^2_{obs} = 31.769$

(Subsection 3.2.5), spatial heterogeneity appears to be predominant in this example.

4.2.2 Point processes and movement

Points are essentially static, and possibly numerous, in Chapter 3 and the rest of this chapter. A point will be mobile and maybe unique hereafter. Movement analysis, of which a brief introduction is given below, is indirectly related to heterogeneity, and goes beyond the links that will be made with the analysis of spatio-temporal aggregated point patterns.

In the same way that state transitions need time to occur, movement needs space and time to be observed because it is characterized (in large sense) by the displacement of a point from (x, y, z, t) to (x', y', z', t') in space-time. Instead of following the point in its displacement, a binary coding can be applied to its path through a grid of spatio-temporal quadrats divided into a number of cells. Once this binary coding is completed, one can count the number of spatio-temporal cells visited by the mobile point in each quadrat. It follows that the beta-binomial

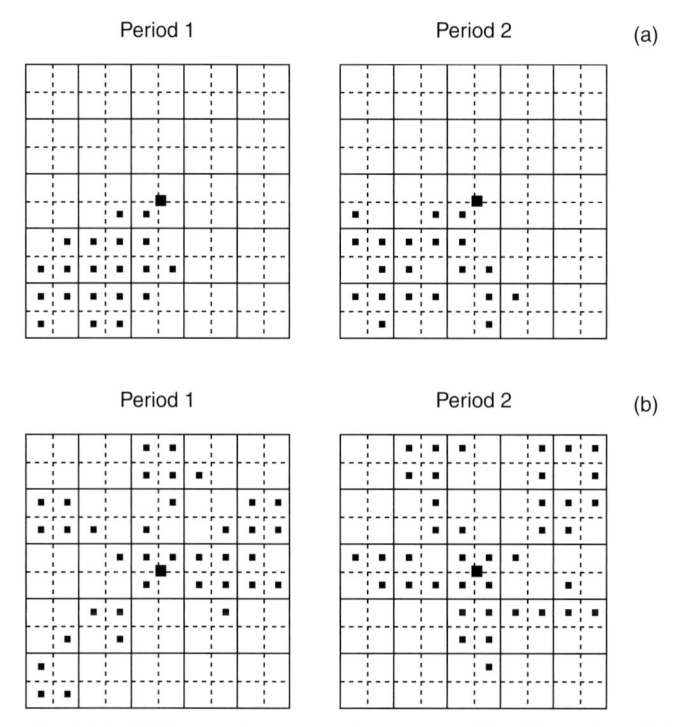

Fig. 4.16. Wolf example in space-time: map of the 25 quadrats divided into 2×2 spatial cells that the alpha female wolf visited or not during her hunting expeditions in two periods, according to scenarios showing (a) more or (b) less similarity in movement of the alpha female wolf between periods. The big dot in the center of the squares indicates the location of the den.

distribution can be used again, not only to measure the spatial aggregation of visits of the mobile point over a given period but also to assess the similarities or differences in the spatial aggregation of visits between periods.

For example, assume the movement of an alpha female wolf is monitored continuously in 2-D space and time. All her displacements from and to the den are recorded, and two sets of records corresponding to two scenarios (i.e., scenario 1 and scenario 2), with two periods (i.e., period 1 and period 2) of 5 days in each scenario, are available for BBD analyses. Under scenario 1, the hunting expeditions of the wolf follow approximately the same path in both periods, indicating some prolonged abundance of prey in a given zone. Under scenario 2, she never follows the same path on any two days because of a generalized abundance of prey or, on the contrary, because of their scarcity. The two scenarios are represented in Fig. 4.16(a) and (b), where a dot in a spatio-temporal cell

Table 4.4. *Summary of BBD outputs for the wolf example of Fig. 4.16, obtained by using 25 quadrats of 2 × 2 spatial cells crossed with two temporal cells under (a) scenario 1 and (b) scenario 2*

(a)				(b)			
$\hat{p} = 0.183$	$SE(\hat{p}) = 0.053$			$\hat{p} = 0.379$	$SE(\hat{p}) = 0.047$		
$\hat{\theta} = 0.992$	$SE(\hat{\theta}) = 0.410$			$\hat{\theta} = 0.140$	$SE(\hat{\theta}) = 0.083$		
x	Observed frequency	BBD frequency	Binomial frequency	x	Observed frequency	BBD frequency	Binomial frequency
0	15	14.76	4.97	0	3	2.03	0.55
1	2	2.78	8.90	1	3	3.83	2.70
2	2	1.69	6.97	2	1	4.75	5.77
3	1	1.27	3.12	3	9	4.74	7.04
4	0	1.05	0.87	4	5	4.01	5.37
5	0	0.92	0.16	5	2	2.89	2.62
6	4	0.84	0.02	6	0	1.74	0.80
7	1	0.81	0.00	7	2	0.80	0.14
8	0	0.89	0.00	8	0	0.22	0.01
BBD goodness-of-fit: $\chi^2_{obs} = 0.012$				BBD goodness-of-fit: $\chi^2_{obs} = 4.856$			
Binomial goodness-of-fit: $\chi^2_{obs} = 1.591$				Binomial goodness-of-fit: $\chi^2_{obs} = 4.488$			

indicates it has been visited by the wolf on at least one occasion (code 1) and no dot means the contrary (code 0). The value of 0.992 for $\hat{\theta}$ (Table 4.4) leaves little doubt about the spatio-temporal aggregation (positive autocorrelation) of the visits of the wolf in scenario 1, and results from the presence of one "cluster" of large size in the bottom-left quarter of the plot in both periods. The binomial distribution is totally eclipsed by the beta-binomial distribution in this scenario, due to the combination of 15 zero counts and four counts of six visits. In fact, such a combination would be unlikely if the probability of visit was constant throughout the plot and over time. The situation is different with scenario 2, for which the observed frequencies of counts are more evenly distributed between 0 and 5 and two counts of seven events are observed. Accordingly, the value of $\hat{\theta}$ drops to 0.140 and the beta-binomial provides a worse goodness-of-fit than the binomial. These numerical results reflect that in scenario 2 the alpha female wolf inconsistently visited the area around the den within and between periods.

Summary: In the distance-based approach to the analysis of spatio-temporal point patterns, the distribution of inter-distances can be used to measure how close two spatial point patterns observed in the same plot at different times are. In the count-based approach, the beta-binomial distribution can be applied to study aggregation in spatio-temporal point patterns and to analyze the movement of points in space and time, which confirms the generality and usefulness of this statistical distribution.

5 · *Heterogeneity analysis of point patterns: process modeling and summary*

Quantification of the heterogeneity contained in an observed point pattern was the topic of Chapters 3 and 4. In this chapter, the modeling of parameter functions of the underlying point process is discussed, and a general summary concerning the pros and cons of the quadrat-based and distance-based approaches available for heterogeneity analysis of point patterns is presented. Specific aspects related to the modeling of the first-order intensity function, second-order moment functions, and the distribution of nearest-neighbor distances are studied in Subsections 5.1.1–5.1.3. Section 5.1 is completed with a discussion of the conditions of application of Ripley's functions, which are involved in the modeling of second-order moments of a point process. The general summary announced above is presented in Section 5.2. Simulated and real data are used in the course of presentation of the material. The modeling aspects covered here hopefully provide a good, simple and clear introduction to the subject, but represent only elements of point process modeling. Complementary readings on the analysis of point patterns are proposed in Section 5.3.

5.1 Elements of point process modeling

In the two previous chapters, the heterogeneity of an observed point pattern was studied in three parts (mean, variance, autocorrelation) in the count-based approach, and was described empirically through the distribution of distances (e.g., nearest-neighbor distances, inter-distances) in the distance-based approach. In this section, a step further is made by using the observed point pattern to infer three main characteristics of the underlying point process: the first-order and second-order intensity functions defined in Section A3.1, and the statistical distribution of nearest-neighbor distances. Previous examples with real data (*Arisaema*, California earthquakes) and a new example provided by polygonal terrain networks on Earth and Mars are used to keep developments concrete.

5.1.1 Modeling of first-order intensity function

In equation (A3.1), the first-order intensity function $\lambda_1(s)$ was introduced as the point density function that, once integrated over some quadrat A, provides $E[N(A)]$, the population mean of $N(A)$ or the number of points to be expected in the quadrat. Thus, $\lambda_1(s)$ is defined at any s; its domain of definition is continuous. One can extract the first-order intensity function from the integral by deriving mathematically both sides of equation (A3.1), that is, by calculating the limit of the ratio of the number of points expected in quadrat A to the size $|A|$ of the quadrat as $|A|$ tends to zero:

$$\lim_{|A|\to 0} \frac{E[N(A)]}{|A|} = \lambda_1(s), \tag{5.1}$$

where quadrat A is assumed to be centered at s. The fact of speaking of a quadrat suggests that the framework is 2-D spatial, but A may be an interval in time or 1-D space or a cell in 3-D space as well.

In applications, the theoretically expected number of points in any quadrat is unknown, and so is $\lambda_1(s)$ at any s. The observed point pattern, via the numbers of points counted in quadrats of a given size, can be used to estimate $\lambda_1(s)$ at the center of each quadrat, by the observed number of points divided by a measure of the size of the quadrat, and the estimates can be mapped over the study area. It is impossible to estimate $\lambda_1(s)$ for each and every s in the study area, but in a study area of 100 m^2, for example, it is possible to calculate more than 100 estimates by working with 1 m $\times$ 1 m quadrats; it suffices to remove the constraint that quadrats be non-overlapping! Quadrats can also be of another shape than square (e.g., circular), and in the estimation less weight can be given to points inside the quadrat but far from the center than to points close to it. This simplified description is that of kernel-based estimators of the first-order intensity function; see Schabenberger and Gotway (2005, pp. 110–112) for a formal description. The choice of the kernel function (e.g., uniform – a full weight of 1 is given to each point within a quadrat – vs. quadratic or negative quadratic exponential) is less important for the quality of the estimation than the choice of the quadrat size, or "bandwidth." With a small bandwidth, the expected value of $\hat{\lambda}_1(s)$ is close to $\lambda_1(s)$ (i.e., the estimator is almost unbiased), but the estimated function $\hat{\lambda}_1(s)$ is highly variable. Conversely, with a large bandwidth, the estimated first-order intensity function is more biased and smoother.

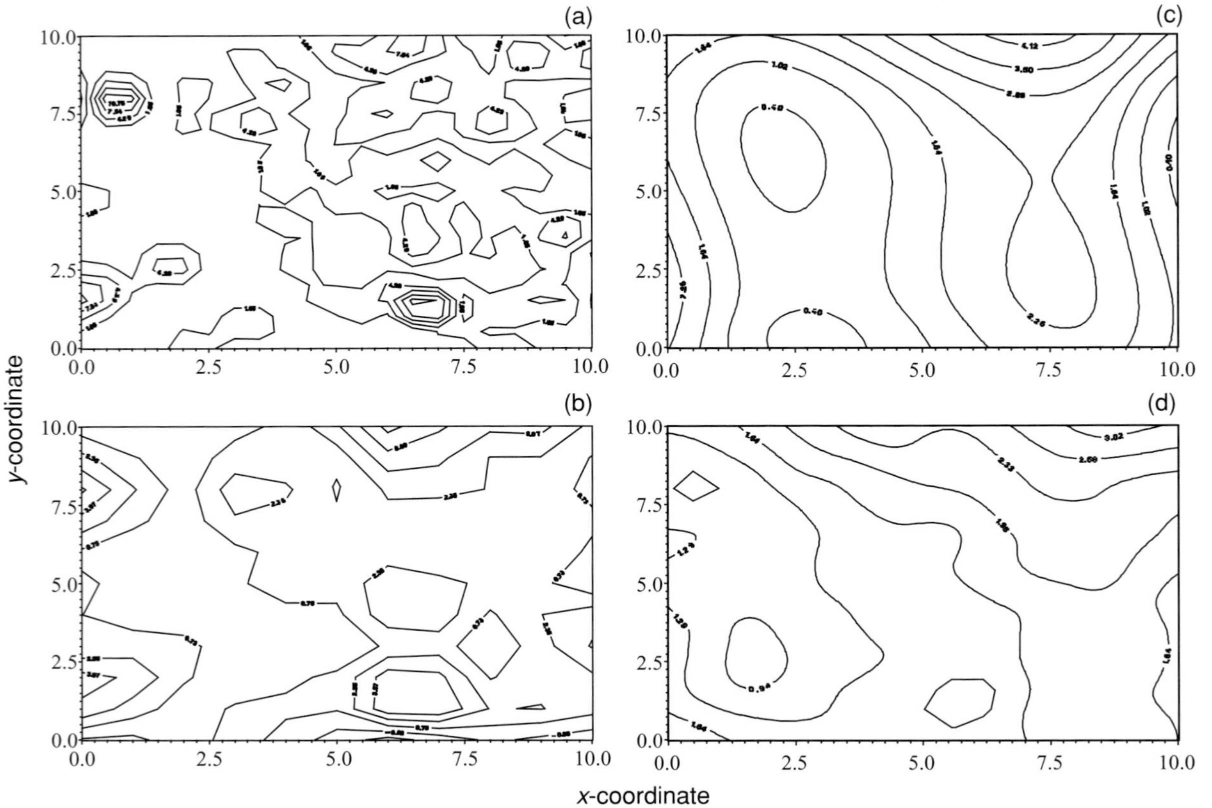

Fig. 5.1. First-order intensity function estimates obtained from the Arisaema 5 point pattern of Fig. 3.8, by using a kernel with (a) a smaller or (b) a larger bandwidth vs. by using (c) a third-degree polynomial in the spatial coordinates x and y or (d) a moving window with a diameter of 5 m, to model the large-scale component of the intensity function.

Below, the Arisaema 5 dataset (*Arisaema* example, Subsection 3.2.5) is used for an illustration.

Since the spatial location of *Arisaema* was recorded at the 0.1 m × 0.1 m scale and each plot was 10 m × 10 m, counts made in quadrats of 0.5 m × 0.5 m (small bandwidth) and 1 m × 1 m (large bandwidth) were used to estimate the first-order intensity function of the underlying point process every 0.25 m along each spatial axis; mapping was completed with a bivariate spline [Fig. 5.1(a) and (b)]. Similar spatial distributions characterized by peaks evenly distributed along the y-axis are observed, except that peaks on the left side of the plot close to the edge and in the right half are of medium size and height (from 4 to 10) with the small bandwidth, and of larger size and intermediate height (between 2 and 4) with the large bandwidth. These results are in accordance with a comment made in view of Fig. 3.8 in Subsection 3.2.5: the observed 2-D spatial point pattern of Arisaema 5 presents a non-monotonic trend in point density from left to right.

In a case like Arisaema 5, it is possible to model the large-scale component of the first-order intensity function of the point process, by fitting a polynomial regression model in spatial coordinates x and y (called "trend surface" in Chapter 7; see also Diggle, 2003, p. 105) to quadrat counts obtained in the classical manner (Subsection 3.2.3).

Key note: The fact of working with quadrat counts in the point-pattern approach allows the application of statistical methods designed for the analysis of surface patterns. The statistical method of this type applied here is coregionalization analysis with a drift (CRAD), due to Pelletier et al. *(2009a, 2009b); see Chapter 7.*

The result of such a fitting to 100 counts in 1 m × 1 m quadrats, using a third-degree polynomial in x and y as regression model and adjusting for spatial autocorrelation between quadrat counts with an appropriate least-squares estimation procedure (i.e., generalized least squares; see Chapter 7), is shown in Fig. 5.1(c). Statistically speaking, the coefficients of all the terms including y or one of its powers are non-significant, while those of x, x^2 and x^3 are of alternating sign (i.e., −, +, −) and significant ($P < 0.05$). The significance reflects the left-to-right trend in point density, while the changes in sign correspond to the absence of *Arisaema* in a main part of the left half of the plot due to unfavorable growth conditions [see the trough in estimates of $\lambda_1(x, y)$], which results in fluctuations of the point density along the x-axis. As an alternative to a global modeling of the large-scale component of $\lambda_1(x, y)$, a reduced trend surface model (e.g., a plane in x and y) can be fitted to same quadrat counts with the same adjustment for spatial autocorrelation between quadrat counts, but within a moving window over the plot. In the case of Arisaema 5, the result [Fig. 5.1(d)] resembles the contour map obtained with the kernel function with large bandwidth [Fig. 5.1(b)].

5.1.2 Modeling of second-order moment functions

Important second-order moments in statistics are generally centered; see the population variance and equation (A2.2), for example. In point pattern analysis, the second-order intensity function can be seen as an exception to this rule. Like $\lambda_1(s)$, it can be written as a limit:

$$\lim_{|A|,|B|\to 0} \frac{\mathrm{E}[N(A)\,N(B)]}{|A|\,|B|} = \lambda_2(s, s'), \tag{5.2}$$

where $A \neq B$ are two quadrats centered at s and s', respectively. But $\mathrm{E}[N(A)N(B)]$ differs from the covariance between counts $N(A)$ and $N(B)$ (i.e., the centered second-order joint moment), by the quantity $\mathrm{E}[N(A)]\,\mathrm{E}[N(B)]$, since $\mathrm{Cov}[N(A),\ N(B)] = \mathrm{E}[\{N(A) - \mathrm{E}[N(A)]\}\{N(B) - \mathrm{E}[N(B)]\}] = \mathrm{E}[N(A)\ N(B)] - \mathrm{E}[N(A)]\ \mathrm{E}[N(B)]$. In words, the numerator on the left-hand side of equation (5.2) is a measure of the degree of association between the numbers of points in quadrats A and B, but it is not a classical one. Therefore, second-order moment functions related to $\lambda_2(s, s')$ but not $\lambda_2(s, s')$ itself are used for modeling in point pattern analysis. This is the case in particular with Ripley's (1981) K and L functions.

Under first-order stationarity and isotropy (Sections 2.3 and A3.1), that is, $\lambda_1(s) = \lambda_1$ for all s and $\lambda_2(s, s') = \lambda_2(h)$ with $h = \|s' - s\|$, Ripley's K function is the anti-derivative of the second-order intensity function, up to a multiplicative constant:

$$\frac{\mathrm{d}K(h)}{\mathrm{d}h} = \frac{2\pi h}{\lambda_1^2}\lambda_2(h). \tag{5.3}$$

For questions of better statistical properties and greater ease of interpretation in applications, the following function is used instead:

$$L(h) - h \text{ with } L(h) = \sqrt{\frac{K(h)}{\pi}}. \tag{5.4}$$

I refer to Schabenberger and Gotway (2005, pp. 102–103) for the estimation of Ripley's K and L functions, with edge correction. Hereafter, L will denote the function $L(h) - h$.

Key note: First-order stationarity and isotropy are important conditions of application of Ripley's functions. First-order stationarity in particular and the implications of it not being satisfied (i.e., when there is heterogeneity of the mean) for the analysis of the L function will be discussed in Subsection 5.1.4.

As a first illustration of the overall behavior of the K and L functions defined above, the three simulated 2-D spatial point patterns of Fig. 3.1 are used again here [Fig. 5.2(a1) and (a2)]. In view of Fig. 5.2(a1), differences between the estimated K functions for the completely random point pattern and the regular point pattern are only apparent for distances h of 7.5 and more. One cannot discern it in the figure, but the long-dashed curve (regular pattern) is systematically below the continuous one (completely random pattern) for distances h smaller than 7.5; recall that the plot is a 10×10 square. More clearly, the short-dashed curve (aggregated pattern) fluctuates above and below the continuous curve; in particular, the former is above the latter over small distances up to the diameter of a cluster and over distances greater than the average distance between cluster centers.

Once multiplied by λ_1 or λ_1^2, $K(h)$ is usually interpreted in terms of the expected number of points within distance h from a point (Ripley, 1981, p. 158; Schabenberger and Gotway, 2005, p. 101). From the estimates of $L(h) - h$ plotted in Fig. 5.2(a2), an interpretation generalizing that of Pearson's r statistic introduced as a measure of autocorrelation between counts in neighboring quadrats in Subsection 3.2.3 can be made. *Note:* Any correlation coefficient is based on a centered second-order joint moment; see Appendix A2. There is no autocorrelation in the completely random point pattern over distances from 0 to 7.5; there is moderate negative autocorrelation over small distances of 2 and less in the regular point pattern; and there seems to be autocorrelation alternating in sign and varying in magnitude over distances from 0 to about 8 in the aggregated point pattern. The latter sine wave behavior of $\hat{L}(h) - h$ results from the well-delimited clusters in the aggregated point pattern of Fig. 3.1(c) and the deterministic nature of heterogeneity in this case. Strictly speaking, the "observed autocorrelation" in this case is spurious and the behavior of the estimated function actually reflects spatial heterogeneity of the mean of patchy type. In the next illustrations, the estimation of L will be restricted to small distances (e.g., 3 in a 10×10 square) far from the maximum distance between two points.

Another aggregated 2-D spatial point pattern provided by a simulated partial realization of a Cox process is displayed in Fig. 5.2(b1). Unlike the

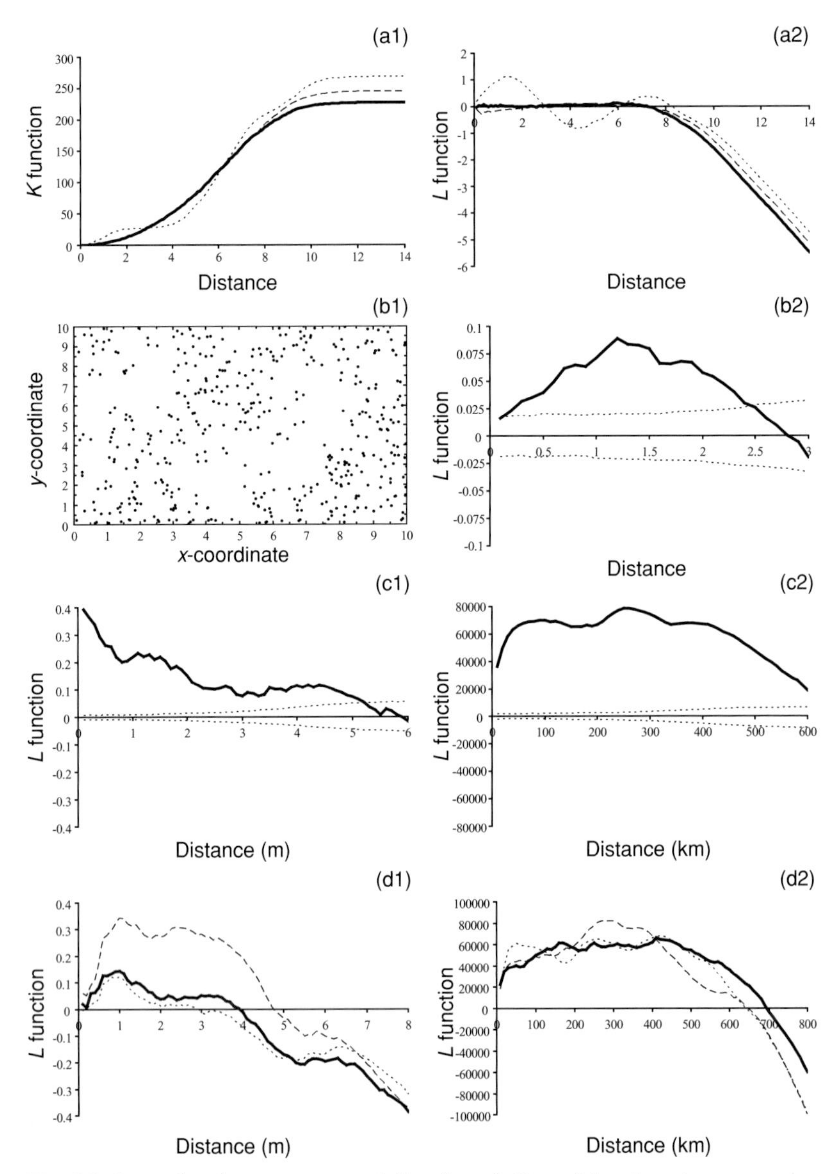

Fig. 5.2. Second-order moment modeling from 2-D spatial and spatio-temporal point patterns. (a1) and (a2) Estimated Ripley's K and L functions for the three simulated 2-D spatial point patterns of Fig. 3.1 (continuous curve: completely random; long-dashed: regular; short-dashed: aggregated). (b1) Autocorrelated 2-D spatial point pattern ($n = 429$) generated from 100 counts simulated on a 10×10 grid by using a spherical variogram model with range 2 and (b2) the estimated L function (continuous curve), with the 2.5th and 97.5th percentile envelopes (dashed) evaluated from 999 independent partial realizations ($n = 429$) of a completely random point process. Estimated L function for (c1) Arisaema 3

simulated 2-D spatial point patterns in Fig. 2.2(d)–(f), for which a 2-D spatial AR(1) process was used, a spherical variogram model with a short range of 2 was used here to generate autocorrelated counts on a 10×10 grid. In all these cases, however, spatial heterogeneity in the point pattern is due to autocorrelation. Compared to Fig. 3.1(c), limits between clusters are fuzzier in Fig. 5.2(b1), resulting in estimates of $L(h) - h$ characterized by positive autocorrelation over distances smaller than 3 [Fig. 5.2(b2)], without sine waving thereafter (not shown). Note that 2.5th and 97.5th percentile envelopes (dashed curves) are plotted in Fig. 5.2(b2), together with the observed L function (continuous), estimated from the data. Such envelopes were produced with a procedure similar to Diggle's randomization testing procedure (Subsection 3.2.4), that is, by simulating 999 independent partial realizations with same n under the null hypothesis of complete randomness for the point process. As expected, the behavior of the estimated L function is significantly different from one that would correspond to complete randomness – the observed curve is not within the envelopes, up to a distance (about 2.5) close to the theoretical value of the range of spatial autocorrelation (i.e., 2).

The *Arisaema* example and the example of California earthquakes, in space (Subsection 3.2.5) and in space-time (Subsection 4.2.1), are revisited below in the light of Ripley's L function analyses. In space, the observed point patterns are provided by Arisaema 3 (Fig. 3.8) and the California earthquakes of magnitude 5 and more over the period 1940–1999 [Fig. 3.12(a)]. In view of Fig. 5.2(c1) and (c2), the following common points and differences can be discussed. In both cases, the estimated L function shows positive autocorrelation up to almost half the length of the diagonal of the plot, and does not behave like a sine wave, which suggests that autocorrelation is true (i.e., from a random component). Values of the two estimated L functions differ by several orders of magnitude,

←

Fig. 5.2. (cont.) (Fig. 3.8) and (c2) the 2-D spatial point pattern of California earthquakes [Fig. 3.12(a)], with the corresponding envelopes. The bivariate form of Ripley's L function estimated (d1) between 2-D spatial point patterns of *Arisaema* in Fig. 4.11 (continuous curve: 1988 with 1989; long-dashed: 1988 with 1990; short-dashed: 1988 with 1991) and (d2) between the 2-D spatial point patterns of California earthquakes in Fig. 4.13 (continuous curve: 1940–1959 with 1960–1979; long-dashed: 1940–1959 with 1980–1999; short-dashed: 1960–1979 with 1980–1999). Note that the estimated Ripley's L functions in this figure are centered with respect to the distance on the horizontal axis; in other words, $\hat{L}(h) - h$ is plotted, instead of $\hat{L}(h)$.

due to huge differences in plot size and the presence of a larger number of points in the smaller plot. More importantly, the L function estimated for Arisaema 3 tends to behave like a negative exponential curve up to distance 6, the highest autocorrelation being observed near the zero distance; the large numbers of *Arisaema* found in the same dm^2 (i.e., the sampling scale) are responsible for this. By comparison, the L function estimated for California earthquakes is strictly positive at very small distances for similar reasons, but rather presents a bump resembling the one in Fig. 5.2(b2) and reaches a peak around 275 km.

In space-time, the point patterns analyzed with Ripley's bivariate L function are those mapped in Fig. 4.11 for *Arisaema* and those of Fig. 4.13 for California earthquakes. I refer to Schabenberger and Gotway (2005, p. 122) for the formal definition of Ripley's K function for a bivariate or spatio-temporal point process and its estimation with edge correction. The L function is then the result of a transformation similar to that defined by equation (5.4), and $L(h) - h$ is interpreted as a function of the inter-distance h and in terms of cross-correlation instead of autocorrelation, between two years (*Arisaema*) or two 20-year periods (California earthquakes). For *Arisaema* [Fig. 5.2(d1)], the cross-correlation between the first and last years is zero at very small distances, and is characterized by the shortest range of positive values (i.e., 3 vs. 4 or 5 for the other pairs of years); the long-dashed curve (i.e., for 1988 and 1990) presents a higher peak essentially because of the lower abundance of *Arisaema* in 1990. For California earthquakes, the general behavior of the estimated L function and the range of positive cross-correlation are similar, whether the 20-year periods are consecutive, or not [Fig. 5.2(d2)].

5.1.3 Modeling of nearest-neighbor distance distributions

In this subsection, we will see that for some regular point patterns it is possible to model the statistical distribution of nearest-neighbor distances, which opens new avenues for the construction of 2.5th and 97.5th percentile envelopes to test hypotheses about the underlying point processes. The 2-D spatial point patterns of two polygonal terrain networks, one Arctic terrestrial and one Martian (Dutilleul *et al.*, 2009), are used for illustrations.

Below, a point is a polygon trough intersection with identifiable spatial coordinates, on Earth in the Arctic [Fig. 5.3(a1)] and at the surface of Mars [Fig. 5.3(a2)]. Both 2-D spatial point patterns are clearly regular. To correct for possible edge effects, 5% cuttings were applied on each side of the frames (see the dashed lines). This resulted in a decrease of the

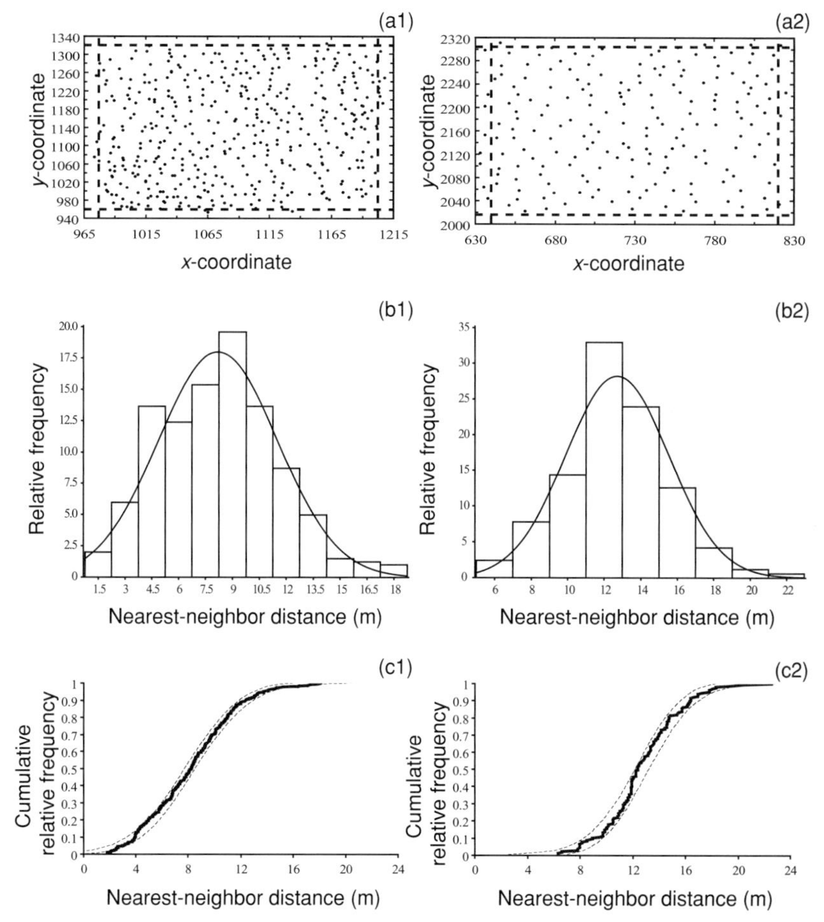

Fig. 5.3. Modeling of the statistical distribution of nearest-neighbor distances from regular 2-D spatial point patterns. (a1)–(c1) and (a2)–(c2) Observed 2-D spatial point patterns; bell-shaped curves fitted to histograms of nearest-neighbor distances; and observed cumulative relative frequency distributions (bold solid curves) of nearest-neighbor distances, with the 2.5th and 97.5th percentile envelopes (dashed curves) evaluated from 999 random samples of n simulated nearest-neighbor distances drawn from a normal distribution with the mean and variance parameters estimated in (b1) and (b2), for Arctic ice wedge intersection points on Earth and trough intersection points at the surface of Mars, respectively.

number n of points actually used for spatial point pattern analysis, from 437 to 403 in (a1)–(c1) and from 192 to 167 in (a2)–(c2).

Histograms in Fig. 5.3(b1) and (b2) resemble the one presented for the simulated regular 2-D spatial point pattern in Fig. 3.5(b), in that they show symmetrical sample distributions of nearest-neighbor distances to which bell-shaped curves representing normal distributions can be fitted (see the continuous curves). The corresponding mean and variance parameter estimates being (8.2, 11.06) and (12.7, 8.00), the mean nearest-neighbor distances (m) for the observed 2-D spatial point patterns of Fig. 5.3(a1) and (a2) are 8.2 $\pm$ 3.3 and 12.7 $\pm$ 2.8, respectively. Recall that, in statistics, the normal distribution is entirely characterized by its mean and variance.

No spatial autocorrelation that would justify a modified estimation of the variance parameter or would prevent a random sampling was found in the observed nearest-neighbor distances. (Variogram analysis, which will be covered in Chapter 7, was used to assess the presence of spatial autocorrelation.) In this case, envelopes can be obtained by sampling the hypothesized distribution of nearest-neighbor distances directly instead of simulating 999 independent partial realizations of the hypothesized model of point process. Unsurprisingly, the observed cumulative relative frequency distributions in Fig. 5.3(c1) and (c2) do not cross the 2.5th and 97.5th percentile envelopes, which were obtained by drawing 999 random samples of 403 and 167 nearest-neighbor distances from normal distributions with the mean and variance values reported above. Strictly speaking, envelopes would be better used, in fact, for purposes of comparison with another observed point pattern. If the other pattern had a number $n' \neq n$ of points, the construction of envelopes could be easily modified; simply, random samples of size n' would be drawn from the hypothesized normal distribution.

Note that: (i) envelopes constructed with the basic version of Diggle's randomization testing procedure for complete randomness would appear, if plotted, on the left of the new ones in Fig. 5.3(c1) and (c2); (ii) envelopes constructed from random samples drawn in the normal distribution of nearest-neighbor distances for the Martian point pattern and adjusted for the number of points in the Artic terrestrial one would appear on the right of those in Fig. 5.3(c1), because of the difference in mean nearest-neighbor distance between the two point patterns.

Summary: The main characteristics of point processes (i.e., the first-order and second-order intensity functions, the statistical distribution of nearest-neighbor distances) can be modeled via the observed point patterns. Such modeling can provide key information about the major axis of change in point density, in relation to heterogeneity of the mean; the range and general behavior of correlations as a function of distance or inter-distance, in relation to heterogeneity due to autocorrelation; and special features of the distribution of points, in direct relation to distances.

5.1.4 Ripley's functions and their conditions of application: a review

As we have seen in Subsection 5.1.2, Ripley's (1981) K and L functions can be very useful tools for performing autocorrelation analysis of point processes from observed point patterns. That said, it is important to be aware of the implications of the assumptions made in their application, starting with the definition of the K function (equation 5.3) and following up with the definition of $L(h)$ from $K(h)$ (equation 5.4) and the approximation of $\sqrt{\frac{K(h)}{\pi}}$ by a linear function of h for small and large distances. This is why in this subsection we will discuss the use of an overall mean density of points per unit of space or time in equation (5.3), illustrate what happens with the L function when such a use is violated by the presence of heterogeneity of the mean, and develop an approach and propose an analytical procedure in order to overcome the issue. Without loss of generality, the framework is 2-D spatial.

Consider the situation of a soil seed bank defining a gradient of increasing density in a site [Fig. 5.4(a1)] and on which a set of seeds dispersed by wind is superimposed, resulting in a point pattern with two components [Fig. 5.4(b1)]. The first component (i.e., the soil seed bank) is essentially deterministic and related to spatial heterogeneity of the mean; the second (i.e., seeds dispersed by wind) is random and could be spatially autocorrelated. To illustrate this situation in Fig. 5.4, a simple spatial structure in the form of a plane with equation $z = 0.5 + 0.5x + 0.5y$ was first created for the number of bank seeds per square quadrat of unit side, using the coordinates of the centre of the quadrat in the equation and rounding the result to the integer part, and within each quadrat the resulting numbers of bank seeds were then completely randomly distributed, using one of the simulation procedures from Section A3.2; the spatially autocorrelated point pattern with range 2 of Fig. 5.2(b1) was used for the second component. In this situation, 1 seed = 1 point.

In the absence of any (true) spatial autocorrelation in the point pattern of Fig. 5.4(a1), the behavior of the estimated L function in panel (a2) is at the least surprising. In fact, if one follows the guidelines of Subsection 5.1.2, the bump observed in the function over distances 0–8 would be indicative of spatial autocorrelation with long range; the plot is a

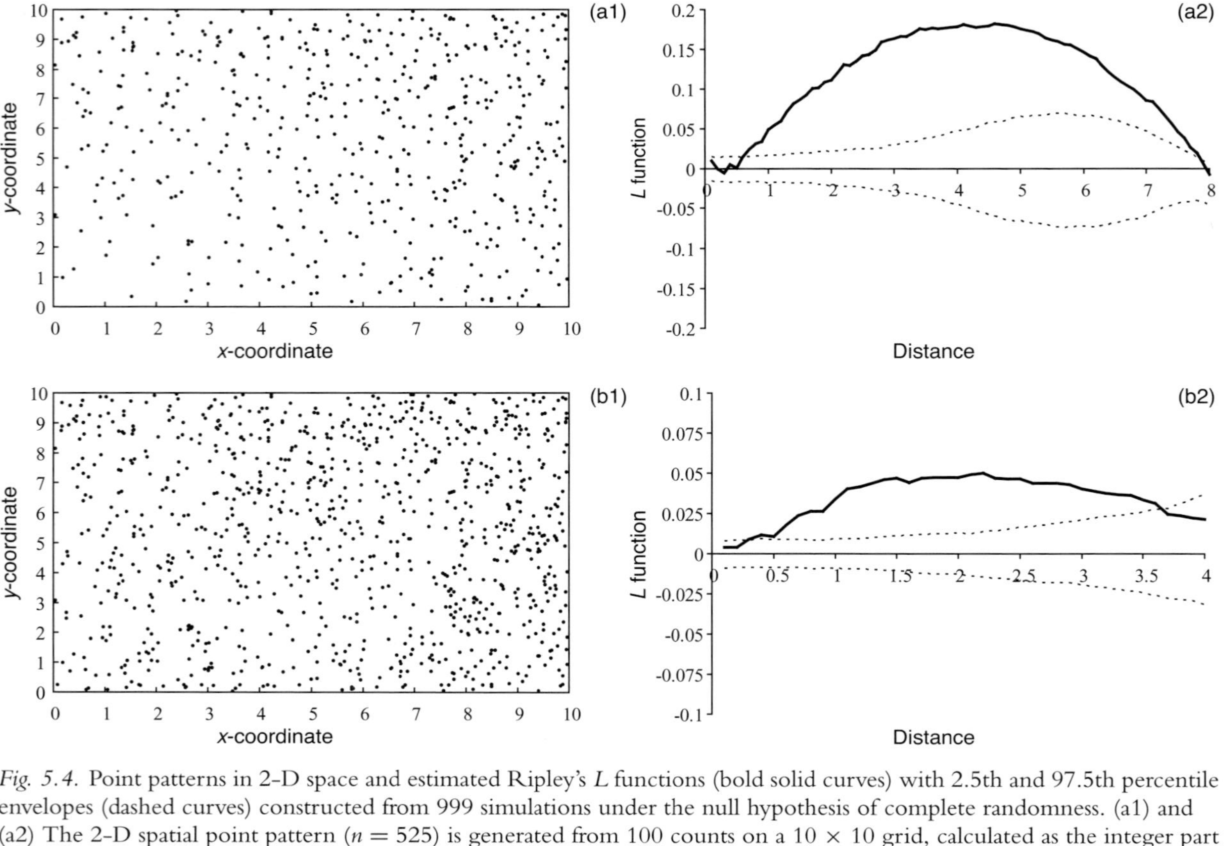

Fig. 5.4. Point patterns in 2-D space and estimated Ripley's L functions (bold solid curves) with 2.5th and 97.5th percentile envelopes (dashed curves) constructed from 999 simulations under the null hypothesis of complete randomness. (a1) and (a2) The 2-D spatial point pattern ($n = 525$) is generated from 100 counts on a 10×10 grid, calculated as the integer part of $\frac{1+x+y}{2}$, where x and y denote the spatial coordinates of a node of the grid. (b1) and (b2) The point pattern ($n = 954$) is the result of the superimposition of point pattern (a1) with the autocorrelated 2-D spatial point pattern of Fig. 5.2(b1).

10×10 square. In this case, the condition of first-order stationarity is clearly not satisfied, so that the use of one overall mean point density in the estimation of the K function, before its use in the estimated L function, is not justified. The main source of the confusing result is there, but the confusion is exemplified by the loss of precision with increasing distance, as shown by the envelopes (which correspond to complete randomness) after distance 4, despite the edge correction. When working with the sum of the two components [Fig. 5.4(b1) and (b2)], which will be the case in practice, the problem due to spatial heterogeneity of the mean pointed out above is attenuated, because patches of seeds dispersed by wind in zones where the seed bank is not so rich somewhat stabilizes the number of points per area unit. But still clearly, the bump observed in the estimated L function [Fig. 5.4(b2)] suggests the presence of spatial autocorrelation within a range of 3.5, which is substantially longer than the true range of 2.

Naturally, the following question then arises: Is it possible to separate the two components of the point pattern in Fig. 5.4(b1), for further statistical analysis in the conditions of application of the method? The answer proposed below is "yes" within reasonable limits. By transforming points into counts made in square quadrats of unit side, it is possible to map those counts using 2-D contours; the result of this is presented in Fig. 5.5(a1), (b1), and (c), for the two true components and their sum, respectively. Using a procedure similar to that used for Fig. 5.1(c), under the model $U(x, y) = \mu(x, y) + \varepsilon(x, y)$, where $U(x, y)$ denotes the number of points counted in the quadrat centered at (x, y), $\mu(x, y) = a + bx + cy$ and $\varepsilon(x, y)$ is spatially autocorrelated following a spherical variogram model with range 2, it is possible to recover, at least for a good part, the first component of the partial realization of the nonstationary point process on the site and, by difference, the second component [Fig. 5.5(a2) and (b2)]. (In practice, both the polynomial model to be used for $\mu(x, y)$ and the variogram model for the spatial autocorrelation of $\varepsilon(x, y)$ will need to be identified.) As in Fig. 5.1(c), spatial autocorrelation was taken into account in the capturing of spatial heterogeneity of the mean. Thereafter, it would be possible to simulate the two components, using the plane fitted to counts as first-order intensity function for the former and the autocorrelation contained in the residuals obtained by difference for the latter (not shown here).

Key note: With the distance-based approach, it does not seem possible to take heterogeneity of the mean in a point pattern into account in order to satisfy the

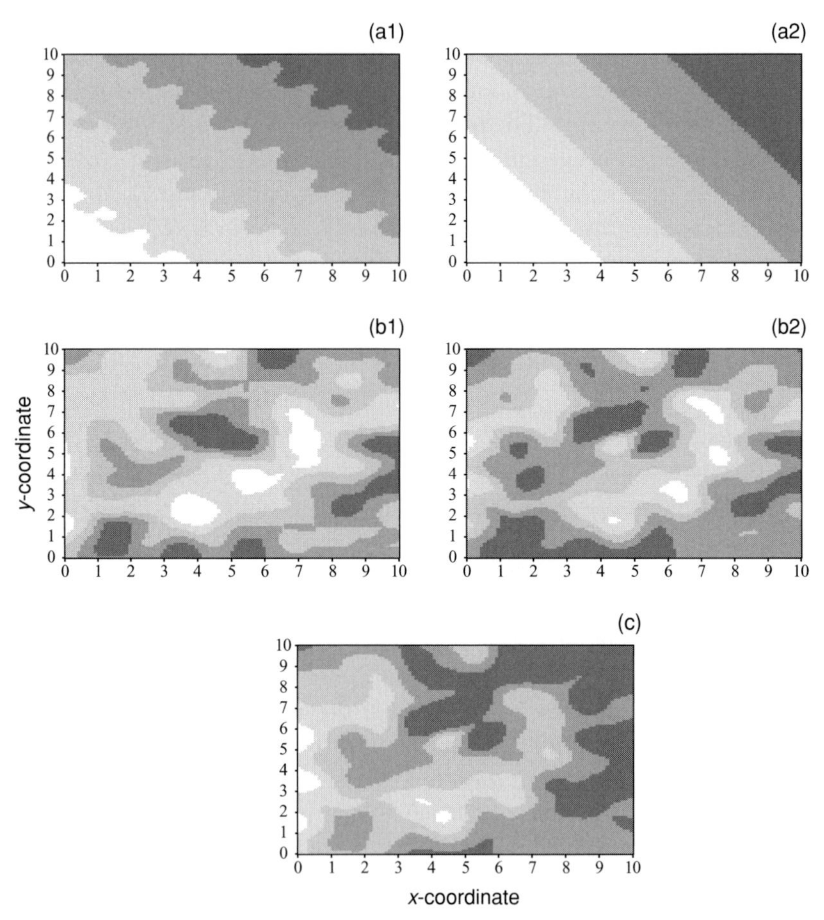

Fig. 5.5. Contour maps produced from 100 counts, observed or estimated at the scale of square quadrats of size 1. (a1), (b1), and (c) The counts are observed, and arise from the 2-D spatial point patterns of Fig. 5.4(a1), Fig. 5.2(b1), and Fig. 5.4(b1), respectively. (a2) and (b2) The estimated counts used to produce the contour map in (a2) were obtained by fitting a plane to the observed counts of Fig. 5.4(b1), while taking spatial autocorrelation into account; the estimated counts (i.e., residuals) used for (b2) were obtained by subtraction; see text for details.

first-order stationarity condition required for applying Ripley's K *and* L *functions. The reason appears simple: heterogeneity of the mean in the context of point process modeling refers to the first-order intensity function* $\lambda_1(\mathrm{s})$*, which, by definition, is a function of the location* s *alone, while distances, by nature, involve pairs of locations* (s, s′)*. With the count-based approach, despite its problem of scaling (i.e., the definition of an "appropriate" quadrat size), it seems possible to do something, at* *the least to separate the two components of a point pattern like the one depicted in Fig. 5.4(b1). To the best of my knowledge, there exists no generalized form of Ripley's* K *and* L *functions that does not require first-order stationarity as a condition of application. Such a modification would start with* $\lambda_1(\mathrm{s})$ *instead of* λ_1 *in the differential equation defining* K(h) *from* $\lambda_2(\mathrm{h})$, *although this might lead to as many* K *functions as there are locations* s . . . *There are several similarities between the situation described here and the autocorrelation analysis of surface patterns, as we will see in Chapters 6 and 7.*

5.2 General summary

By nature and definition, points are central to point pattern analysis. When one attaches heterogeneity aspects to it and the analysis becomes that of heterogeneity of point patterns, one discovers after some time that both counts and distances have a role to play in this because there is not one approach that unilaterally dominates the other and could "do the job" alone. In other words, neither can cover the three key aspects of heterogeneity (mean, variance, autocorrelation) in point patterns, completely and perfectly.

The success and popularity of the distance-based approach may be, for a part, the artificial consequence of the imposing of the isotropy condition on the second-order intensity function [i.e., $\lambda_2(s, s') = \lambda_2(h)$] in some methods of point pattern analysis, in Ripley's *K* and *L* functions in particular. More practical reasons are the greater availability of equipment for positioning points of interest and the dependency of results of the count-based approach on the definition of quadrat size. It is important to bear in mind, however, that Ripley's functions are designed for autocorrelation analysis (Subsection 5.1.2), which represents only one aspect of heterogeneity in point patterns; as a proof, their conditions of application include first-order stationarity (i.e., homogeneity of the mean). In other words, such functions cannot deliver all the information contained in a point pattern, and if misused (e.g., in the presence of heterogeneity of the mean), they can produce incorrect results (Subsection 5.1.4). Among all distances, the nearest-neighbor distances provide an excellent basis for discriminating point patterns and modeling point processes via frequency plots, cumulative or not (Chapters 3 and 4 and Subsection 5.1.3). This is the case with Diggle's randomization testing procedure (introduced in Subsection 3.2.4), for the main types of point patterns and point processes: completely random or simple Poisson; aggregated (including various modifications of the simple Poisson process; see Fig. 3.2, Section A3.2 and below); and regular. Concerning regular point patterns, the modeling of nearest-neighbor distances with a normal distribution (Subsection 5.1.3) may somehow recall the pioneering work of Clark and Evans (1954) (Subsection 3.2.4), but it is different because it involves the fitting of a curve to the histogram [Fig. 5.3(b1) and (b2)].

When heterogeneity of the mean is of interest, quadrat counts (i.e., numbers of points observed in parts of a plot in 2-D space) have a key role to play because the first-order intensity function $\lambda_1(s)$ (i.e., the potential source of heterogeneity of the mean) is a function of location instead of distance. The analytical procedure proposed as an alternative to kernel-based estimators in Subsection 5.1.1 and illustrated in Fig. 5.1(c) and (d) has the advantage of incorporating the autocorrelation among counts in the capturing of heterogeneity of the mean. This procedure finds its roots in surface pattern analysis (see Chapter 7 in 2-D space), via the transformation of points to quadrat counts. It does not mean that there are no nonstationary models for point processes. It suffices to think of the generalized and compound Poisson processes (Cliff and Ord, 1981, pp. 88–89), but the population mean of a Poisson distribution is also its population variance and in these conditions it seems difficult to model the two components of heterogeneity separately. For a similar reason, the sample variance and the sample mean of quadrat counts are linked in the variance-to-mean ratio (introduced in Subsection 3.2.3), which is useful within reasonable limits in point pattern analysis even though only one variance measure is used in it.

A number of elements of point process modeling were presented here, with more emphasis on heterogeneity aspects than usual. Mathematically more demanding references, on the fitting of models in Diggle's randomization testing procedure and for Ripley's K and L functions, are given in Section 5.3. Definitively, the topic (including the modeling of heterogeneity of the mean in combination with autocorrelation, and vice versa) represents fertile ground for research work by statisticians.

In Chapters 3 and 4 and Section 5.1, the demonstration has been made, through examples, that point patterns are encountered in many situations in the biological and environmental sciences. In particular, the *Arisaema* example in plant ecology and the example of California earthquakes in the earth sciences were followed in space, time and space-time, which allowed us to see how the heterogeneity aspects were changing depending on the framework and how the model, the analytical procedure and the interpretation of results were adjusted accordingly. These two main examples were completed by others, used to illustrate the method of time-dependent transition probabilities (chewing behavior of sheep, development of mosquito larvae; Subsection 4.1.3) and the application of the beta-binomial distribution to study movement in space-time (hunting expeditions of wolf, Subsection 4.2.2), to name a few. In each case, at least one heterogeneity aspect of the point pattern was studied and discussed in statistical and non-statistical terms.

5.3 Recommended readings

I recommend the following books and articles as complements to my writings.

A comprehensive summary of the various quadrat methods and indices of dispersion (e.g., Morisita's I_δ and Lloyd's *IMC* and *IP*) is given by Upton and Fingleton (1985, Chapters 1, 2, and 4). Cliff and Ord (1981, Chapter 4) discuss the differences between true and false contagion and between generalized and compound Poisson processes in space; Daley and Vere-Jones (1988) elaborate on the theoretical aspects of point processes; and Cressie (1993, p. 620) gives, among other things, the mathematical explanations concerning the derivation of the Poisson distribution from the simple Poisson process. Details about the introduction of the beta-binomial distribution in plant pathology and its use in the spatial framework can be found in Hughes and Madden (1993), Madden and Hughes (1994), and Madden *et al.* (1995), while Smith (1983) presents the algorithm for the maximum likelihood estimation of the beta-binomial distribution. From the ecological perspective, Hurlbert (1990) provides a constructive critique of unjustified practices and claims in the analysis of spatial point patterns. That being said, I am still unclear about the existence of montane unicorn populations, the example used throughout this article . . .

Diggle's (2003) *Statistical Analysis of Spatial Point Patterns* (first edition, 1983) represents a reference that cannot be ignored, and was the inspiration for the distance-based approach developed in the temporal and spatio-temporal frameworks in Sections 4.1 and 4.2. Chapter 8 in Ripley's (1981) *Spatial Statistics* provides a clear introduction to the popular K and L functions used to model the second-order moments of point processes in Subsection 5.1.2. Marcoux *et al.* (2010) present a nice application of Ripley's functions in the temporal framework and propose modifications for marked temporal point patterns, with behavioral data on marine mammals for illustration. Chapters 6 and 7 in Diggle (2003) are about least-squares and likelihood-based estimation procedures for model fitting, based on the cumulative relative frequency distribution of nearest-neighbor distances and Ripley's K function; given the continuous nature of space and time in point pattern analysis, both types of procedures involve integrals and require a good mathematical background.

Brillinger (1994) very clearly describes the differences between temporal point patterns and time series (to be studied in Chapter 6), and rightly calls the marked temporal point processes "hybrids." In Dutilleul *et al.* (2000a), the interested reader will find more references on the use of transition probabilities in the animal behavior sciences as well as further details on the method of time-dependent transition probabilities. Finally, Turchin's (1998) *Quantitative Analysis of Movement* is devoted to the theory and practice of examining movement in plant and animal populations.

6 · *Heterogeneity analysis of time series*

In this chapter and the next two, models, methods, and procedures are presented to analyze the heterogeneity of surface patterns statistically. In the temporal framework, such patterns are called "time series," and Chapter 6 focuses on these in particular. After a preamble (Section 6.1), where the main differences with the heterogeneity analysis of point patterns are highlighted, heterogeneity due to temporal autocorrelation is studied solely, under the assumption of weak stationarity (i.e., stationarity at orders 1 and 2), in Section 6.2. Thereafter, we will see how to study the heterogeneity of the mean of time series in the absence of temporal autocorrelation vs. in its presence, in both cases in the absence of heteroscedasticity (Section 6.3). In subsections of Sections 6.2 and 6.3, some statistical functions are defined and analyzed in a domain other than time, that is, the frequency or spectral domain. In Section 6.4, the emphasis is on the identification of heteroscedastic patterns in the presence of the two other types of heterogeneity. In Section 6.5, the statistical methods and procedures of previous sections are compared, and differences and common points between them are discussed. Bivariate analyses of time series, in the time domain as well as in the frequency domain, are presented in Section 6.6. With the exception of Section 6.2, in which simulated data are mainly used to help establish guidelines, real datasets from various fields are used in examples in other sections. Complementary readings on the vast topic of time series analysis are proposed in Section 6.7. Details about simulation procedures and more mathematical aspects are grouped in Appendix A6. The MATLAB® and SAS codes implementing the analytical procedures are available on the CD-ROM accompanying the book.

6.1 Preamble

In this section, priority is given to time series, namely, series of data on continuous quantitative responses indexed by discrete values of time,

$\{u(t)|t = 1, \ldots, n\}$, which are collected as partial realizations of discrete-time stochastic processes, $\{U(t)|t \in \mathbb{Z}\}$, with $U(t) \in \mathbb{R}$ for any given t. Even though space, as an index of a stochastic process, tends to be continuous and more than one-dimensional (Section 2.1), this preamble will be very helpful for further developments in Chapters 7 and 8, allowing one to focus on questions specific to spatial and spatio-temporal data in those chapters.

First of all, the link between the equation of a model and the stochastic process underlying an observed surface pattern in general and an observed time series in particular is more direct than for a point process and a point pattern. The reason for this is very simple. There is no need to pass by probabilities, intensity functions, or counts to interpret the fundamental equation (2.1), $U(s) = \mu(s) + \varepsilon(s)$, in the case of surface patterns: the indexed random variable $U(s)$ is equal to its mean or expected value $\mu(s)$ plus a random deviation $\varepsilon(s)$ from it, characterized by a zero mean, a variance and a possible correlation with the random deviation $\varepsilon(s')$ of $U(s')$ from $\mu(s')$, with $s \neq s'$. Therefore, models will be incorporated to the presentation of methods and procedures in the heterogeneity analysis of surface patterns, instead of keeping modeling aspects apart and treating them separately as for point processes (Chapter 5). Among the temporal variables $U(t)$ (i.e., $s = t$) that will be studied, are air and soil temperatures, atmospheric CO_2 concentration, and wood characteristics (ring width, fiber length, microfibril angle).

The most general version of equation (2.1) in the case of time series is

$$U(t) = \mu(t) + \varepsilon(t), \tag{6.1}$$

without any constraint on $\mu(t)$ and the variance and correlation parameters of $\varepsilon(t)$. The following particular cases are related to the examples with real datasets:

- under weak stationarity (Section 2.3),

$$U(t) = \mu + \varepsilon(t), \tag{6.2}$$

  with $\mathrm{Var}[U(t)] = \sigma^2$ for all t and $\mathrm{Corr}[U(t), U(t')] = \rho(t' - t)$ for any $t \neq t'$;
- under stationarity restricted at order 2 (Section 2.3),

$$U(t) = \mu(t) + \varepsilon(t), \tag{6.3}$$

  with $\mathrm{Var}[U(t)] = \sigma^2$ for all t and $\mathrm{Corr}[U(t), U(t')] = 0$ (absence of temporal autocorrelation) or $\rho(t' - t)$ (presence of temporal autocorrelation) for any $t \neq t'$. Equation (6.3) reads the same as equation (6.1), but there is an important difference between them: there is no constraint on $\mu(t)$ and $\varepsilon(t)$ in equation (6.1), but the conditions of stationarity at order 2 apply to $\varepsilon(t)$ in equation (6.3); recall that $\mathrm{Var}[U(t)] = \mathrm{Var}[\varepsilon(t)]$ and $\mathrm{Corr}[U(t), U(t')] = \mathrm{Corr}[\varepsilon(t), \varepsilon(t')]$ (Section 2.2).

Equations (6.1)–(6.3) are aimed at representing one temporal stochastic process by assuming that a single time series is observed. In these conditions, a model of heterogeneity of the variance (i.e., a heteroscedastic variance–covariance structure) can be postulated and even tested, but it is submitted to restrictions in terms of number of parameters. The collection of $K > 1$ replicated time series from the same temporal stochastic process (i.e., temporal repeated measures)

$$U_k(t) = \mu(t) + \varepsilon_k(t) \quad (k = 1, \ldots, K; t = 1, \ldots, n) \tag{6.4}$$

allows the broadening of the domain of inference for variance–covariance parameters as well as the calculation of a time series of averages with reduced variability:

$$\bar{U}(t) = \mu(t) + \bar{\varepsilon}(t) \quad (t = 1, \ldots, n), \tag{6.5}$$

with $\bar{U}(t) = \frac{1}{K}\sum_{k=1}^{K} U_k(t)$ and $\bar{\varepsilon}(t) = \frac{1}{K}\sum_{k=1}^{K} \varepsilon_k(t)$ for a given t, and $\{\bar{U}(t) | t \in \mathbb{Z}\}$ and $\{\bar{\varepsilon}(t) | t \in \mathbb{Z}\}$ are the processes corresponding to the time series of averages and the associated random deviations from expected values, respectively.

Temporal repeated measures can also be collected for different levels of a classification factor (e.g., the response is observed over time on individuals belonging to different experimental groups), which leads to the models:

$$U_{ik}(t) = \mu_i(t) + \varepsilon_{ik}(t) \quad (k = 1, \ldots, K; t = 1, \ldots, n); \tag{6.6}$$

$$\bar{U}_i(t) = \mu_i(t) + \bar{\varepsilon}_i(t) \quad (t = 1, \ldots, n), \tag{6.7}$$

where $i = 1, \ldots, I$ denotes the level of a fixed classification factor and the definitions of $\bar{U}_i(t)$ and $\bar{\varepsilon}_i(t)$ are similar to those of $\bar{U}(t)$ and $\bar{\varepsilon}(t)$ in equation (6.5).

Several variables observed at the same times can be analyzed jointly as a multiple time series $\{(u_1(t), \ldots, u_p(t))' | t = 1, \ldots, n\}$. The temporal version of equation (2.12) written below, where $\boldsymbol{u}(t)$, $\boldsymbol{\mu}(t)$, and $\boldsymbol{\varepsilon}(t)$ are the vector notations for $U(t)$, $\mu(t)$, and $\varepsilon(t)$, corresponds to this situation:

$$\boldsymbol{u}(t) = \boldsymbol{\mu}(t) + \boldsymbol{\varepsilon}(t). \tag{6.8}$$

Under weak stationarity, it becomes [cf. equation (6.2)]:

$$\boldsymbol{u}(t) = \boldsymbol{\mu} + \boldsymbol{\varepsilon}(t), \tag{6.9}$$

where $\mathrm{Var}[U_j(t)] = \mathrm{Var}[\varepsilon_j(t)] = \sigma_j^2$ for all t and $j = 1, \ldots, p$, and the cross-correlations $\mathrm{Corr}[U_j(t), U_{j'}(t')] = \mathrm{Corr}[\varepsilon_j(t), \varepsilon_{j'}(t')] = \rho_{jj'}(t' - t)$ for $j \neq j'$ and $t \neq t'$ are population parameters of special interest (Section A2.2).

Model equations (6.2)–(6.9) will be used in the following sections. For example, the mean function $\mu(t)$ in equation (6.3) will be modeled differently in Subsections 6.3.1 and 6.3.2, and the identification of a heteroscedastic variance–covariance structure from data collected under equation (6.6) will be the object of Section 6.4.

6.2 Temporal autocorrelation analysis

A primary objective of this section is to familiarize our eyes with the various forms of heterogeneity in time series due to autocorrelation, because it is possible to recognize and distinguish graphically weakly vs. strongly autocorrelated time series and negatively vs. positively autocorrelated time series. Of course, such an eye training exercise has its limits, which we will draw before proceeding to quantification and estimation.

In this exploratory stage, we proceed by simulation in order to know with certainty the model of the stochastic process underlying the observed time series; the simulation procedures are presented in Section A6.1. As the first time series model, consider the classical discrete-time process that we encountered in Chapter 2 (Figs. 2.3 and 2.4).

AR(1), or first-order autoregressive

With $\mu(t) = \mu = 0$, its general equation is

$$U(t) = \varepsilon(t) = \phi\varepsilon(t-1) + \eta(t), \tag{6.10}$$

where $-1 < \phi < 1$, and $\eta(t)$ follows a normal distribution with a zero mean and a variance of 1 for all t and $\mathrm{Corr}[\eta(t), \eta(t')] = 0$ for all $t \neq t'$; a temporal stochastic process $\{\eta(t) | t \in \mathbb{Z}\}$ with such characteristics is called "white noise."

Thus, partial realizations of discrete-time AR(1) processes were simulated for values of the autoregressive parameter ϕ ranging from –0.9

to 0.9 by steps of 0.2 (Fig. 6.1, left). Autocorrelation is the only source of heterogeneity in all ten time series, but their fluctuations are very different from $\phi = -0.9$ to ± 0.1 and from $\phi = \pm 0.1$ to 0.9. In Subsection 6.2.1, we will see that (i) negative values of ϕ in equation (6.10) correspond to autocorrelation for $U(t)$ that alternates in sign depending on the oddness or evenness of the lag $|t' - t|$, starting with the negative sign when $|t' - t| = 1$, and (ii) the theoretical autocorrelation function of a discrete-time AR(1) process is positive at all lags when ϕ is positive. When $\phi = -0.9$, we observe a "yo-yo effect" in the time series [Fig. 6.1(a), left], which reflects the strong negative autocorrelation at lag 1; in other words, the correlation between successive observations is strong and negative, so that a high value at one time follows and is followed by a low value at the previous and next times. When $\phi = \pm 0.1$, no distinct or consistent type of fluctuation can be discerned in the time series [Fig. 6.1(e) and (f), left], because the autocorrelation is weak, close to the total absence of autocorrelation [Fig. 2.1(a)]. In between, for $\phi = -0.7, -0.5$ and -0.3 [Fig. 6.1(b)–(d), left], there is a decreasing gradient in the amplitude of fluctuations (vertically) and the fluctuations themselves are less and less systematic, because the autocorrelation is still negative at odd lags (1, 3, etc.), but weaker than for $\phi = -0.9$. At the other extreme, when $\phi = 0.9$, the fluctuations of the times series are very smooth and long (horizontally); when increasing (decreasing), the values of $U(t)$ tend to increase (decrease) over a long period of time [cf. Fig. 2.3(b)]. From $\phi = 0.3$ to 0.5 and 0.7, the smoothness of fluctuations is developing and their length over time is increasing, reflecting the increase in positive temporal autocorrelation.

First-order autoregressive processes are useful for pedagogical purposes, as we have just seen. Whether temporal or spatial, these processes are also practical in simulation studies where the statistical objective consists in assessing the efficiency of estimation procedures or the validity of testing procedures, because the results obtained for AR(1) processes generally provide insight for the results to be obtained for other processes (e.g., Diggle, 1990, pp. 87–92; Alpargu and Dutilleul, 2006). That being said, several other discrete-time processes exist and may be more appropriate than the AR(1) in biological and environmental studies. The model equations for four of them are written below, all with $\mu(t) = \mu = 0$ and the same definition for $\eta(t)$ as in equation (6.10). The parameter values used for simulation are reported in parentheses after each model equation, and the simulated partial realizations are plotted in Fig. 6.2(left).

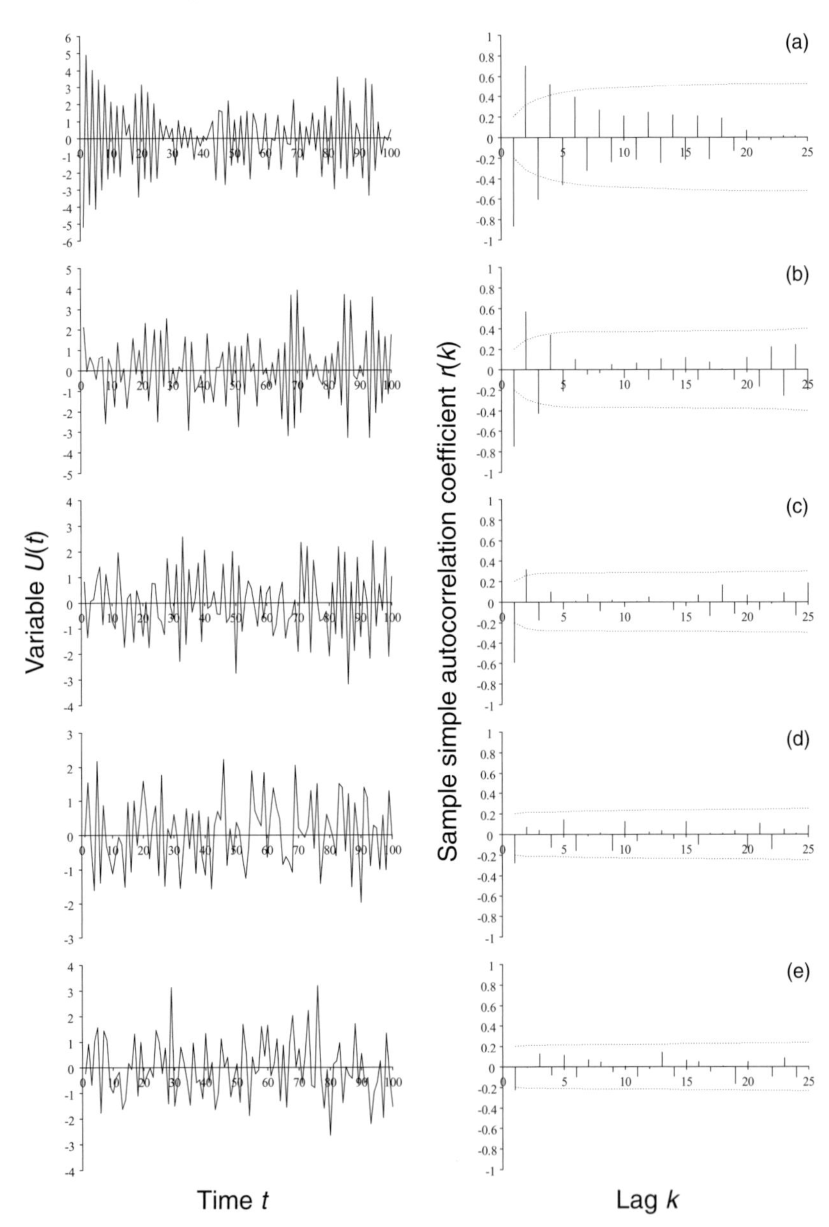

Fig. 6.1. Sample simple autocorrelation coefficients (right) and the corresponding partial realization (left) of discrete-time AR(1) processes [$\mu(t) = 0$ for all t; $\varepsilon(t) = \phi\varepsilon(t-1) + \eta(t)$ for any t, with $\text{Var}[\eta(t)] = 1$ for all t, so $\sigma^2(t) = \frac{1}{1-\phi^2}$ for all t, and $\rho(t, t') = \phi^{|t'-t|}$ for any $t \neq t'$], for values of ϕ ranging from (a) –0.9 to (j) 0.9, by steps of 0.2; $n = 100$ for the ten time series.

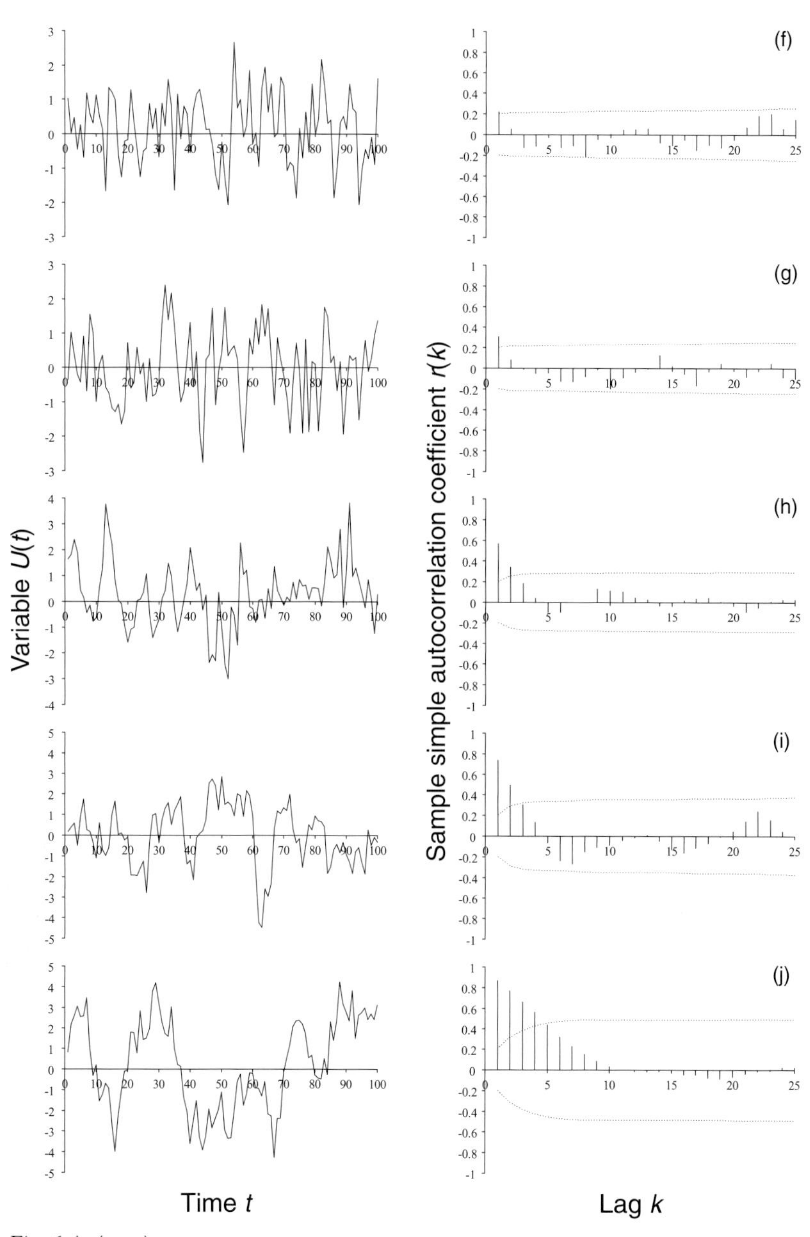

(f)
(g)
(h)
(i)
(j)
Variable $U(t)$
Sample simple autocorrelation coefficient $r(k)$
Time t
Lag k

Fig. 6.1. (cont.)

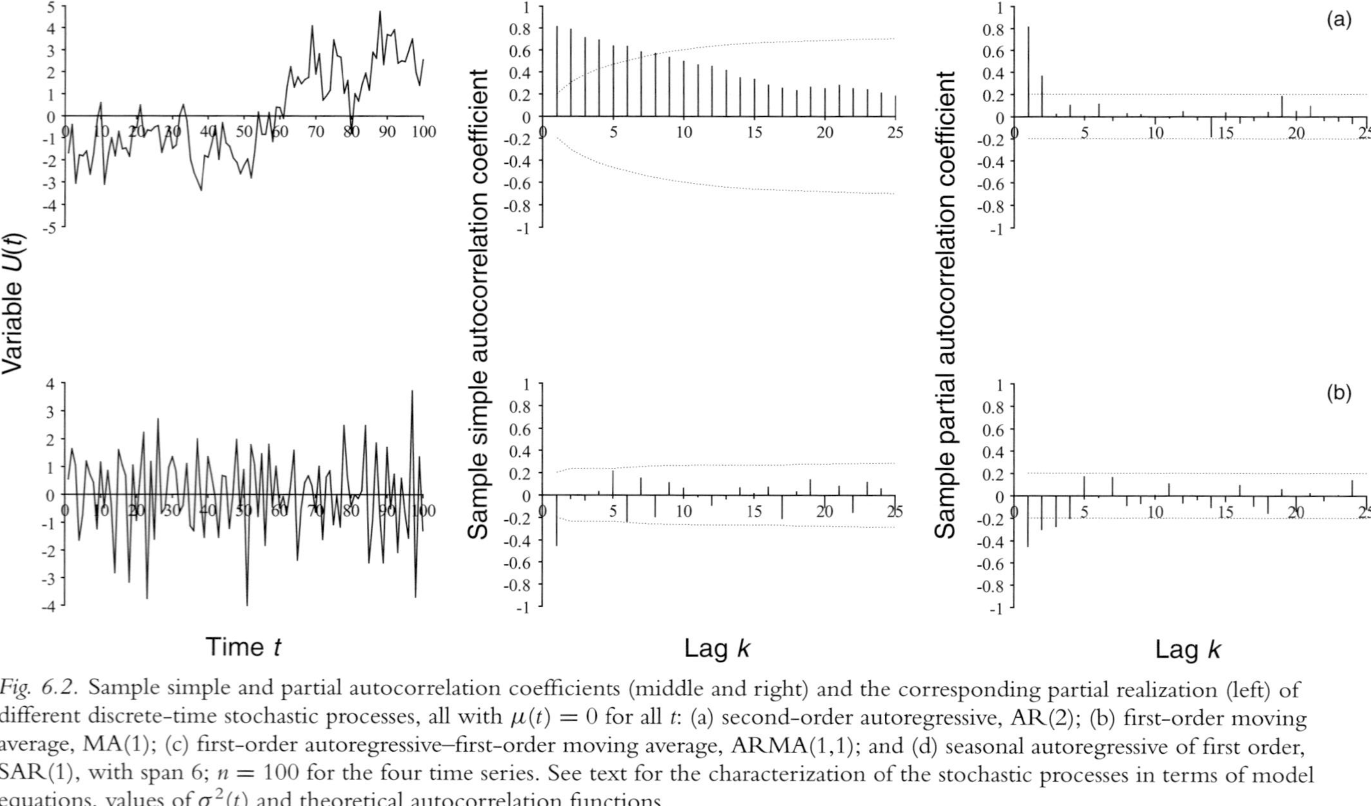

Fig. 6.2. Sample simple and partial autocorrelation coefficients (middle and right) and the corresponding partial realization (left) of different discrete-time stochastic processes, all with $\mu(t) = 0$ for all t: (a) second-order autoregressive, AR(2); (b) first-order moving average, MA(1); (c) first-order autoregressive–first-order moving average, ARMA(1,1); and (d) seasonal autoregressive of first order, SAR(1), with span 6; $n = 100$ for the four time series. See text for the characterization of the stochastic processes in terms of model equations, values of $\sigma^2(t)$ and theoretical autocorrelation functions.

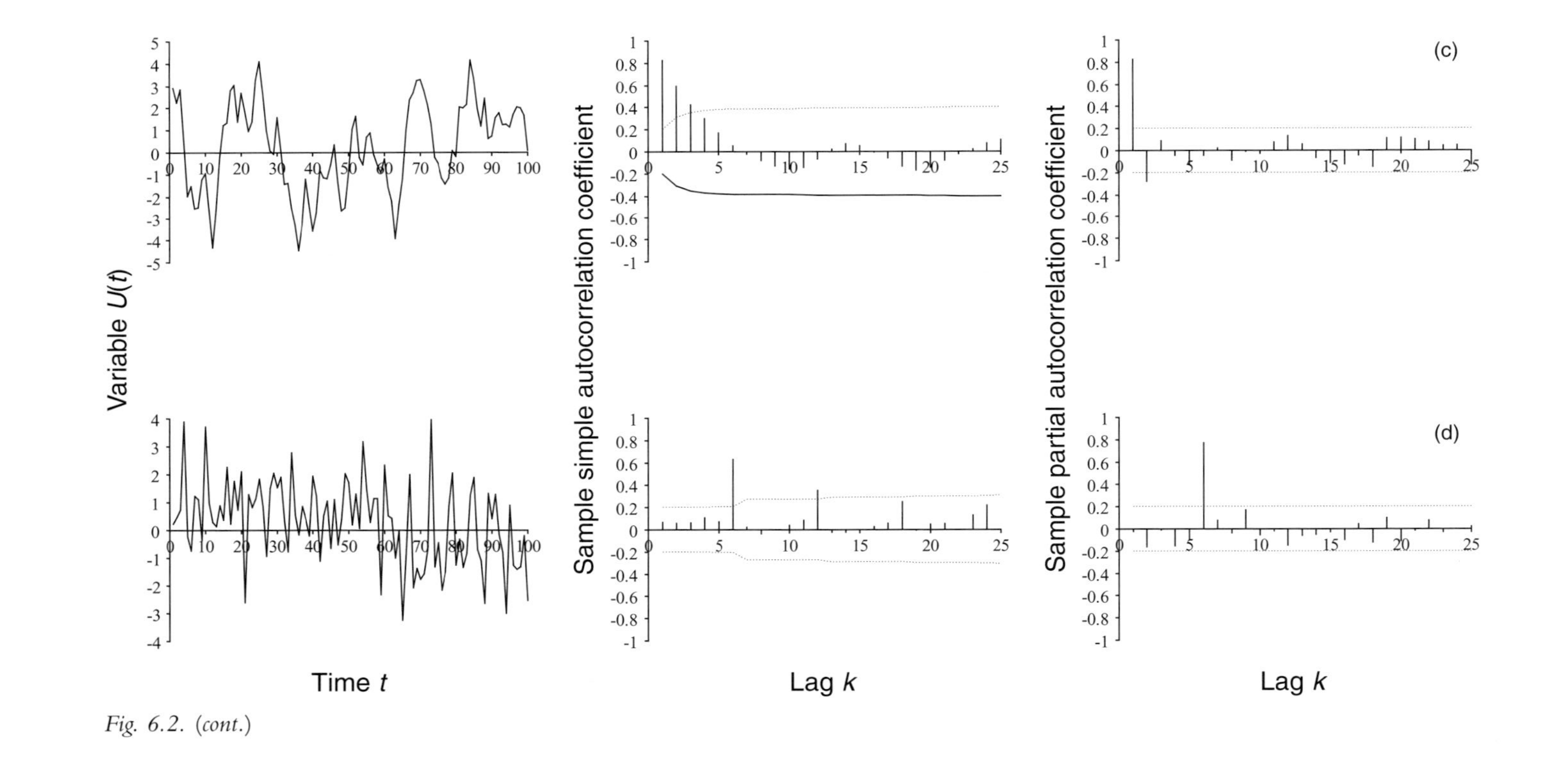

Variable $U(t)$
Time t
Sample simple autocorrelation coefficient
Lag k
Sample partial autocorrelation coefficient
Lag k
(c)
(d)

Fig. 6.2. (*cont.*)

AR(2), or second-order autoregressive

$$U(t) = \varepsilon(t) = \phi_1\varepsilon(t-1) + \phi_2\varepsilon(t-2) + \eta(t), \qquad (6.11)$$

where $\phi_1 + \phi_2 < 1$, $\phi_2 - \phi_1 < 1$, and $-1 < \phi_2 < 1$ ($\phi_1 = 0.6$, $\phi_2 = 0.3$).

MA(1), or first-order moving average

$$U(t) = \varepsilon(t) = \eta(t) - \theta\eta(t-1), \qquad (6.12)$$

where $-1 < \theta < 1$ ($\theta = 0.8$).

ARMA(1,1), or first-order autoregressive–first-order moving average

$$U(t) - \phi U(t-1) = \varepsilon(t) - \phi\varepsilon(t-1) = \eta(t) - \theta\eta(t-1), \qquad (6.13)$$

where $-1 < \phi < 1$ and $-1 < \theta < 1$ ($\phi = 0.7$, $\theta = -0.5$).

SAR(1), or seasonal autoregressive of first order, with span 6

$$U(t) = \varepsilon(t) = \Phi\varepsilon(t-6) + \eta(t), \qquad (6.14)$$

where $-1 < \Phi < 1$ ($\Phi = 0.8$).

In the partial realizations of discrete-time AR(2) and ARMA(1,1) processes considered here [Fig. 6.2(a) and (c), left], autocorrelation appears to be positive and strong because fluctuations are smooth and long. In the MA(1) partial realization [Fig. 6.2(b), left], it should be negative at first lag and relatively strong because of brief but wide up-and-down fluctuations. The SAR(1) partial realization [Fig. 6.2(d), left] shows a rather consistent succession of peaks every six observations (i.e., the value of the span), with decreasing or increasing tendencies from peak to peak. With the exception of the SAR(1) partial realization, the other three do not show fluctuations and hence forms of temporal heterogeneity due to autocorrelation, very different from some of the AR(1) partial realizations plotted in Fig. 6.1(left). Therefore, guidelines for quantitative analyses that can be used to identify more accurately the model of each of those temporal stochastic processes are presented in the next two subsections.

6.2.1 Estimation in the time domain

It is natural to estimate autocorrelation (i.e., the correlation between a continuous quantitative random variable U and itself for different index values) in the domain where the observations are made, whether this domain is time (here) or space (Chapter 7). In Subsection 6.2.1, autocorrelation is estimated in the time domain under the assumption of weak stationarity (Section 2.3), which includes constant population mean (i.e., homogeneity of the mean) and constant population variance (i.e., homogeneity of the variance). The model used for this is defined by equation (6.2), of which equations (6.10)–(6.14) are particular cases. In Subsection 6.2.2, the same model (equation 6.2) will be used to analyze autocorrelation in the frequency domain.

Key note: Weak stationarity is not an absolute requirement for the estimation of autocorrelation, and in different conditions autocorrelation will be analyzed differently; see Sections 6.3 and 6.4.

Below, the quantification of temporal heterogeneity due to autocorrelation is performed in four steps:

(i) the calculation of autocorrelation statistics, called "sample autocorrelation coefficients," from the observed time series – there will be two families of such coefficients;
(ii) the identification of a model for the underlying temporal stochastic process, based on the calculated coefficients – theoretical autocorrelation functions are directly related to the type of model and the values of its parameters;
(iii) the estimation of parameters for the postulated model;
(iv) the estimation of the variance of the autocorrelated process relative to the generating white noise [see $\eta(t)$ in equations (6.10)–(6.14)].

Assume a time series of length n, $\{u(t)|t = 1, \ldots, n\}$, has been observed. The "sample simple autocorrelation coefficient" (SSAC) at lag $k = 1, \ldots, [\frac{n}{4}]$, where $[\frac{n}{4}]$ denotes the integer part of $\frac{n}{4}$ (i.e., a common upper bound for lags), is then calculated by

$$r(k) = \frac{\frac{1}{n}\sum_{t=1}^{n-k} \{u(t) - \bar{u}\}\{u(t+k) - \bar{u}\}}{\hat{\sigma}^2} \tag{6.15}$$

(Priestley, 1981, pp. 323 and 330) as an estimate of $\rho(t, t+k)$ [equations (A2.3) and (A2.6)] under weak stationarity, that is,

$$\rho(t, t+k) = \rho(|t+k-t|) = \rho(k) = \frac{\mathrm{E}[\{U(t) - \mu\}\{U(t+k) - \mu\}]}{\sigma^2}. \tag{6.16}$$

The application of the weak stationarity assumption in equation (6.15) is apparent through the use of (i) one mean estimate, $\bar{u} = \frac{1}{n}\sum_{t=1}^{n} u(t)$, for one population mean parameter, μ; (ii) one variance estimate, $\hat{\sigma}^2 = \frac{1}{n}\sum_{t=1}^{n}\{u(t) - \bar{u}\}^2$, for one population variance parameter, σ^2; and (iii) all $n - k$ pairs of observations made at intervals of k time units. Such calculation of estimates is also related to ergodicity (Section 2.3).

To facilitate and improve the identification of a time series model, a second family of autocorrelation statistics is used: the "sample partial autocorrelation coefficients" (SPAC), which are usually calculated at lags $k = 1, \ldots, [\frac{n}{4}]$ also. In general terms, and simply put, the partial correlation between two random variables, X and Y, given a third one, z (i.e., conditionally to a fixed value of Z), is defined as the simple correlation between the former two variables after removing the linear effect of the third one by regression; its theoretical expression is

$$\rho_{XY.z} = \frac{\rho_{YZ} - \rho_{XZ}\,\rho_{YZ}}{\sqrt{(1-\rho_{XZ}^2)(1-\rho_{YZ}^2)}}, \tag{6.17}$$

where ρ_{XY}, ρ_{XZ} and ρ_{YZ} denote the population simple correlations between the different pairs of random variables. By substituting $U(t)$, $U(t+2)$ and $U(t+1)$ to X, Y and Z, one obtains, under weak stationarity,

$$\rho_{02.1} = \frac{\rho(2) - \rho(1)^2}{1 - \rho(1)^2} = \xi(2). \tag{6.18}$$

The definition of theoretical partial autocorrelation coefficients $\xi(k)$ extends to lags greater than 1; the expression just becomes more complicated to write, as the random variable Z above is then replaced by a vector of random variables, so ρ_{XZ} and ρ_{YZ} are no longer scalar quantities (see, e.g., Box *et al.*, 1994, pp. 64–67). In practice, SPACs are calculated by using the required SSACs. For example, by replacing $\rho(1)$ and $\rho(2)$ by $r(1)$ and $r(2)$ in equation (6.18), one obtains the SPAC at lag 2, $\hat{\xi}(2)$. Since there is no observation between $u(t)$ and $u(t+1)$ in discrete time, where 1 denotes one time unit, $\rho(1) = \xi(1)$ and $r(1) = \hat{\xi}(1)$.

We can now look at Figs. 6.1 and 6.2 again, focusing on the SSAC and SPAC plots; no SPAC plot is presented in Fig. 6.1 (i.e., for discrete-time AR(1) processes) for a reason given below.

Key note: The sample autocorrelation coefficients plotted in Fig. 6.1 (right) and Fig. 6.2 (middle and right) are, by definition, autocorrelation estimates, but they reliably reproduce the behavior of the theoretical autocorrelation functions.

The objective here is to develop guidelines for time series model identification in applications; an example with real data is presented at the end of this subsection.

AR(1), Fig. 6.1 (right)

In this case, the theoretical simple autocorrelation function is

$$\rho(k) = \phi^k \text{ for } k = 1, 2, \ldots \tag{6.19}$$

That is, simple autocorrelations decrease exponentially with increasing lags when $\phi > 0$, and decrease exponentially, but in absolute value only, when $\phi < 0$ – for negative values of ϕ, $\rho(k)$ alternates in sign, being negative at odd lags and positive at even lags. This is what we may observe in Fig. 6.1 (right), besides small discrepancies inherent to the use of partial realizations to calculate sample autocorrelation coefficients.

Concerning partial autocorrelations, it is easy to see from equations (6.18) and (6.19), that $\xi(2) = 0$ for a discrete-time AR(1) process, and actually that $\xi(k) = 0$ for $k = 2, 3, \ldots$ for this process. This is why SPACs are not plotted in Fig. 6.1, because they would have looked tiny starting at lag $k = 2$, with $\hat{\xi}(1) = r(1)$.

In short, one may thus say that a discrete-time AR(1) process is characterized by an exponential decrease of its simple autocorrelations (in absolute value or not) and a cut-off of its partial autocorrelations after lag 1.

AR(2), Fig. 6.2 (a, middle and right)

The autoregression is expanded by one term, $\phi_2\varepsilon(t-2) = \phi_2 U(t-2)$ in equation (6.11), compared to the AR(1). This addition may seem anodyne, but it is directly related to the absence of a friendly general equation for $\rho(k)$ in the case of the AR(2) (Box *et al.*, 1994, p. 60). For the choice of ϕ_1 and ϕ_2 values made here (i.e., 0.6 and 0.3), simple autocorrelations behave like a mixture of slowly decreasing exponentials. In all cases, discrete-time AR(2) processes are characterized by a cut-off of their partial autocorrelations after lag 2 (i.e., their order): $\xi(k) = 0$ for $k = 3, 4, \ldots$

MA(1), Fig. 6.2 (b, middle and right)

In the discrete-time moving average process of first order [equation (6.12)], the current value of the white noise process, $\eta(t)$, and the previous value, $\eta(t-1)$, are "averaged," with respective weights of 1 and $-\theta$. Compared with the AR(1) and AR(2), the behavior of SSACs and SPACs is reversed. There is a cut-off in simple autocorrelations after lag 1: $\rho(1) = \frac{-\theta}{1+\theta^2}$ with $\rho(k) = 0$ for $k = 2, 3, \ldots$, and there is an exponential decrease of partial autocorrelations in the negative values when $\theta > 0$.

ARMA(1,1), Fig. 6.2 (c, middle and right)
Because of the mixture of one autoregressive component with one moving average component [equation (6.13)], there is no cut-off in either autocorrelation function for discrete-time ARMA(1,1) processes. Like the AR(2), the addition of one term in the model equation, compared to the pure AR(1) and the pure MA(1), is responsible for complications in obtaining friendly general expressions for $\rho(k)$ and $\xi(k)$. For a positive value of ϕ (e.g., 0.7) and a negative value of θ (e.g., –0.5), the decrease of SSACs is exponential and resembles that of a pure AR(1) with same value of ϕ [Fig. 6.1 (i, right)], but the SPAC at lag 2 exceeds two standard errors, indicating it is significantly different from 0 [Fig. 6.2 (c, right)].

SAR(1) with span 6, Fig. 6.2 (d, middle and right)
In equation (6.14), autoregression is performed on $U(t - 6)$, that is, on the variable six time units behind (i.e., the span length). Therefore, autocorrelation is restricted to be different from zero at lags that are multiples of 6 (i.e., 6, 12, 18, etc.), following a decreasing exponential pattern over those lags and creating "stochastic periodicity" (Dutilleul, 1995). Like the AR(1), but in a seasonal version, partial autocorrelations are different from zero only at one multiple of the span, that is, at lag 6.

The estimation options are essentially least squares, unconditional (ULS) or conditional (CLS), and maximum likelihood (ML); the three options are available in SAS PROC ARIMA, statement ESTIMATE. Theoretically, the estimation methods of least squares and maximum likelihood are equivalent when the random errors in linear models are normally distributed (Searle, 1971, p. 87). In the examples here, ML was used if normality was found to be an acceptable hypothesis, or when there seemed to have been a convergence problem with ULS or CLS otherwise.

Key note: Models are fitted to the data, and not to the SSACs and SPACs, which are used to identify the time series model to fit.

Concerning variances, the five discrete-time processes considered above are weakly stationary. As part of the condition of stationarity at order 2, their population variances are thus constant over time:

$\sigma^2(t) = \sigma^2$ for all t. As was the case with theoretical autocorrelation functions, population variances are functions of the model parameters. Their expressions below assume a variance of 1 for the white noise process, or must be regarded as ratios, relative to the variance of the white noise process. Numbers in parentheses correspond to the population variance values for the processes of Fig. 6.2. In applications, variance estimates are obtained by replacing model parameters by estimates in the expressions below.

AR(1): $\sigma^2 = \dfrac{1}{1-\phi^2}$ – the value of σ^2 ranges from 5.26 (a) to 1.01 (e) and from 1.01 (f) to 5.26 (j) in Fig. 6.1.

AR(2): $\sigma^2 = \dfrac{1-\phi_2}{(1+\phi_2)\{(1-\phi_2)^2-\phi_1^2\}}$ (4.14).

MA(1): $\sigma^2 = 1+\theta^2$ (1.64).

ARMA(1,1): $\sigma^2 = \dfrac{1+\theta^2-2\phi\theta}{1-\phi^2}$ (3.82).

SAR(1) with a span of 6: $\sigma^2 = \dfrac{1}{1-\Phi^2}$ (2.78).

The example below is the first of a series, in this chapter and the next two, using temporal, spatial and spatio-temporal data collected at Gault Nature Reserve in Mont-Saint-Hilaire (Québec, Canada). Hopefully, these "MSH examples" will be an inspiration for quantitative analyses in conservation projects.

First MSH air–soil temperature example

The specific objectives of this first MSH example are to apply the guidelines for the identification of a time series model presented above and to propose complementary diagnostic rules for the validation of the model after it has been fitted to the time series. Note that, in the applications of the AR(1) and other time series models under weak stationarity, the population mean of $U(t)$, $\mu(t) = \mu$, does not have to be equal to zero; with $\mu \neq 0$, it suffices to write the model equation as $U(t) = \mu + \varepsilon(t)$, with $\varepsilon(t) = \phi\varepsilon(t-1) + \eta(t)$ in the case of the AR(1).

Air and soil temperatures are important environmental variables in plant ecology studies, especially in the spring when adult plants exit dormancy and new seedlings emerge. Once identified, their autocorrelation might be included in a model with other variables. The time series

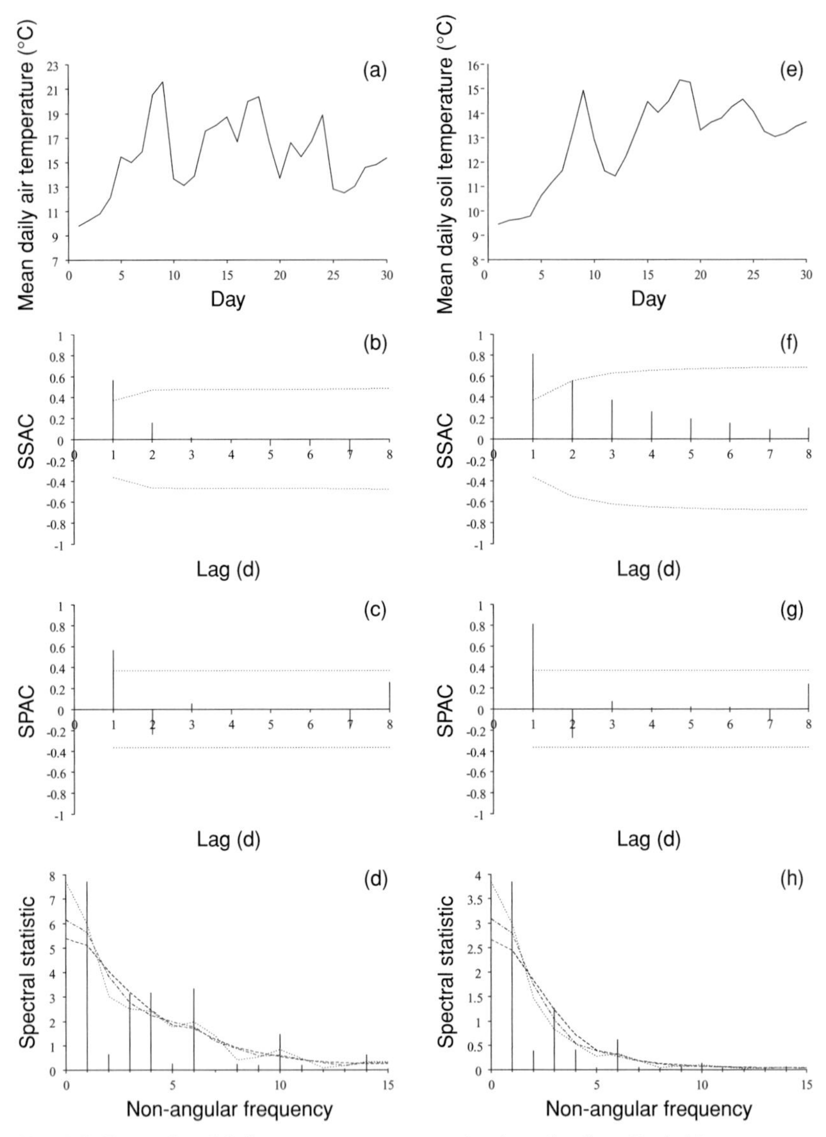

Fig. 6.3. Example of daily mean temperatures in air and soil at Gault Nature Reserve in Mont-Saint-Hilaire (Québec, Canada) over the period June 1–30, 2004: (a) and (e) the time series ($n = 30$); (b) and (f) sample simple autocorrelation coefficients (SSAC); (c) and (g) sample partial autocorrelation coefficients (SPAC); and (d) and (h) spectral statistics: periodogram (bars) and spectral density function estimates, from the least smoothed (short-dashed) to the most smoothed (long-dashed).

analyzed here consist in daily mean temperatures in air and soil, covering the period June 1–30, 2004 (i.e., late spring and early summer), so $n = 30$. In view of Fig. 6.3(a) and (e), the assumption of weak stationarity seems reasonable. Actually, the smooth fluctuations observed resemble those of the short time series with strong positive autocorrelation, or "false trend," plotted in Fig. 2.3(b). Accordingly, SSACs and SPACs were calculated for both time series [Fig. 6.3(b)–(c) and (f)–(g)]. Both daily mean temperatures are positively autocorrelated, but the one in soil is more strongly autocorrelated. In both cases, however, the statistical significance of coefficients is difficult to reach because of large standard errors due to the small value of n.

Based on the identification guidelines, the AR(1) might be a candidate model for the time-discrete processes of both time series, but it would be wise to try other models as well, like the MA(1) for daily mean air temperature and the AR(2) and ARMA(1,1) for daily mean soil temperature. Therefore, the four models were fitted to both time series and diagnostic statistics are reported in Table 6.1.

Among the diagnostic statistics, the information criteria were not defined above. As they will be more extensively used in Section 6.4, their detailed presentation is delayed until there. In short, information criteria are related to the residual variance computed for the model fitted to the time series, so the rule is: the smaller, the better. The main difference between Akaike's and Schwarz's criteria lies in the use of a more severe penalty in the latter, to take into account the number of parameters of the fitted model and apply the parsimony principle. As for parameter estimates, the rule is: the more significant (i.e., the smaller the probability of significance), the better. The SSACs calculated on residuals reflect the capacity of the model to incorporate the autocorrelation contained in the time series, so the lesser autocorrelation left in the residuals (i.e., the greater the probability of significance), the better.

For daily mean air temperature [Table 6.1(a)], the residual SSACs discard none of the four models, but the information criteria and the parameter estimates give a strong advantage to the AR(1) and MA(1) models, compared to the AR(2) and ARMA(1,1). Eventually, the preference is given to the AR(1) model because of slightly better statistics overall, compared to the MA(1). The story is different for daily mean soil temperature [Table 6.1(b)]. First of all, the MA(1) must be discarded immediately because of the values of information criteria and residual SSACs obtained for this model. Then, the need for a second parameter in the model arises from the values of information criteria, which

Table 6.1. *Example of daily mean temperatures in (a) air and (b) soil at Gault Nature Reserve in Mont-Saint-Hilaire (Québec, Canada) over the period June 1–30, 2004: information criterion values, parameter estimates, and sample simple autocorrelation coefficients of residuals (with probabilities of significance in parentheses) for four time series models fitted*

	Akaike's criterion	Schwarz's criterion	Parameter estimates		Sample simple autocorrelation coefficients of residuals at lags 1–6					
(a)										
AR(1)	142.7	145.5	$\hat{\phi} = 0.6172$ (0.0001)		0.094 (0.7590)	–0.220	–0.090	–0.003	–0.036	0.090
AR(2)	143.2	147.4	$\hat{\phi}_1 = 0.7396$ (0.0001)	$\hat{\phi}_2 = -0.2266$ (0.2213)	–0.007 (0.8692)	–0.094	0.024	0.040	–0.079	0.126
MA(1)	143.1	145.9	$\hat{\theta} = -0.5919$ (0.0002)		0.139 (0.8117)	0.146	–0.078	0.049	–0.111	0.054
ARMA(1,1)	143.0	147.2	$\hat{\phi} = 0.3914$ (0.1365)	$\hat{\theta} = -0.3695$ (0.1709)	–0.025 (0.8917)	–0.048	–0.052	0.036	–0.093	0.116
(b)										
AR(1)	80.86	83.66	$\hat{\phi} = 0.8920$ (0.0001)		0.252 (0.2293)	–0.225	–0.226	–0.097	–0.070	0.132
AR(2)	79.31	83.52	$\hat{\phi}_1 = 1.1813$ (0.0001)	$\hat{\phi}_2 = -0.3447$ (0.0492)	0.042 (0.3626)	–0.217	–0.094	0.039	–0.081	0.226
MA(1)	96.35	99.15	$\hat{\theta} = -0.7645$ (0.0001)		0.473 (0.0006)	0.507	0.162	0.274	0.046	0.210
ARMA(1,1)	78.15	82.35	$\hat{\phi} = 0.8050$ (0.0001)	$\hat{\theta} = -0.4247$ (0.0200)	–0.013 (0.3493)	–0.088	–0.169	0.045	–0.154	0.233

AR(1) = first-order autoregressive; AR(2) = second-order autoregressive; MA(1) = first-order moving average; and ARMA(1,1) = first-order autoregressive–first-order moving average are smaller for the AR(2) and ARMA(1,1) models than for the AR(1). Eventually, the smallest values of information criteria and the highest significance of parameter estimates give the advantage to the ARMA(1,1), as the model best representing the heterogeneity due to autocorrelation in the time series of daily mean soil temperatures.

6.2.2 Estimation in the frequency domain

The frequency domain offers a "new world" to analyze the autocorrelation contained in a time series and, through this, the temporal stochastic process that has generated the observed heterogeneity. Compared with the time domain, this world corresponds to an alternative space (in the mathematical sense), where autocorrelation can theoretically be analyzed in an equivalent manner in the sense that no information is lost or gained in it. To get an intuitive idea of this space, assume n observations have been made for U at equally spaced times. Seen as a function in discrete time, the observed time series $\{u(t)|t = 1, \ldots, n\}$ can be rebuilt exactly from the coefficients of its Fourier series development, that is, from its projections in the space of cosine and sine waves at frequencies $\frac{2\pi p}{n}$ for $p = 0, 1, \ldots, [\frac{n}{2}]$. (Strictly speaking, a distinction must be made between the cases of odd n vs. even n: the sine wave for $p = [\frac{n}{2}]$ is required in the Fourier series development when n is odd, while it is not used when n is even because its value is then zero. The sine wave for $p = 0$ is never used because its value is always zero, whether n is odd or even.)

In this subsection, we continue to work under weak stationarity. Accordingly, we are going to explore one facet of spectral analysis: that of "continuous spectra," so called because they are characterized by a spectral density function continuously defined over the interval $[-\pi, \pi]$; discrete and mixed spectra will be analyzed in Subsection 6.3.2. For a given frequency $\omega \in [-\pi, \pi]$, the spectral density function of a weakly stationary discrete-time process is obtained from the theoretical simple autocorrelation function by

$$f(\omega) = \frac{1}{2\pi} \sum_{k=-\infty}^{\infty} \rho(k) e^{-i\omega k}, \tag{6.20}$$

with $e^{-i\omega k}$ the complex exponential $\cos(\omega k) - i \sin(\omega k)$. Hereafter, nonangular frequencies will be used for ease of interpretation, because they are expressed in number of cycles over the time series instead of radians.

From equation (6.20), it follows that $f(\omega)$, via $\rho(k)$, is a function of the model parameters (i.e., ϕ, ϕ_1, ϕ_2, θ, and Φ) for the AR(1), AR(2), MA(1), ARMA(1,1), and SAR(1) processes. The explicit expressions of $f(\omega)$ for the different processes above are given in several books (Anderson, 1971; Priestley, 1981; Box *et al.*, 1994). Here, they are plotted in Fig. 6.4(a)–(h) (see continuous curves). The behavior of a spectral density function depends on the sign and strength of autocorrelation, and can be interpreted in terms of "energy" conveyed by the time series (i.e., fluctuations, oscillations, seasonality). When the autocorrelation is predominantly negative [see the AR(1) processes with $\phi = -0.9$ and -0.5 and the MA(1) process with $\theta = 0.8$ in Fig. 6.4(a), (b) and (f)], the highest values of $f(\omega)$ are in the highest frequencies; the time series oscillate a lot and frequently. When the autocorrelation is non-seasonal and restricted to be positive [see the AR(1) processes with $\phi = 0.5$ and 0.9 and the AR(2) and ARMA(1,1) in Fig. 6.4(c)–(e) and (g)], the highest values of $f(\omega)$ are in the lowest frequencies; the time series fluctuate smoothly and even more if autocorrelation is stronger. When the autocorrelation is seasonal [see the SAR(1) process with a span of 6 in Fig. 6.4(h)], there are several peaks in $f(\omega)$, located at multiples of $\frac{n}{6}$; there is stochastic periodicity in the time series. The spectral density function of a white noise process (i.e., not presented in Fig. 6.4) would be a horizontal continuous straight line if represented, thus justifying the name of the process; when passing through a prism, white light decomposes into all frequencies equally.

Schuster's (1898) classical periodogram plays a central part in spectral analysis. At the Fourier frequencies $\frac{2\pi p}{n}$ for $p = 0, 1, 2, \ldots, [\frac{n}{2}]$, which correspond to integer numbers of cycles over the time series and periods of ∞, n, $\frac{n}{2}$, etc., the value of this uni-frequential spectral statistic is equal, up to the multiplicative constant $\frac{n}{2}$, to the squared amplitude of the cosine wave with phase λ_p, $A_p \cos\left(\frac{2\pi p}{n} + \lambda_p\right)$, fitted to the time series by ordinary least squares:

$$I^C\left(\frac{2\pi p}{n}\right) = \frac{2}{n}\left[\left\{\sum_{t=1}^{n} u(t)\cos\left(\frac{2\pi p}{n}t\right)\right\}^2 + \left\{\sum_{t=1}^{n} u(t)\sin\left(\frac{2\pi p}{n}t\right)\right\}^2\right]. \tag{6.21}$$

That is, $I^C(\frac{2\pi p}{n}) = \frac{n}{2}\hat{A}_p^2$. At non-Fourier frequencies, the latter equality is approximate, and equation (6.21) is used. In some statistical packages, the value of Schuster's periodogram above is divided by 4π.

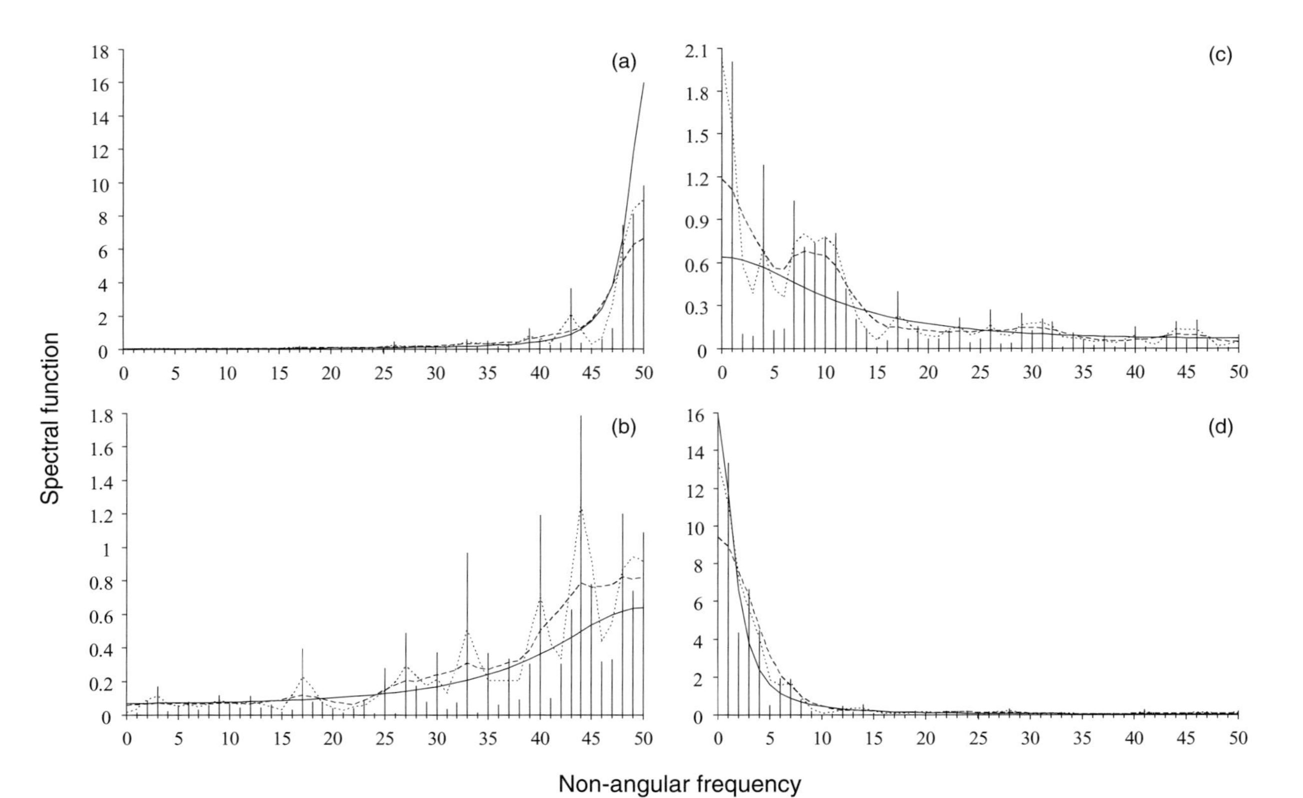

Fig. 6.4. Periodogram (bars) and other spectral functions evaluated for (a)–(d) the partial realizations of discrete-time AR(1) processes in Fig. 6.1 with $\phi = -0.9, -0.5, 0.5$, and 0.9, and (e)–(h) the four partial realizations of discrete-time stochastic processes in Fig. 6.2. The theoretical spectral density function (continuous curve) is also represented for each process, while the smoothed estimates are dashed, from short (least smoothed) to long (most smoothed).

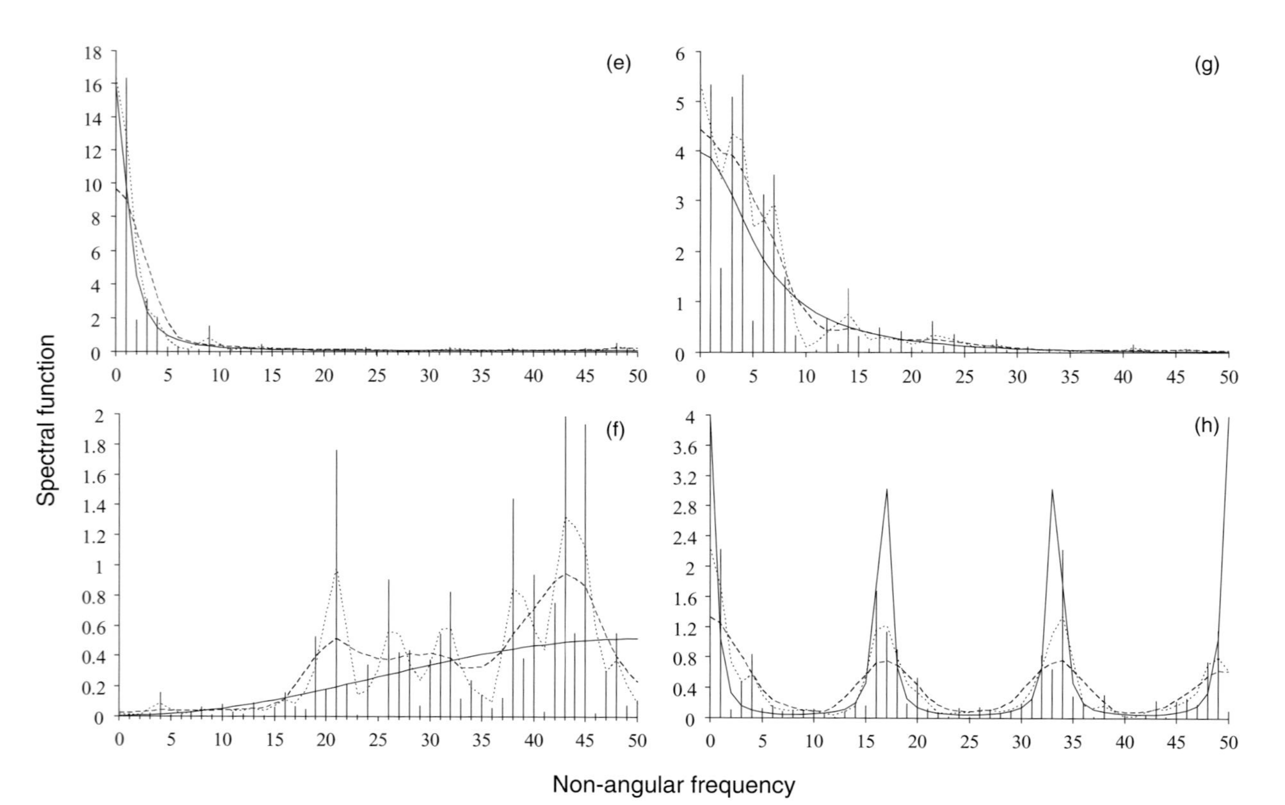

Fig. 6.4. (*cont.*)

Asymptotically (i.e., for time series of infinite length or $n \to \infty$), Schuster's periodogram is an unbiased estimator of the spectral density function, but its variance does not vanish, so it is not consistent. One family of better spectral density function estimators is obtained by smoothing Schuster's periodogram within a triangular window

$$\hat{f}\left(\frac{2\pi p}{n}\right) = \frac{1}{4\pi}\sum_{\ell=0}^{2L} w_\ell I^C\left[\frac{2\pi(p - L + \ell)}{n}\right], \tag{6.22}$$

where w_ℓ ($\ell = 0, 1, \ldots, 2L$) are the smoothing weights, with $w_L > w_{L-1} = w_{L+1} > w_{L-2} = w_{L+2}$, etc. Such spectral density function estimators are available in SAS PROC SPECTRA, but several other families of estimators have been proposed to overcome the lack of consistency of Schuster's periodogram in the analysis of continuous spectra (e.g., autoregressive, maximum entropy, and with tapered data; Brillinger, 1981; Priestley, 1981). Figure 6.4(a)–(h) (see dashed curves) helps us appreciate the effect of smoothing in the performance of spectral density function estimators, with the use of triangular windows of width 3 (short-dashed) vs. 9 (long-dashed).

Back to the first MSH air–soil temperature example, two remarks arise from Fig. 6.3(d) and (h). First, the energy conveyed by the daily mean soil temperatures is more concentrated in the low frequencies than the daily mean air temperatures. Second, the behavior of spectral statistics depicted in Fig. 6.3(d) and (h) is in agreement with the choice of models made in Subsection 6.2.1 from the SSACs, SPACs, and diagnostic statistics in Table 6.1.

6.3 Analysis of temporal heterogeneity of the mean

In this section, the weak stationarity assumption is partly relaxed in that the population mean of $U(t)$, $\mu(t)$, is allowed to change with t. In other words, the temporal stochastic process under observation is no longer stationary at order 1, and there is heterogeneity of the mean to quantify in the observed time series. Equations (6.3) and (6.5) provide appropriate models for this. Temporal heterogeneity of the mean will be non-periodic in Subsection 6.3.1, and periodic, possibly in combination with a non-periodic component, in Subsection 6.3.2. Depending on the example used for discussion, there will be absence or presence of temporal autocorrelation, meaning there may be one or two types of heterogeneity to analyze; two types = heterogeneity of the mean and heterogeneity due to autocorrelation. The presentation of statistical models and methods will be incorporated into examples, which are aimed at highlighting specific aspects of the analysis of temporal heterogeneity of the mean. In all the examples, there will be no evidence of temporal heterogeneity of the variance, leaving this topic for Section 6.5.

6.3.1 Polynomial models without, vs. with, autocorrelated errors

Non-periodic temporal heterogeneity of the mean arises when there is a quasi-monotonic increase, or decrease, of the values from beginning to end of the observed time series, resulting in a true large-scale trend like those displayed in Figs. 6.5(a) and 6.6(a); see also Figs. 2.3(a) and 2.4(a)–(c), which were produced from simulated data. Polynomials, and especially orthogonal polynomials, are recommended to model the mean function $\mu(t)$ in such cases.

Depending on the complexity of the trend, polynomials with increasing degree can be used:

- degree 1, linear: $\mu(t) = a + bt$;
- degree 2, quadratic: $\mu(t) = a + bt + ct^2$;
- degree 3, cubic: $\mu(t) = a + bt + ct^2 + dt^3$.

Only rarely are higher-degree polynomials used, for the true trend to remain large scale.

The construction of orthogonal polynomials is mathematical, and its details are given in Section A6.2. In simple terms, the monoms t, t^2 and t^3 in the equations above are replaced by the polynomials $P_{O1}(t)$, $P_{O2}(t)$ and $P_{O3}(t)$, to be used as new regressors in the model fitting; working with orthogonal polynomials has the advantage that coefficients a, b, c, d are estimated independently, so the removal of one orthogonal polynomial from the model has no effect on the coefficients estimated for the others.

For example, for $n = 15$ (see the first forestry example):

- $P_{O1}(t) = t - 8$;
- $P_{O2}(t) = t^2 - 16(t - 8) - 82.67 = t^2 - 16t + 45.33$;
- $P_{O3}(t) = t^3 - 24(t^2 - 16t + 45.33) - 225.4(t - 8) - 960$
  $= t^3 - 24t^2 - 158.6t - 244.72$.

The two examples below differ in the autocorrelation structure found for the errors $\varepsilon(t)$. This difference will have important implications for

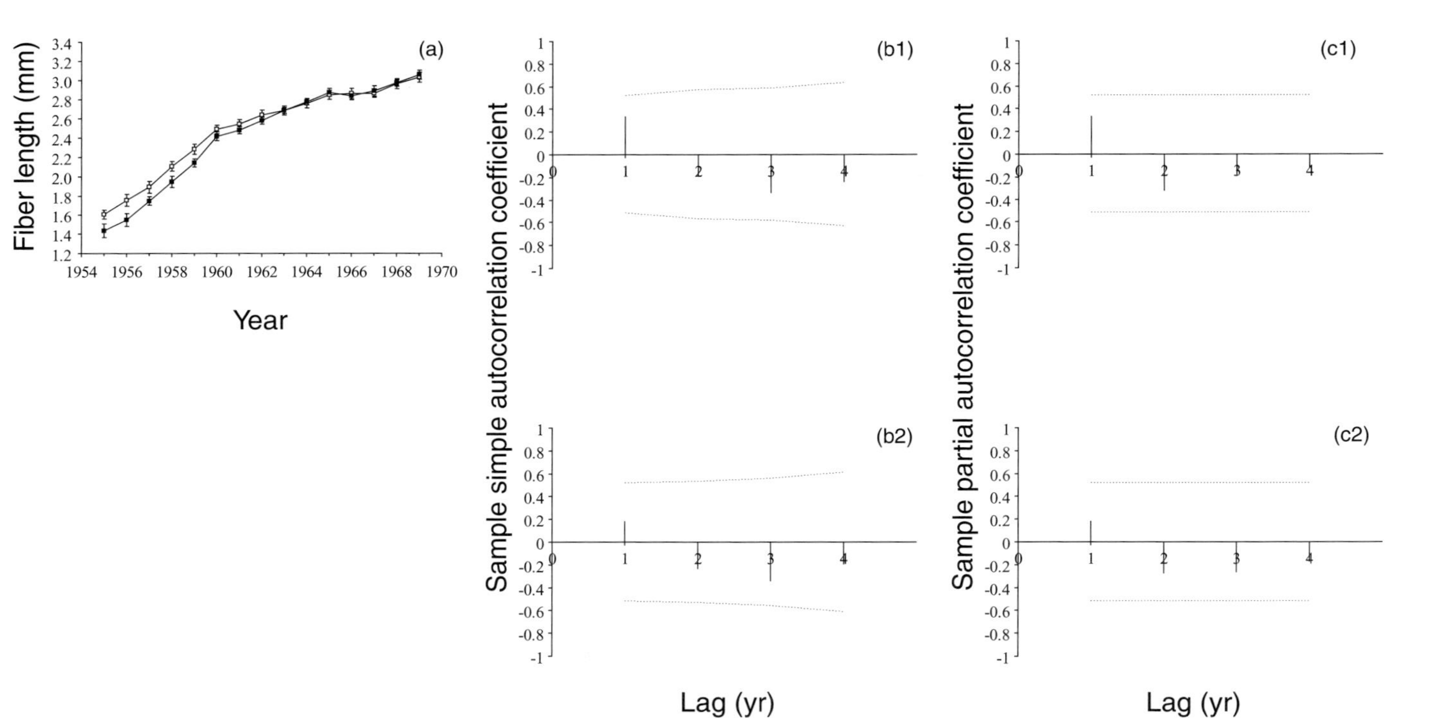

Fig. 6.5. Forestry example with yearly repeated measures of fiber length on Norway spruces (*Picea abies* (L.) Karst.) before first thinning: (a) the time series ($n = 15$) of averages with the corresponding standard errors calculated for 20 fast-grown trees (open squares) and 20 slow-grown trees (filled squares); (b1) and (b2) sample simple autocorrelation coefficients (±2 standard errors) computed from the residuals of a second-degree polynomial model fitted by ordinary least squares (OLS) to the time series of fast-grown trees and slow-grown trees, respectively; and (c1) and (c2) sample partial autocorrelation coefficients (±2 standard errors) computed from the same OLS residuals as in (b1) and (b2). Note that the OLS method of estimation assumes the homogeneity of variance and the absence of autocorrelation.

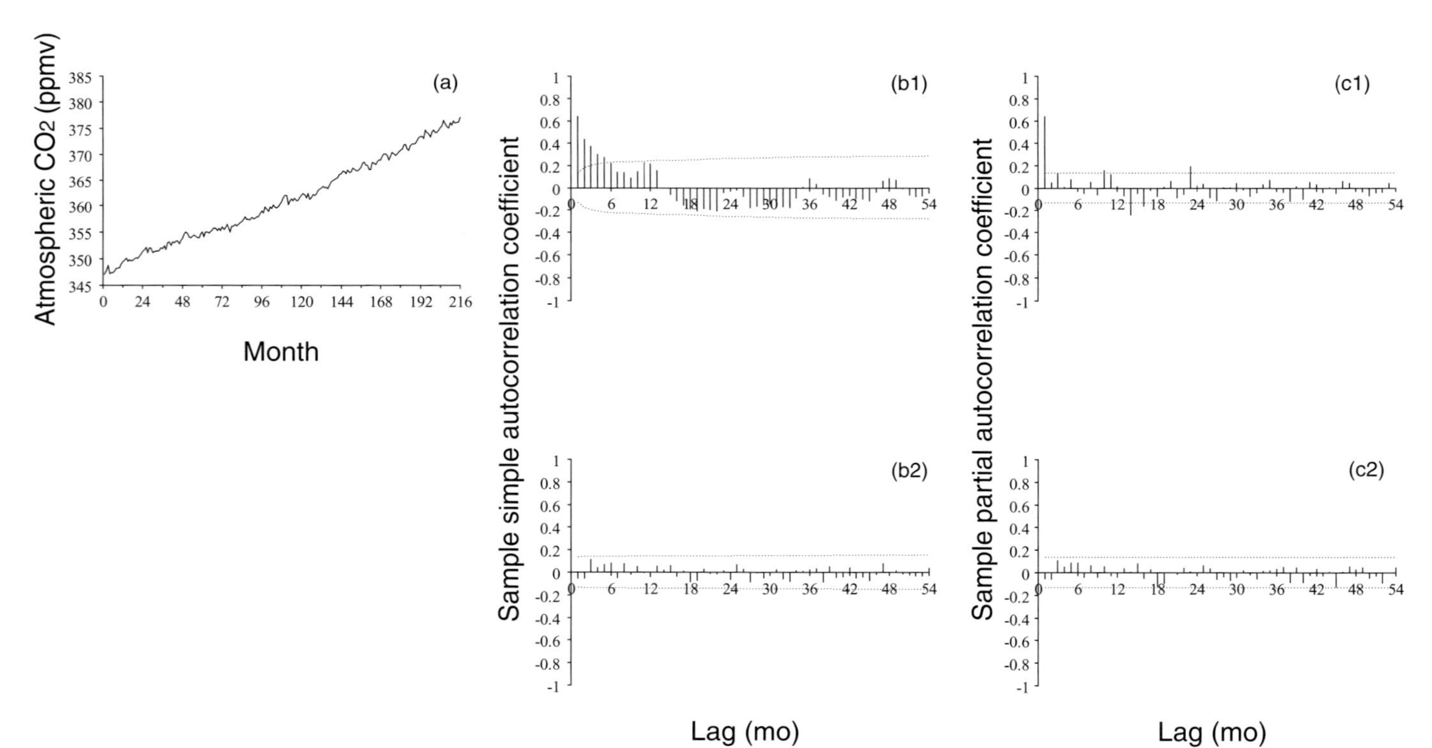

Fig. 6.6. Example of atmospheric CO_2 concentration in Samoa Islands (Polynesia) over the period January 1987–December 2004: (a) the time series ($n = 216$); (b1) and (b2) sample simple autocorrelation coefficients (±2 standard errors) computed from the residuals of the orthogonal polynomial models of second degree fitted to the time series by OLS and by estimated generalized least squares (EGLS), respectively; and (c1) and (c2) sample partial autocorrelation coefficients (±2 standard errors) computed from the same residuals as in (b1) and (b2). Here, the EGLS method of estimation helps relax the assumption of absence of autocorrelation of the OLS method.

the statistical treatment and the estimation of parameters of the mean function.

First forestry example

In this simple example, 15 yearly repeated measures of fiber length were made on 20 fast-grown and 20 slow-grown Norway spruces (*Picea abies* (L.) Karst.) before first thinning (Herman *et al.*, 1998). In other words, time series data were collected under model (6.6), with growth category as the classification factor. For each growth category, standard errors of yearly averages are almost constant from beginning to end of the time series [Fig. 6.5(a)]. Thus, the model defined by equation (6.5) can be used for analyzing temporal heterogeneity of the mean in each time series of yearly averages, with $\mathrm{Var}[\bar{U}(t)] = \frac{\sigma^2}{20}$ for $t = 1, \ldots, 15$. Classical multiple regression (i.e., by ordinary least squares, OLS) of $\bar{U}(t)$ on $a + b\,P_{O1}(t) + c\,P_{O2}(t) + d\,P_{O3}(t)$ led to non-significance of the estimated slope for $P_{O3}(t)$, meaning a second-degree polynomial model is sufficient to describe each trend. It also led to OLS residuals that were analyzed for autocorrelation [Fig. 6.5(b) and (c)], but no SSAC or SPAC was significantly different from zero.

In conclusion, temporal heterogeneity of the mean is dominant in both time series, with $R^2 = 0.99$. Concerning the underlying process, one can reasonably assume that $\mathrm{Corr}[\bar{U}(t), \bar{U}(t')] = 0$ for $t' \neq t = 1, \ldots,$ 15, with $\mathrm{Var}[\bar{U}(t)] = \frac{\sigma^2}{20}$.

First atmospheric CO_2 concentration example

The series of 216 monthly values of atmospheric CO_2 concentration in Samoa Islands (Polynesia) over the period January 1987–December 2004 (Keeling and Whorf, 2005) is used in this example. This time series was analyzed under equation (6.3), because of a large-scale trend of sustained increase of atmospheric CO_2 concentration [Fig. 6.6(a)]. In a preliminary step, classical multiple regression of $U(t)$ on $\mu(t) = a + b\,P_{O1}(t) + c\,P_{O2}(t) + d\,P_{O3}(t)$ was performed. From the OLS fitting, it appears that the estimated slope of the cubic term is close to the significance level of 5% ($P = 0.0599$) [Table 6.2(a)] and the residuals are autocorrelated [Fig. 6.6(b1) and (c1)]. Actually, the behavior of the SSACs and SPACs of OLS residuals mainly resembles that of a partial realization of a discrete-time AR(1) process with a positive and relatively high value of ϕ (Subsection 6.2.1). These SSACs and SPACs also show traces of seasonal autocorrelation around lags 12 and 24. Since the OLS estimation method assumes the absence of autocorrelation of the errors

Table 6.2. *Example of atmospheric CO_2 concentration in Samoa Islands (Polynesia) over the period January 1987–December 2004: slope estimates and the corresponding standard errors and probabilities of significance for the third-degree orthogonal polynomial model fitted to the time series by (a) ordinary least squares (OLS) and (b) estimated generalized least squares (EGLS)*

	(a)			(b)		
	Slope estimates	Standard errors	Probabilities of significance	Slope estimates	Standard errors	Probabilities of significance
Linear term	0.1334	0.0006621	0.0001	0.1336	0.001332	0.0001
Quadratic term	0.0001828	0.00001187	0.0001	0.0001740	0.00002360	0.0001
Cubic term	0.0000004101	0.0000002168	0.0599	0.0000005010	0.0000004265	0.2414

in addition to the homogeneity of their variance, an alternative fitting procedure is required to perform the multiple regression.

A reasonable alternative fitting procedure is called "estimated generalized least squares" (EGLS); the matrix algebra aspects involved in it are detailed in Section A6.3. In simple terms, EGLS consists in "prewhitening" the time series, that is, removing autocorrelation from the time series by applying a linear transformation to it, based on the autocorrelation structure identified; applying the same transformation to the regressors [$P_{O1}(t)$, $P_{O2}(t)$, and $P_{O3}(t)$]; and regressing the prewhitened time series on the transformed regressors by OLS. If the errors are autocorrelated and the autocorrelation structure is correctly specified, EGLS provides unbiased standard errors for estimated slopes, whereas OLS standard errors are biased downwards when autocorrelation is positive [Table 6.2(b) vs. (a); see also Diggle (1990, pp. 87–92) and Alpargu and Dutilleul (2001)]. Accordingly, the EGLS estimated slope of $P_{O3}(t)$ is not declared significantly different from zero ($P = 0.2414$) and a second-degree orthogonal polynomial model is retained to analyze the temporal heterogeneity of the mean. In view of Fig. 6.6(b2) and (c2), there is no remaining autocorrelation in the EGLS residuals.

In conclusion, two types of temporal heterogeneity were found in the series of atmospheric CO_2 concentration in Samoa Islands (Polynesia): in the mean (non-periodic), accounting for 97.8% of the total variation, and in the autocorrelation of random deviations, accounting for a mere 1.9%.

6.3.2 Purely trigonometric models and extensions

Temporal heterogeneity of the mean at intermediate scale is observed when some type of fluctuation is repeated a number of times over the series, and this number is sufficiently large so that a polynomial model would not fit, but is not too large for temporal heterogeneity not to be at small scale. Examples of such periodicity are shown with real data in Figs. 6.7–6.10; see also Fig. 2.6(a)–(c), which was produced from simulated data. In some of these cases, temporal heterogeneity of the mean is solely periodic [Figs. 6.7(a) and 6.8(a)]; in the others, it is combined with non-periodic temporal heterogeneity of the mean at large scale [Figs. 6.9(a) and 6.10(a)]. In all these cases, cosine and sine waves will be at the core of the models used for statistical analyses of the data. More specifically, equations (6.3) and (6.5) will be used with

$$\mu(t) = a + b\cos(\omega_{01}t) + c\sin(\omega_{01}t) + d\cos(\omega_{02}t) + e\sin(\omega_{02}t)$$

in the case of two periodic components and no large-scale trend, or

$$\mu(t) = a + b\,P_{O1}(t) + c\,P_{O2}(t) + d\,P_{O3}(t) + e\cos(\omega_0 t) + f\sin(\omega_0 t)$$

in the case of one main periodic component combined with some large-scale trend.

Key note: A periodic signal need not be a pure cosine or sine wave to be analyzed with a trigonometric model. Furthermore, trigonometric models apply as well to periodic signals with varying amplitude and pseudo-periodic signals with unevenly spaced peaks.

In some biological and environmental studies, frequencies corresponding to natural periodicities (e.g., 24 hours and 12 months for the circadian rhythm and the annual cycle) and fractions of them (e.g., 12, 8 hours; 6, 4 months) can be postulated, but in most studies the frequencies ω_{01}, ω_{02}, etc. will need to be estimated because unknown. Dutilleul's (1990) multi-frequential periodogram offers a flexible tool for this. For a given vector of frequencies $(\omega_1, \ldots, \omega_F)$, the statistic, denoted $I^M(\omega_1, \ldots, \omega_F)$, is basically defined as the sum of squares of the trigonometric model

$$U(t) = a + \sum_{f=1}^{F}\{b_f\cos(\omega_f t) + c_f\sin(\omega_f t)\} + \varepsilon(t), \qquad (6.23)$$

fitted to the time series by OLS. Frequency estimates are obtained by maximization of $I^M(\omega_1, \ldots, \omega_F)$ with respect to $(\omega_1, \ldots, \omega_F)$. Mathematical details are given in Section A6.3. The statistical method has proved appropriate in the absence of temporal autocorrelation as well as in its presence, namely, for the analyses of discrete and mixed spectra (Dutilleul, 2001). In Section 6.1 of the same reference, the author proposed a joint trend and periodicity analysis, using (i) a pre-specified model for the trend; (ii) unknown frequencies (i.e., to be estimated) for the periodicities; and (iii) the sum of squares of the complete model (see below) as statistic, denoted $I^{M*}(\omega_1, \ldots, \omega_F)$,

$$U(t) = a + \sum_{q=1}^{Q} b_q P_{O_q}(t) + \sum_{f=1}^{F} \{c_f \cos(\omega_f t) + d_f \sin(\omega_f t)\} + \varepsilon(t). \tag{6.24}$$

This extension will be particularly useful for the second atmospheric CO_2 concentration example and the second MSH air–soil temperature example here.

Key note: In periodicity analysis in the presence of a large-scale trend, it is not recommended to fit the trend by OLS and limit the spectral analysis to the resulting OLS residuals, without updating them into EGLS residuals, for example (Thomas and Wallis, 1971; Wallis, 1974).

Wistar rat example

In a chronobiological experiment involving the Wistar rat (Schelstraete *et al.*, 1992), two groups of 12 mother rats were kept in individual cages with their young, and observed continuously during 5 days. One experimental group (EXP) was submitted to an atypical synchronizer (i.e., shifted light and temperature regimes), while the other group (CTL) was in control conditions of light and temperature. The indexed random variable is the duration of a given maternal behavior (i.e., care of the young) within a 2-h interval (i.e., in discrete time), so the length of time series is $n = 60$. Thus, the data (i.e., replicated time series) were obtained under equation (6.6) and after calculation of averages over replicates, the statistical analysis was performed under equation (6.5) for each group separately. The basic hypothesis tested in this experiment was that the atypical synchronizer should affect the 24-h circadian rhythm of maternal behavior in the EXP group. Here, it is assessed via the autocorrelation analysis of average series combined with the analysis of

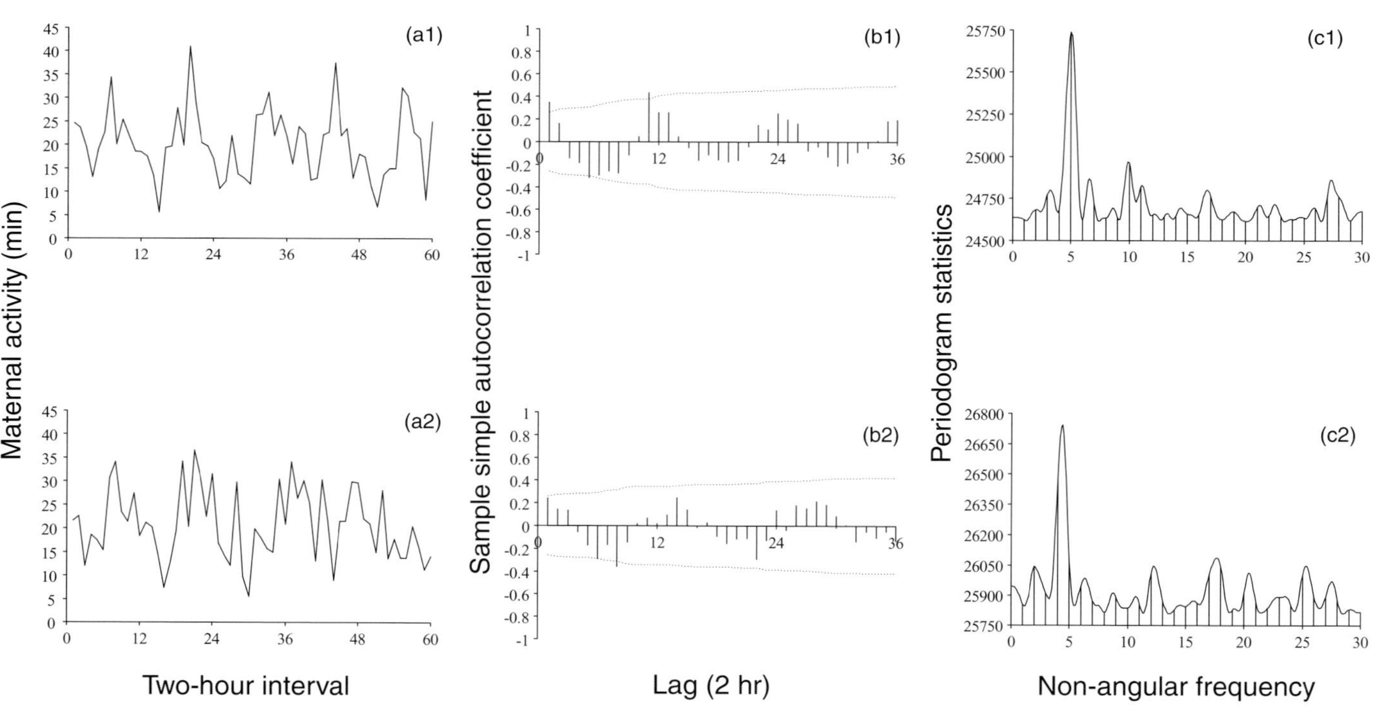

Fig. 6.7. Wistar rat example, with the duration of a maternal activity within 2 h as the variable of observation: (a1) and (a2) the time series of averages during 5 successive days ($n = 60$) for a control group of mother rats and a group of mother rats submitted to an atypical synchronizer, respectively; (b1) and (b2) sample simple autocorrelation coefficients for each series (raw data); and (c1) and (c2) the uni-frequential version of Dutilleul's periodogram for each series; the bars correspond to Schuster's classical periodogram evaluated at integer non-angular frequencies.

their heterogeneity of the mean, using sample autocorrelation coefficients and periodograms.

The average series for the groups CTL and EXP are plotted in Fig. 6.7(a1) and (a2). Visual inspection indicates differences in the fluctuations of the two time series, with five relatively clear peaks in the CTL series and seemingly less than five more diffuse periods of activity in the EXP series. These preliminary observations are refined below.

Autocorrelation analysis provides a first quantification of the information contained in the two time series [Fig. 6.7(b1) and (b2)]. Besides the first SSACs, which are positive for both series, other positive coefficients are around lags 12, 24, and 36 for the CTL series and over lags 13–15 and 24–30 for the EXP series. Results for lags 12, 24, and 36, which correspond to 1, 2, and 3 days, were expected, but not the others. Note that SSACs were calculated for lags greater than $[\frac{n}{4}] = 15$, to cover at least two days and preferably three.

The analysis of uni-frequential periodograms $I^C(\omega)$ and $I^M(\omega)$ provides complementary information [Fig. 6.7(c1) and (c2)]; the evaluation of the former periodogram statistic is restricted to the Fourier frequencies $\frac{2\pi p}{n}$ for $p = 0, 1, \ldots, [\frac{n}{2}] = 30$ here. For the CTL series, the two dominant peaks are centered at, or very close to, the non-angular frequencies of 5 and 10 cycles, whereas the dominant peak is between 4 and 5 cycles and there is a secondary peak between 17 and 18 cycles for the EXP series. Finally, the maximization of the bi-frequential periodogram $I^M(\omega_1, \omega_2)$ provided frequency estimates of 4.4 and 17.7 for the EXP series and of 5.0 and 10.0 for the CTL series. In conclusion, the atypical synchronizer did affect the way mother rats cared their young in the EXP group, which could in turn affect the later development and behavior of the young.

Sunspot number example

The yearly mean sunspot numbers analyzed here were retrieved from the Website of the National Geophysical Data Center (NGDC) in Boulder (Colorado, USA). The time series covers the period 1749–1944 (i.e., $n = 196$), because the other data available (i.e., 1700–1748 and 1945–present) have been recorded with different procedures. Compared to the previous example, the specific objectives of this one are to illustrate the bi-frequential periodogram analysis and explain how to build models to make predictions. On the environmental side, solar activity is known to influence terrestrial activities, including plant growth (see, e.g., Dutilleul and Till, 1992 and the references therein).

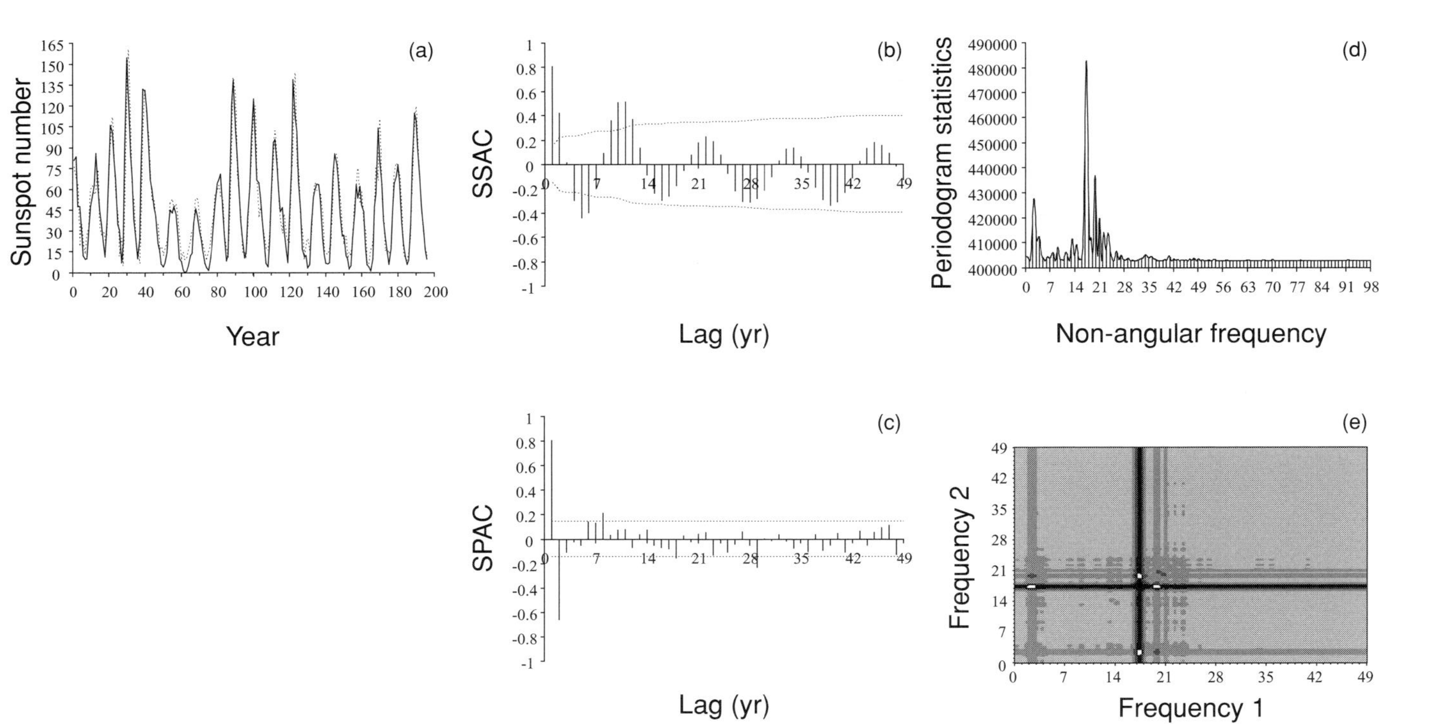

Fig. 6.8. Example of yearly mean sunspot numbers over the period 1749–1944: (a) the observed (continuous curve) and predicted (dashed) time series ($n = 196$) – a multiple regression model with two pairs of cosine and sine waves at the non-angular frequencies derived from (e) and AR(2) errors [see (b) and (c)] was used for prediction; (b) and (c) sample simple and partial autocorrelation coefficients, respectively; (d) the uni-frequential version of Dutilleul's periodogram; and (e) the bi-frequential version of Dutilleul's periodogram – the four spots in white indicate the highest values of the statistic, and the plot is limited to $[0, 49] \times [0, 49]$ because the information contained in the rest of the frequency domain is ancillary.

There is clearly a periodicity in the observed time series [see the continuous curve in Fig. 6.8(a)], but this periodicity does not appear to have a constant amplitude (vertically). Looking more closely, the length of the time interval separating two successive peaks is not constant either (horizontally), as it ranges from about 9 to 13 yr (i.e., 11 yr on average). In what follows, we will see how to define a trigonometric model with the appropriate cosine and sine waves in order to capture this kind of "periodicity."

Results of autocorrelation analysis [Fig. 6.8(b) and (c)] indicate that there is autocorrelation at small scale in the time series, in addition to some periodicity of about 11 yr (see the positive SSACs around lags 11, 22, 33, and 44), and that the autocorrelation at small scale could be modeled with a discrete-time AR(2) process (Subsection 6.2.1). For the rest, the behavior of SSACs does not reflect the varying amplitude of the periodicity or the uneven spacing between maximum sunspot numbers, since the SSACs tend to peak at multiples of 11 yr.

The uni-frequential periodograms $I^C(\omega)$ and $I^M(\omega)$ [Fig. 6.8(d)] confirm the presence of positive autocorrelation at small scale (see the secondary peak, extreme left), with a main peak between 17 and 18 cycles (i.e., period of >11 yr) and a secondary peak between 19 and 20 cycles (i.e., 9–11 yr), which we should pay attention to. In fact, in the bi-frequential periodogram $I^M(\omega_1, \omega_2)$ [Fig. 6.8(e)], four peaks (see white spots), or two pairs of peaks by symmetry, are observed. Putting aside the peaks involving one very short frequency (because of the positive autocorrelation at small scale), there remains the pair of frequency estimates 17.4 and 19.8, which correspond to periods of 11.3 and 9.9 yr. The combination of two cosine and two sine waves with close frequencies like these is capable of producing a pseudo-periodic signal with varying amplitude [see the dashed curve in Fig. 6.8(a)]. To be complete, the corresponding trigonometric model fitted to the time series by OLS (i.e., with white noise errors) explains 43.7% of the total variation, whereas its fitting by EGLS (i.e., with AR(2) errors) provides an explanation of 83.8%, stressing the importance of autocorrelation in this case; SAS PROC AUTOREG was used for this.

Second atmospheric CO_2 concentration example

The atmospheric CO_2 concentration data used for this example come from the same source as those analyzed in the first example presented in Subsection 6.3.1 (Keeling and Whorf, 2005). They are monthly values for Mauna Loa (Hawaii) and cover the period January 1965–December 2004,

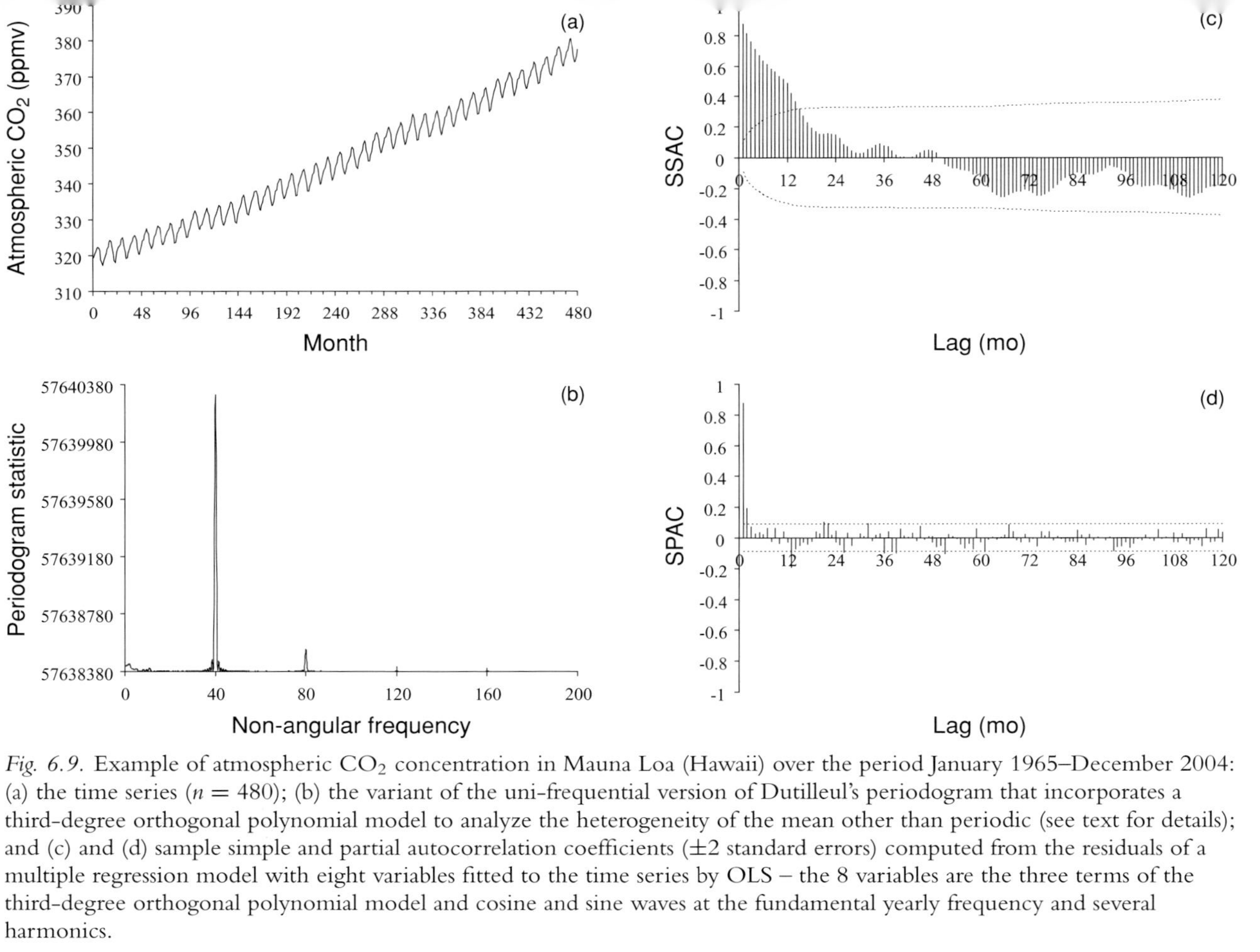

Fig. 6.9. Example of atmospheric CO_2 concentration in Mauna Loa (Hawaii) over the period January 1965–December 2004: (a) the time series ($n = 480$); (b) the variant of the uni-frequential version of Dutilleul's periodogram that incorporates a third-degree orthogonal polynomial model to analyze the heterogeneity of the mean other than periodic (see text for details); and (c) and (d) sample simple and partial autocorrelation coefficients (± 2 standard errors) computed from the residuals of a multiple regression model with eight variables fitted to the time series by OLS – the 8 variables are the three terms of the third-degree orthogonal polynomial model and cosine and sine waves at the fundamental yearly frequency and several harmonics.

so $n = 480$. Clearly, a large-scale trend is superimposed to the annual periodicity [Fig. 6.9(a)]. Therefore, $I^{M*}(\omega)$ or its multi-frequential version is required, and $I^{C}(\omega)$ or $I^{M}(\omega)$ would be inappropriate. The pre-specified model for the trend includes $P_{O1}(t)$, $P_{O2}(t)$ and $P_{O3}(t)$.

From the plot of $I^{M*}(\omega)$ [Fig. 6.9(b)], it appears that the dominant peak is, unsurprisingly, at 40 cycles (i.e., the number of years), with one secondary peak at the first harmonic of 6 months (i.e., 80 cycles) and another on the extreme left, for positive autocorrelation at small scale. Other peaks exist, but are very tiny, at the second and third harmonics of 4 and 3 months (i.e., 120 and 160 cycles). The corresponding sine waves were included and kept in the model for $\mu(t)$ because their estimated coefficients were found to be significant; similarly for the first harmonic of 6 months.

The OLS fitting of a joint polynomial–trigonometric model, with a total of eight regressors including the cosine and sine waves for the annual periodicity, provided residuals that were strongly and positively autocorrelated at small scale [Fig. 6.9(c) and (d)]. Accordingly, the errors $\varepsilon(t)$ were modeled with a discrete-time AR(2) process (Subsection 6.2.1).

In conclusion, temporal heterogeneity was found in the mean, at large scale (non-periodic) and at intermediate scale (periodic), and in the autocorrelation of random deviations. As in the first atmospheric CO_2 concentration example, the heterogeneity of the mean at large scale accounts for about 98% of the total variation, even without $P_{O3}(t)$ that was discarded again following the EGLS fitting.

Second MSH air–soil temperature example

This example can be seen as a follow-up to the first MSH air–soil temperature example of Section 6.2, except that the data here are hourly means instead of daily means and the time series cover the period June 10–18, 2004 (i.e., $n = 216$), instead of the entire month. Consequently, there is periodicity in both time series [Fig. 6.10(a) and (e)], but the daily oscillations in them are of different shapes: more marked and sharp for hourly mean air temperature vs. smoother and flat for hourly mean soil temperature. There is also a large-scale trend of seemingly intermediate importance in each series. Below, we will see that similar statistical tools can be used to analyze the two time series.

In Fig. 6.10(b) and (f), where $I^{M*}(\omega)$ is plotted, one can see that for each series there is positive autocorrelation at small scale – this is more important for hourly mean soil temperature – and there is a fundamental peak at 9 cycles, without strong evidence for harmonics. One can also see

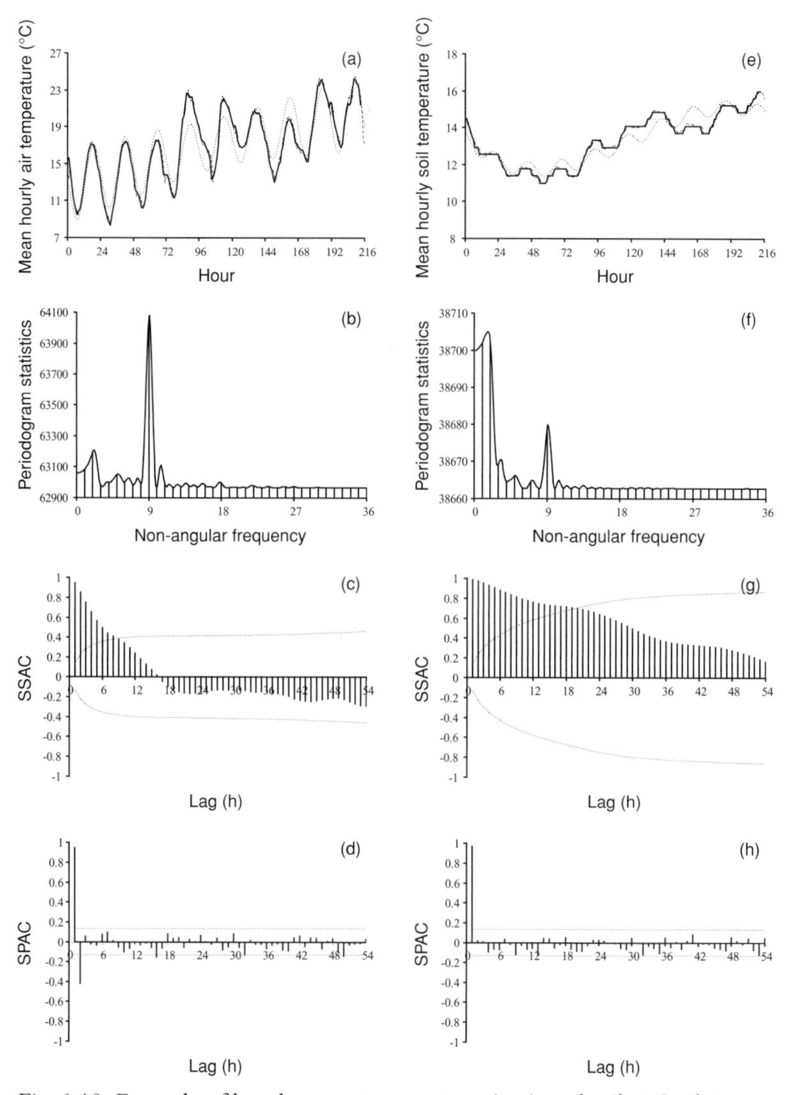

Fig. 6.10. Example of hourly mean temperatures in air and soil at Gault Nature Reserve in Mont-Saint-Hilaire (Québec, Canada) over the period June 10–18, 2004: (a) and (e) the observed time series (continuous curve) and two predicted time series ($n = 216$) – the two predicted time series differ by the temporal autocorrelation structure used for the random errors in the multiple regression model, independence (short-dashed) vs. AR(2) in (a) or AR(1) in (e) (long-dashed) [see (c)–(d) and (g)–(h)]; (b) and (f) the variant of the uni-frequential version of Dutilleul's periodogram that incorporates a polynomial trend – the plot is limited to [0, 36] because the information contained in the rest of the frequency domain is ancillary; (c) and (g) sample simple autocorrelation coefficients (SSAC) of the residuals of the multiple regression model with a polynomial and a trigonometric part (see text for details) fitted to the observed time series by OLS; and (d) and (h) sample partial autocorrelation coefficients (SPAC) of the same OLS residuals as in (c) and (g).

that for hourly mean air temperature, there is a secondary peak between 10 and 11 cycles (i.e., close to 9). The maximization of $I^{M*}(\omega_1, \omega_2)$ with an orthogonal polynomial model of second degree provided frequency estimates of 9.1 and 10.2 cycles for hourly mean air temperature, while the maximization of $I^{M*}(\omega)$ with an orthogonal polynomial model of third degree provided a frequency estimate of 9.1 cycles for hourly mean soil temperature.

The OLS fitting of joint polynomial–trigonometric models provided positively autocorrelated residuals [Fig. 6.10(c)–(d) and (g)–(h)]. The corresponding errors were modeled with a discrete-time AR(2) process for hourly mean air temperature and a discrete-time AR(1) process for hourly mean soil temperature. Despite the second order of autoregression for air, it is the autocorrelation for soil that is stronger, the value of $\hat{\phi}$ being close to its possible maximum of 1; this makes sense, since we are speaking of soil temperature on an hourly basis.

Clearly, EGLS predictions are better than OLS predictions [Fig. 6.10(a) and (e)]. In fact, each joint polynomial–trigonometric model explains 81.3% (air) and 87.1% (soil) of the total variation when fitted with white noise errors, and the percentages increase to 98.8% (air) and 99.4% (soil) with AR(2) and AR(1) errors, respectively. To be complete, the percentages of explained variation for the pure polynomial and pure trigonometric models fitted by OLS are 44.5% (polynomial) and 44.1% (trigonometric) for air and 82.9% (polynomial) and 5.5% (trigonometric) for soil; note that, since the polynomial model is orthogonal, these percentages of explained variation for the two pure models would have perfectly added up to the percentage for the joint model if the frequency estimates had been Fourier frequencies.

6.4 Analysis of temporal heterogeneity of the variance

In this section, the three types of heterogeneity (i.e., mean, variance, autocorrelation) are studied jointly, but the focus is on temporal heterogeneity of the variance, or heteroscedasticity. We will see how to infer the variance–covariance structure from temporal repeated measures, under the mixed model defined by equation (6.6); "mixed," in regard to its deterministic part $\mu_i(t)$ and its random part $\varepsilon_{ik}(t)$. With a single time series, a heteroscedastic variance–covariance structure can be postulated and tested using equation (6.3), but the model of temporal heterogeneity of the variance is submitted to restrictions in terms of number of parameters, as the number of variance parameters cannot be equal to the time series length n. Note that with temporal repeated measures initially collected according to equation (6.6), separate heterogeneity analyses might be performed under equation (6.4) if there is evidence for differences in the variance–covariance structure between classification levels.

Below, a number of homoscedastic variance–covariance structures differing by their modeling of temporal autocorrelation, together with their heteroscedastic version, are presented for the vectors of random errors $\varepsilon_{ik}(t)$ ($t = 1, \ldots, n$) in equation (6.6). Thereafter, "information criteria" due to Akaike (1974a) and Schwarz (1978) are defined; these two criteria are used to choose an appropriate variance–covariance structure for a given dataset. Finally, an application is provided by a second forestry example.

Through $\mu_i(t)$ (i.e., a function of t and i), equation (6.6) allows the presence of temporal heterogeneity of the mean in the data and possible differences in this heterogeneity between classification levels. Through the variance–covariance matrix of the vectors of random errors $\varepsilon_{ik}(t)$ ($t = 1, \ldots, n$), temporal heterogeneity of the variance and temporal autocorrelation are allowed. Hereafter, we will concentrate on that matrix, denoted $\Sigma = (\sigma(t, t'))$; $\sigma(t, t')$, the entry in row t and column t', is the population variance $\mathrm{Var}[U(t)]$ if $t = t'$ and the population covariance $\mathrm{Cov}[U(t), U(t')]$ otherwise.

The seven variance–covariance structures defined here are comprised of three structures in their homoscedastic and heteroscedastic versions, plus the most general permissible structure. They are listed in decreasing order of restrictions on temporal autocorrelation.

Compound symmetry (CS in SAS PROC MIXED; SAS Institute Inc., 2009)

$$\begin{aligned} \sigma(t, t') &= \sigma^2 \text{ if } t = t'; \\ &= \rho\sigma^2 \text{ for any } t \neq t'. \end{aligned} \tag{6.25}$$

This homoscedastic variance–covariance structure with two parameters assumes equal correlation between observations over time, whether these are close to one another or far apart. It has a theoretical interest in statistics, but is not very practical in time series applications.

Heterogeneous compound symmetry (CSH)

$$\begin{aligned} \sigma(t, t') &= \sigma^2(t) \text{ if } t = t'; \\ &= \rho\sigma^2(t) \text{ for any } t \neq t'. \end{aligned} \tag{6.26}$$

This structure with $n + 1$ parameters is the heteroscedastic version of the previous one, with same restrictions on temporal autocorrelation.

First-order autoregressive (AR(1))

$$\sigma(t, t') = \sigma^2 \text{ if } t = t'; \\ = \phi^{|t'-t|}\sigma^2 \text{ for any } t \neq t'. \quad (6.27)$$

See equation (6.19) in Subsection 6.2.1; two parameters.

Heterogeneous first-order autoregressive (ARH(1))

$$\sigma(t, t') = \sigma^2(t) \text{ if } t = t'; \\ = \phi^{|t'-t|}\sigma^2(t) \text{ for any } t \neq t'. \quad (6.28)$$

This heteroscedastic structure possesses $n + 1$ parameters and the same temporal autocorrelation as the previous one.

Toeplitz (TOEP)

$$\sigma(t, t') = \sigma^2 \text{ if } t = t'; \\ = \rho(|t' - t|)\,\sigma^2 \text{ for any } t \neq t'. \quad (6.29)$$

The notation $\rho(|t' - t|)$ above means that the correlation between two observations is a function of the lag separating them in time, all pairs of observations separated by the same lag being correlated the same. Thus, there are $n - 1$ "strips" with the same covariance value above and below the main diagonal in Σ, without specific constraint on the way covariance values vary with the lag. Consequently, the number of parameters of this homoscedastic structure is n.

Heterogeneous Toeplitz (TOEPH)

$$\sigma(t, t') = \sigma^2(t) \text{ if } t = t'; \\ = \rho(|t' - t|)\,\sigma^2(t) \text{ for any } t \neq t'. \quad (6.30)$$

This heteroscedastic structure possesses $2n - 1$ parameters and the same temporal autocorrelations as TOEP.

Unstructured (UN)
This general structure possesses $\frac{n(n+1)}{2}$ parameters because all parameters $\sigma(t, t')$ for $t' \geq t$ are possibly different, with $\sigma(t, t') = \sigma(t', t)$. There is no constraint on Σ other than it is a permissible (i.e., positive definite) variance–covariance matrix, including $\sigma^2(t) > 0$ for all t.

The suitability of the variance–covariance structures above for a given dataset can be assessed on the basis of the values taken for them by Akaike's and Schwarz's information criteria. Both criteria are built on the natural logarithm of the residual variance, denoted $\hat{\sigma}_\eta^2$ below. In simple terms, the residual variance is the portion of variation remaining unexplained after the model has been fitted, by including the modeled errors in the fitting.

If q denotes the number of parameters of a given variance–covariance structure, Akaike's and Schwarz's information criteria are defined by

$$\mathrm{AIC}(q) = \log(\hat{\sigma}_\eta^2) + 2q \text{ (Priestley, 1981, p. 373)} \tag{6.31}$$

and

$$\mathrm{SBC}(q) = \log(\hat{\sigma}_\eta^2) + q \log(n) \text{ (Priestley, 1981, p. 376).} \tag{6.32}$$

One now sees better the reason for the rule: the smaller, the better, introduced on page 181, and the more severe penalty used in Schwarz's criterion to account for the number of parameters of the model fitted; see $q \log(n)$ vs. $2q$ in $\mathrm{AIC}(q)$.

(The use of $\mathrm{AIC}(q)$ and $\mathrm{SBC}(q)$ in statistics goes beyond the mixed models; see, for example, our use of them in Subsection 6.2.1 and the content of Table 6.1. Depending on the book, $\mathrm{AIC}(q)$ and $\mathrm{SBC}(q)$ may be defined with a different sign. The two information criteria were even defined with different signs in Versions 8 and 9 of the SAS software. Fortunately, to avoid any confusion, the rule "the smaller the better" is now indicated in the output of SAS PROC MIXED in Version 9.)

Second forestry example

This example is in two parts: (a) a follow-up to the first forestry example of Subsection 6.3.1, except that the random variable U is ring width instead of fiber length (Herman *et al.*, 1998); and (b) a similar experiment by the same authors (Herman *et al.*, 1999), with temporal repeated measures of microfibril angle at 11 sampling sites within the growth ring of 1959 in seven fast-grown and seven slow-grown Norway spruces. Temporal heterogeneity of the variance is clearly visible in the two pairs of average series [Fig. 6.11(a) and (b)]. However, standard errors tend to decrease from Year 1 to Year 15 in (a), but are smaller at the extremes and larger at the central sites in (b).

From the values of Akaike's and Schwarz's criteria (Table 6.3), it appears that the first choice of variance–covariance structure is heterogeneous first-order autoregressive for ring width and heterogeneous compound symmetry for microfibril angle, and the second choice is heterogenerous

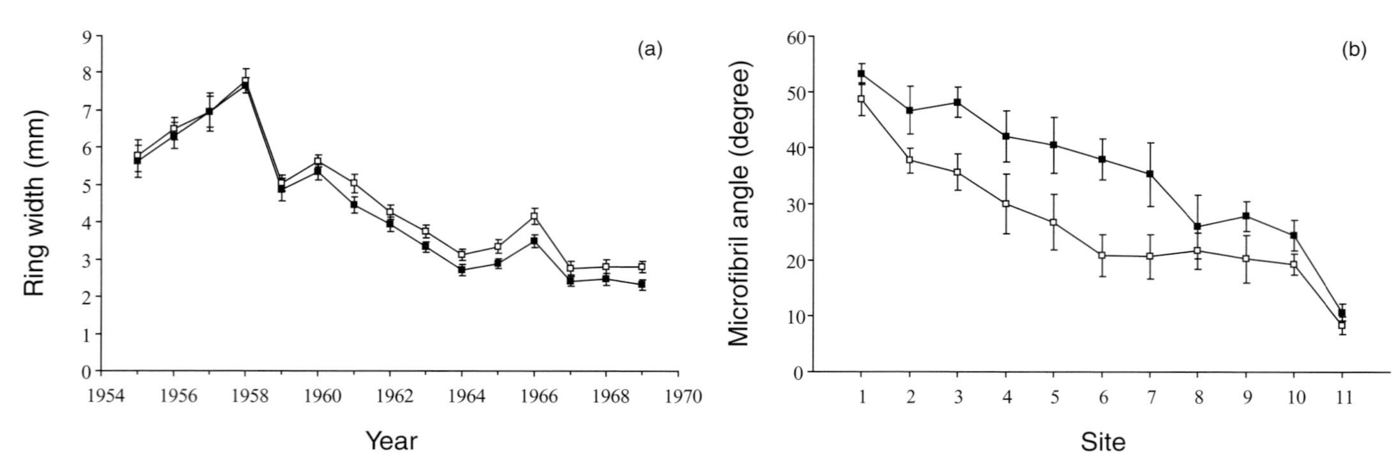

Fig. 6.11. Forestry example with heteroscedastic temporal repeated measures of (a) ring width before first thinning ($n = 15$ years) and (b) microfibril angle within the 1959 growth ring ($n = 11$ intra-ring sites) on fast-grown (open squares) and slow-grown (filled squares) Norway spruces (*Picea abies* (L.) Karst.); averages and the corresponding standard errors calculated from 20 (a) and 7 (b) trees per growth category are plotted.

Table 6.3. *Forestry example with heteroscedastic temporal repeated measures (Fig. 6.11) of (a) ring width before first thinning (n = 15 years) and (b) microfibril angle within the 1959 growth ring (n = 11 intra-ring sites) on Norway spruces (Picea abies (L.) Karst.): information criterion values for various variance–covariance structures used in mixed linear models to take into account the heterogeneity of variance and that due to autocorrelation in the statistical analysis of the data*

	(a)		(b)	
	Akaike's criterion	Schwarz's criterion	Akaike's criterion	Schwarz's criterion
Compound symmetry (CS)	880.5	882.2	481.6	482.2
Heterogeneous CS (CSH)	760.0	773.5	470.6	474.4
First-order autoregressive (AR(1))	808.9	810.5	484.0	484.7
Heterogeneous AR(1) (ARH(1))	681.5	695.0	478.8	482.6
Toeplitz (TOEP)	815.5	828.2	482.7	486.2
Heterogeneous TOEP (TOEPH)	682.2	706.7	472.5	479.2
Unstructured (UN)	685.0	786.3	484.8	505.9

The abbreviations in parentheses are used in the SAS procedure MIXED. See text for details about the definition of variance–covariance structures.

Toeplitz for both variables; the two criteria agree on these choices. The selection of CSH for microfibril angle may surprise, but it must be related to the relatively small number of sampling sites and the greater facility of temporal correlation to be almost constant in this case.

The analysis of mixed models with the ARH(1) structure for ring width and the CSH structure for microfibril angle provided the following results: (a) the variance parameter estimates almost monotonically decrease from 4.22 (Year 1) to 0.341 (Year 15), $\hat{\phi} = 0.704$, and the Year main effects are highly significant ($P < 0.01$); (b) variance estimates are, respectively, 40.7, 183.0 and 18.2 at Sites 1, 7 and 11 $\hat{\rho} = 0.572$, and the Site main effects are highly significant ($P < 0.01$). SAS PROC MIXED was used for this.

In summary, there is heteroscedasticity, positive autocorrelation and heterogeneity of the mean in both sets of temporal repeated measures. Differences between growth categories, in the form of main effects and through interactions with Year and Site, will be studied in other examples presented in Chapter 9 (Subsection 9.3.2).

6.5 Relationships between analytical methods

This section provides a summary of the statistical methods and procedures presented for heterogeneity analysis of time series in previous sections. The objective is to make links between the conditions of application of the different methods and to emphasize the implications of the presence of one type of temporal heterogeneity for the analysis of the others.

Autocorrelation analysis based on sample autocorrelation coefficients (i.e., the SSACs and SPACs) was proposed to quantify temporal heterogeneity due to autocorrelation under weak stationarity (i.e., stationarity at orders 1 and 2). Guidelines for visual inspection of the coefficients and their assessment based on standard errors exist to identify the model of a discrete-time process, with diagnostic rules (e.g., information criteria). When appropriate, the method can be applied to raw data (Subsection 6.2.1) or residuals of fitted regression models (Section 6.3). Spectral analysis based on the spectral density function pursues the same goal under the same stationarity assumption (Subsection 6.2.2).

Under stationarity at order 2 only, the analysis of temporal heterogeneity of the mean is based on multiple regression with fixed explanatory variables: orthogonal polynomials at large scale (Subsection 6.3.1) and cosine and sine waves at intermediate scale (Subsection 6.3.2). The presence of a large-scale trend prevents the use of classical periodogram analysis, but a trend component can be incorporated into the computation of uni-frequential and multi-frequential periodogram statistics in the context of joint polynomial–trigonometric models (Dutilleul, 2001). When present, autocorrelation intervenes in the application of multiple regression in that the model fitting is then recommended to be EGLS instead of OLS. Forgetting this may lead to mistakes like retaining a polynomial term in the model because it is significant in the OLS fitting, while it is non-significant in the EGLS fitting; see the case of $P_{O3}(t)$ in the first atmospheric CO_2 concentration example in Subsection 6.3.1. Thus, the link between autocorrelation and heterogeneity of the mean is double: the analysis of the former precludes the presence of the latter and the analysis of the latter requires a statistical adjustment for the former.

Last but not least, mixed models allow one to include heteroscedasticity in the heterogeneity analysis of time series, together with the two other types of heterogeneity (Section 6.4). These models seem to be more useful when replicated time series or temporal repeated measures collected for different levels of a classification factor are available for analysis. Information criteria help the user identify an appropriate variance–covariance structure; any a priori choice would require tremendous information from the part of the experimenter. With a single time series, preliminary analyses based on OLS polynomial regression or periodograms should help define the fixed part of the model.

6.6 Temporal cross-correlation analysis

In this section, the focus will not be on temporal cross-correlation itself as much as on the effects that temporal heterogeneity of the mean, of the variance and due to autocorrelation can have on the outcome of temporal cross-correlation analysis. Therefore, the continued forestry example with autocorrelated and heteroscedastic temporal repeated measures will be used for estimation in the time domain in Subsection 6.6.1. As for estimation in the frequency domain, stationary time series from the first MSH air–soil temperature example and nonstationary time series from the second MSH air–soil temperature example will be analyzed in Subsection 6.6.2. At the beginning of each subsection, the relevant theoretical function (i.e., the cross-spectral density function in the frequency domain) will be defined under weak stationarity, which includes the absence of temporal heterogeneity of the mean and the variance.

6.6.1 Estimation in the time domain

Consider a situation in which two random variables U_j and $U_{j'}$ have been observed in the same experimental conditions at times $t = 1, \ldots, n$. In other words, a bivariate time series of length n $\{\boldsymbol{u}(t) | t = 1, \ldots, n\}$, with $\boldsymbol{u}(t) = (U_j(t), U_{j'}(t))$, is available for statistical analysis. In model (6.9), each of the two underlying discrete-time processes is stationary at order 1 because $\boldsymbol{\mu}(t) = \boldsymbol{\mu}$ for all t. Assume furthermore that the two discrete-time processes are jointly stationary at order 2; that is, each of them is stationary at order 2 (Section 2.3) and their population cross-correlations [equations (A2.14) and (A2.15)] depend only on the lag k (including its sign), which separates observations in time

$$\begin{aligned} \rho_{jj'}(t, t+k) &= \rho_{jj'}(t+k-t) = \rho_{jj'}(k) \\ &= \frac{\mathrm{E}[\{U_j(t) - \mu_j\}\{U_{j'}(t+k) - \mu_{j'}\}]}{\sqrt{\sigma_j^2 \sigma_{j'}^2}}. \end{aligned} \tag{6.33}$$

In practice, $\rho_{jj'}(k)$ for $k = -[\frac{n}{4}], \ldots, -1, 0, 1, \ldots, [\frac{n}{4}]$ is estimated by the "sample cross-correlation coefficient" (SCCC) at lag k. Under joint weak stationarity, this coefficient is calculated by

$$r_{jj'}(k) = \frac{1}{n} \sum_{t=1}^{n-k} \frac{\{u_j(t) - \bar{u}_j\}\{u_{j'}(t+k) - \bar{u}_{j'}\}}{\sqrt{\hat{\sigma}_j^2 \hat{\sigma}_{j'}^2}}. \qquad (6.34)$$

Clearly, equations (6.33) and (6.34) are, respectively, the extension of equations (6.16) and (6.15) used in temporal autocorrelation analysis (Subsection 6.2.1).

The presence of autocorrelation in each series poses a problem for the cross-correlation analysis between series. More specifically, temporal autocorrelation induces a bias in the standard errors of SCCCs, downwards when positive in both series and upwards when dominantly negative (Jenkins and Watts, 1968, pp. 338–340). In the third forestry example below, we will see that the heterogeneity of the mean poses a different type of problem, in the evaluation of the correlation itself, and the heterogeneity of the variance also requires a statistical adjustment. The solution traditionally put forward for the problem of autocorrelation in cross-correlation analysis with time series consists in using the autocorrelation structure identified for one series to prewhiten the two series; usually, it is recommended that one of the two time series analyzed for cross-correlation be a white noise, even if this can be discussed (Dutilleul, 2008). Then, SCCCs are calculated between prewhitened series. Prewhitening here is based on a linear transformation similar to that used for EGLS estimation of slopes in multiple regression with autocorrelated errors (Subsection 6.3.1). Other solutions based on modified tests of significance exist (e.g., Dutilleul, 1993b), and will be presented in the spatial framework (Chapter 7).

Third forestry example

Autocorrelated yearly repeated measures of ring width and fiber length for fast-grown Norway spruces after first thinning [i.e., $n = 18$; Fig. 6.12(a1) and (a2)] are used for this example, as a follow-up to the first and second (part a) forestry examples. There is temporal heterogeneity of the mean in the data, in the form of decreasing and increasing large-scale trends for ring width and fiber length, respectively (see the short-dashed curves). Heteroscedasticity is moderate and essentially in ring width. Since correlation analysis between the two variables will be performed on average series for a given growth category, model (6.8) will be applied with $\boldsymbol{u}(t) = (\bar{U}_1(t), \bar{U}_2(t))$, where $\bar{U}_1(t)$ denotes the average ring width at time t, and $\bar{U}_2(t)$, the average fiber length at same time.

Depending on the material used for analysis, the following results were obtained. On raw data [Fig. 6.12(b1)], Pearson's sample linear correlation coefficient is –0.800; no probability of significance is reported because the value of the coefficient is biased by the presence of large-scale trends. The EGLS mean estimates obtained with a linear model for ring width and a second-degree orthogonal polynomial model for fiber length [Fig. 6.12(b2)] provide a Spearman's rank-based correlation coefficient of –0.986 ($P < 0.001$). On non-prewhitened differences [Fig. 6.12(b3)], Pearson's coefficient is –0.248 ($P = 0.321$). Note that: (i) the sample variance–covariance matrix calculated from the replicated time series collected for individual trees was used for the linear transformation defining the EGLS estimation of means; (ii) non-prewhitened differences correspond to "stationarized series," after removal of the estimated trends.

Differences between raw data and EGLS mean estimates were autocorrelated [Fig. 6.12(c1)–(c4)], and an AR(1) autocorrelation structure was chosen to prewhiten them prior to cross-correlation analysis. The SCCCs calculated on prewhitened differences [Fig. 6.12(d)] show a relatively strong, negative instantaneous correlation of –0.522 between ring width and fiber length, but nothing comparable to the spurious correlation of –0.800 obtained with raw data and the almost perfect correlation of –0.986 between EGLS mean estimates. Positive SCCCs are observed at lags ± 3 and ± 4; these might be related to the planning of thinning applications over time. In conclusion, the source of the negative association between ring width and fiber length in fast-grown Norway spruces after first thinning lies primarily in large-scale trends and secondarily in random deviations.

6.6.2 Estimation in the frequency domain

Under the joint weak stationarity assumption for two discrete-time processes, the theoretical cross-spectral density function is defined by

$$f_{jj'}(\omega) = \frac{1}{2\pi} \sum_{k=-\infty}^{\infty} \rho_{jj'}(k) e^{-i\omega k}. \tag{6.35}$$

That is, the extension of equation (6.20) defining the spectral density function $f(\omega)$ for one process, where the autocorrelation function $\rho(k)$ is

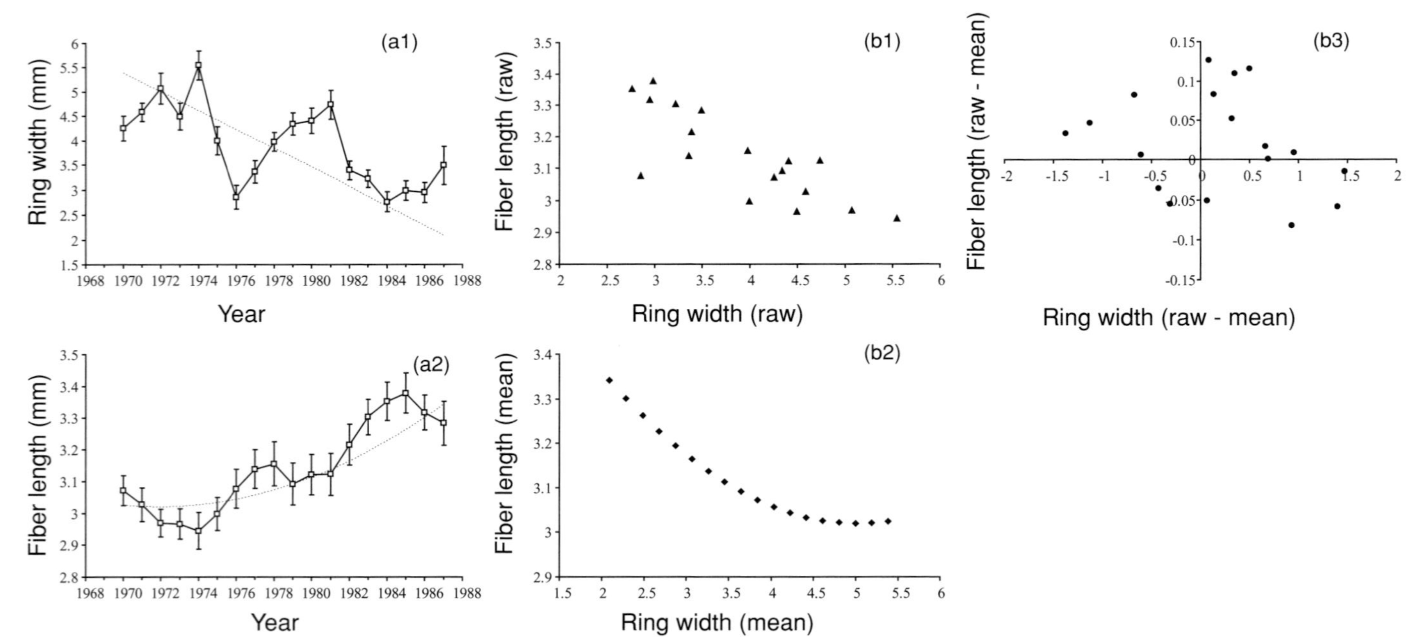

Fig. 6.12. Forestry example with yearly repeated measures of ring width and fiber length on fast-grown Norway spruces (*Picea abies* (L.) Karst.) after first thinning: (a1) and (a2) the observed time series ($n = 18$; continuous curve) of averages with the corresponding standard errors calculated for 20 fast-grown trees and the best EGLS polynomial fitting (short-dashed); (b1)–(b3) scatter plots of fiber length against ring width, using the raw data, the EGLS mean estimates, and differences between the two, respectively; (c1)–(c4) sample simple (SSAC) and partial (SPAC) autocorrelation coefficients calculated on the differences plotted in (b3); and (d) sample cross-correlation coefficients (SCCC) computed for prewhitened differences.

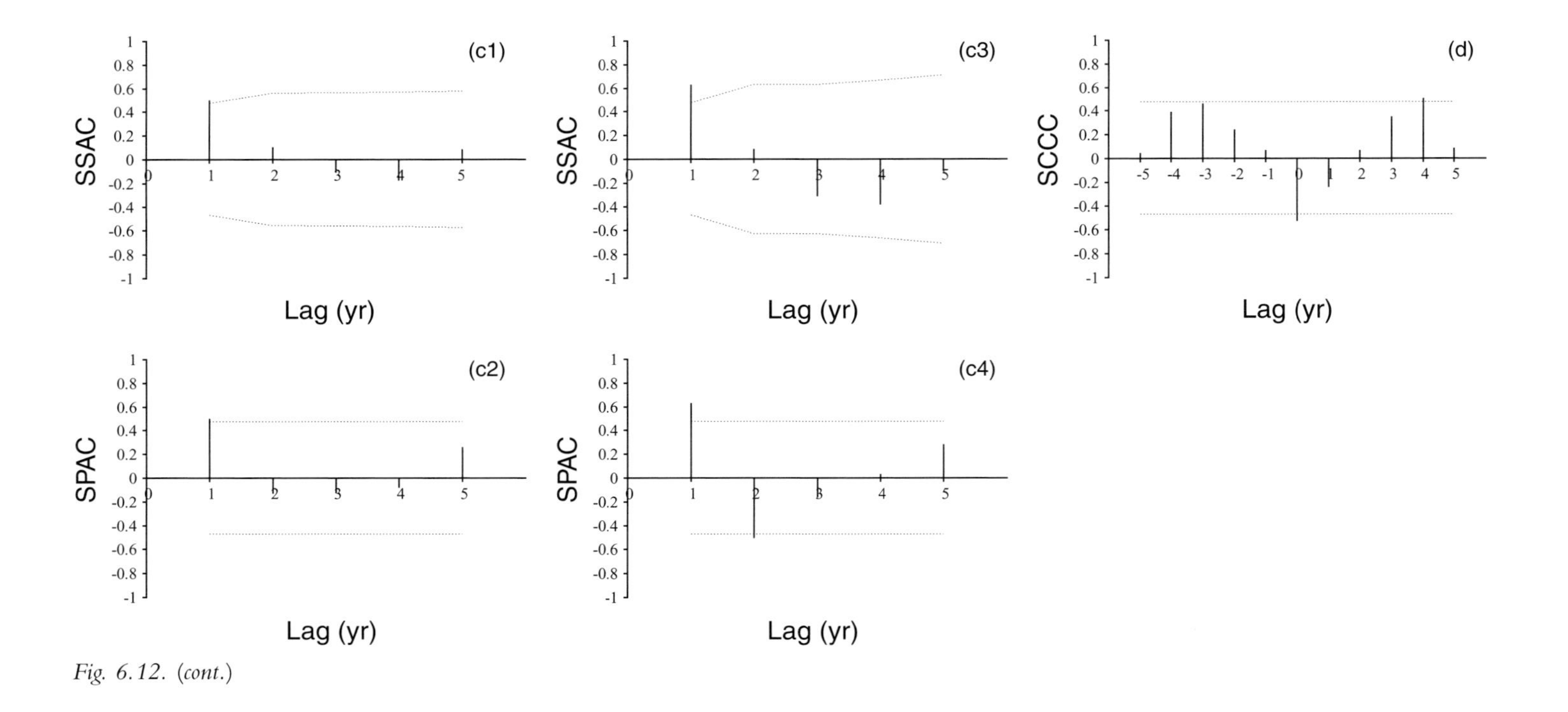

(c1)
SSAC
Lag (yr)
(c3)
SSAC
Lag (yr)
(d)
SCCC
Lag (yr)
(c2)
SPAC
Lag (yr)
(c4)
SPAC
Lag (yr)

Fig. 6.12. (cont.)

replaced by the cross-correlation function. Numerically, this substitution makes a big difference because $\rho_{jj'}(k)$, contrary to $\rho(k)$, is not an even function (in the mathematical sense); that is, $\rho_{jj'}(k)$ is generally different from $\rho_{jj'}(-k)$, while $\rho(k)$ is always equal to $\rho(-k)$. Consequently, $f_{jj'}(\omega)$ is generally complex-valued and its analysis is indirect, via its squared modulus divided by $f_j(\omega)f_{j'}(\omega)$, called the "coherency spectrum," and the arctangent of its imaginary part divided by its real part, called the "phase spectrum." (The squared modulus of the complex number $a + bi$ is $a^2 + b^2$.) By construction, the coherency spectrum is non-negative, with 1 as maximum value, and the unit of the phase spectrum is the radian. Note that the division by $f_j(\omega)f_{j'}(\omega)$ in the definition of the coherency spectrum involves the autocorrelation functions of the two processes.

The coherency spectral statistic can be interpreted as a measure of association (i.e., a coefficient of determination) between two processes in the frequency domain, without the sign; values close to 0 (1) mean that the processes are unrelated (strongly related) over the interval of frequencies considered. As for the phase spectrum, values close to 0 mean that the two processes are in phase over the interval of frequencies – that is, cosine waves fitted to the two time series reach their peaks and troughs about at the same times; values of $-\pi$ and π mean that the two processes are out of phase – that is, when the cosine waves fitted to one series are on a peak, those fitted to the other series at the same frequencies are in a trough. Note that the coherency and phase statistics defined above are for continuous spectra, and by the end of this subsection we will see other coherency and phase statistics for discrete and mixed spectra.

Concerning estimation, $f_{jj'}(\omega)$ can be estimated by smoothing Schuster's cross-periodogram

$$I^C_{jj'}(\omega) = \frac{1}{2\pi} \sum_{k=-(n-1)}^{n-1} r_{jj'}(k)e^{-i\omega k}, \tag{6.36}$$

leading to

$$\hat{f}_{jj'}\left(\frac{2\pi p}{n}\right) = \frac{1}{4\pi} \sum_{\ell=0}^{2L} w_\ell I^C_{jj'}\left[\frac{2\pi(p - L + \ell)}{n}\right], \tag{6.37}$$

with same notations as in equations (6.20)–(6.22).

In a different context, assume that the two discrete-time processes are not stationary at order 1 due to large-scale trends and that each process has a discrete or mixed spectrum (i.e., they have periodic components with white noise or autocorrelated random deviations). In such a case, the cross-spectral density function and, through it, the cross-correlation function cannot be analyzed on raw data using statistics (6.34) and (6.37). Instead, Dutilleul's cross-periodogram $I_{jj'}^{M*}(\omega)$ can be used very simply as follows. In the same way that $I^{M*}(\omega)$ in Subsection 6.3.2 represents the sum of squares of a joint polynomial–trigonometric model fitted to one time series, $I_{jj'}^{M*}(\omega)$ represents the scalar product of the projections of the two time series, each on its own trend and both on the cosine and sine waves at frequency ω. A spectral correlation coefficient (including the sign), or coherency spectrum, is then given by

$$-1 \leq \frac{I_{jj'}^{M*}(\omega)}{I_{j}^{M*}(\omega)\, I_{j'}^{M*}(\omega)} \leq 1, \tag{6.38}$$

and the arccosine of this spectral correlation provides a phase spectrum; the interpretation of correlation statistics as projection angles has a long tradition in statistics (Scheffé, 1959, p. 376). All the equations are given in Section A6.3.

Continued MSH air–soil temperature examples

The spectra in the MSH air–soil temperature examples of Sections 6.2 and 6.3 are, respectively, continuous [Fig. 6.3(d) and (h)] and mixed [Fig. 6.10(b) and (f)]. Therefore, the coherency and phase spectra derived from the estimated cross-spectral density function $\hat{f}_{jj'}(\omega)$ are presented for the first example [Fig. 6.13(a)], whereas those directly computed from $I_{jj'}^{M*}(\omega)$ are displayed for the second example [Fig. 6.13(b)].

The highest level of smoothing in Fig. 6.13(a), which is based on a triangular window with weights 1, 2, 3, 4, 3, 2, 1, provides a suitable continuous curve, compared to the two others with weights 1, 2, 1 and 1, 2, 3, 2, 1. Based on the continuous curve, the coherency varies between about 0.5 and 0.85, indicating an intermediate to strong association between the series of daily mean temperatures in the frequency domain. Moreover, the phase is consistently non-negative with the exception of the highest frequencies, which are not very important in this case because little of the variation of the two time series is associated with them [Fig. 6.3(d) and (h)]. The numerous non-negative phases mean that

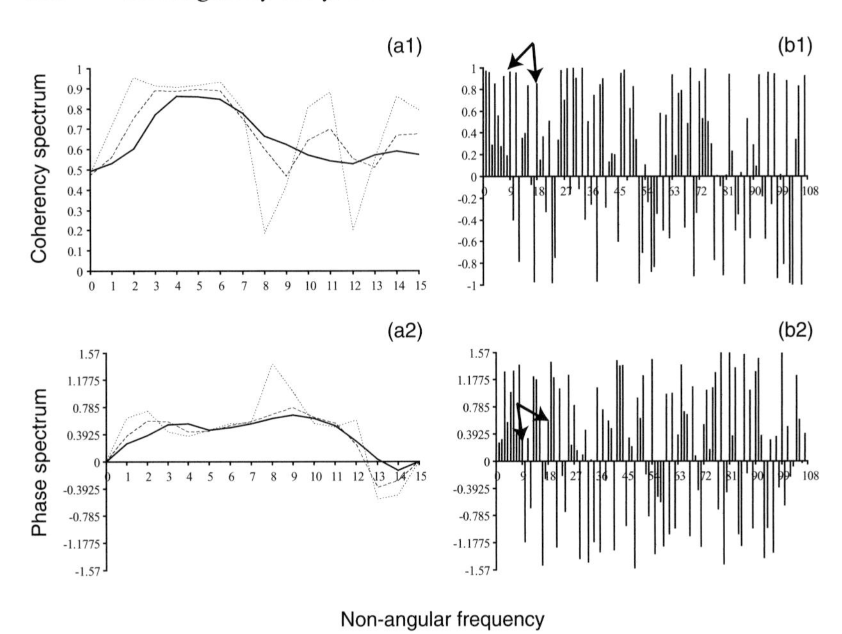

Fig. 6.13. Continued examples of daily [see (a1) and (a2); $n = 30$] and hourly [see (b1) and (b2); $n = 216$] mean temperatures in air and soil at Gault Nature Reserve in Mont-Saint-Hilaire (Québec, Canada): coherency and phase spectra computed (a1) and (a2) by smoothing Schuster's cross-periodogram of raw data to various degrees (low: short-dashed; high: continuous), and (b1) and (b2) from Dutilleul's uni-frequential cross-periodogram of EGLS residuals (i.e., after fitting polynomial terms to model the heterogeneity of the mean at large scale while taking temporal autocorrelation into account) – bars are used here at integer non-angular frequencies to ease reading.

changes in air temperature generally precede changes in soil temperature. To be convinced of the "precede," it would suffice to fit the cosine and sine waves for one of the lower frequencies to each of the series and locate the peaks and troughs in the fitted curves.

Arrows in Fig. 6.13 point to coherencies (panel b1) and phases (panel b2) at two frequencies of particular interest: the fundamental frequency of 9 cycles over the series of hourly mean temperatures, which corresponds to the 24-h periodicity, and its first harmonic of 18 cycles or 12 h. At both frequencies, coherency is strong (i.e., >0.8) and the phase is positive; peaks and troughs in hourly mean air temperature precede those in hourly mean soil temperature. A similar behavior of the coherency and phase spectra is observed at the lowest frequencies, which correspond to the continuous part of the spectra and positively autocorrelated random deviations in the two series [Fig. 6.10(b)–(d) and (f)–(h)].

In conclusion, air and soil temperatures in Mont-Saint-Hilaire (Québec, Canada), and probably in many other parts of the world, tend to be strongly associated in the frequency domain, both among and within days, with changes in air temperature preceding those in soil temperature. This phase shift makes much sense because of the major physical differences between the two media. Despite the presence of temporal heterogeneity of the mean at large scale in the time series of hourly mean temperatures, a bivariate spectral analysis has been made possible thanks to $I_{jj'}^{M*}(\omega)$.

6.7 Recommended readings

The following references will provide complements of information on the theory and practice of time series analysis to the interested reader (in order of presentation of the material here). Priestley (1981, Chapter 5) discusses the statistical properties of the two main estimators of population autocovariances (i.e., with $\frac{1}{n}$ vs. $\frac{1}{n-k}$ as divisor). Box *et al.* (1994, Chapter 3) give a detailed presentation of the identification rules for discrete-time AR, MA, and ARMA processes and their characteristics (including variance and spectrum), while Chatfield (2004) offers a more applied perspective on these topics. Dutilleul (1995) provides a good review of autocorrelation analysis in the chronobiological context. Priestley (1981, Chapters 4 and 6–8) describes the Fourier series development, the statistical properties and analysis of Schuster's periodogram statistic, and the main estimators of the spectral density function. For readers with the appropriate mathematical background, Anderson (1971) and Brillinger (1981) provide a deeper treatment of the subject. Diggle (1990) is a more applied book covering the time-domain and frequency-domain approaches. Dutilleul (2001) is the reference for multi-frequential periodogram analysis, while Jardon *et al.* (2003) report one of the dendrochronological applications and Dutilleul *et al.* (2011) present the version of the method for temporal series with irregularly spaced data, with an application in paleolimnology. The very mathematical Flandrin (1999) shows the links between spectral analysis and non-parametric wavelet analysis. Searle (1971) elaborates on estimation methods available for the analysis of linear models, including OLS and EGLS. Diggle *et al.* (1996, Chapter 5) discuss variance–covariance structures in mixed models as well as the estimation method of restricted maximum likelihood used for their analysis (pp. 64–68). The complete forestry example of correlation analysis with autocorrelated and heteroscedastic temporal repeated measures is available in Dutilleul *et al.* (1998). Jenkins and Watts (1968, Chapters 8 and 9) cannot be ignored for their sections on cross-correlation and bivariate spectral analyses, including their example of two independent AR(1) partial realizations (pp. 338–340). Finally, Priestley (1981, Chapter 9) also discusses the case of bivariate spectral analysis in a thorough manner.

Appendix A6: Simulation procedures and mathematical developments

A6.1 Simulation of partial realizations of discrete-time processes

The simulation procedure of Dutilleul and Legendre (1992) was followed and extended here (Section 6.2). Equations (6.10)–(6.14) are at the center of simulation codes for the discrete-time AR(1), AR(2), MA(1), ARMA(1,1), and SAR(1) processes. These codes were written in SAS, using the function RANNOR to generate pseudo-random numbers from a normal distribution with a zero mean and a variance of 1 (SAS Institute Inc., 2009). All complete SAS codes (including autocorrelation and spectral analyses) can be found on the CD-ROM accompanying the book. The simulation codes for the ARMA(1,1) and SAR(1) processes are copied below for discussion because the former is an extension of the AR(1) and MA(1), and the latter (with a span of 4 instead of 6 to reduce the number of code lines here) is seasonal.

Besides the call of RANNOR with a given seed (e.g., 1) to generate white noise observations (saved in WN in ARMA11), the two codes have the following features in common: (i) the length of simulated time series is expected to be in a previous DATA step named "n"; (ii) a number of commands in a DO-END loop are executed $100 + n + 1$ times, in order to "forget" the starting point of the simulation (i.e., the first white noise observation) and wait sufficiently long (see the IF-THEN command) to ensure a suitable autocorrelation structure for the simulated time series; and (iii) times used for simulation are equally spaced, $t = 1, \ldots, n$. Specific to SAR1_4 or for any SAR(1) process with span S, intermediate observations between $u(t)$ and $u(t - S)$ need to be stored in memory, until $u(t + 1), \ldots, u(t + S - 1)$ are simulated from them, together with newly generated white noise observations.

```
DATA ARMA11;
SET n;
phi = 0.7;
teta = –0.5;
U = RANNOR(1);
WN = RANNOR(1);
WN_1 = WN;
DO t = –100 TO n BY 1;
WN = RANNOR(1);
U = phi*U + WN – teta*WN_1; /* translation of equation (6.13) in
  SAS */
WN_1 = WN;
IF (t > 0) THEN OUTPUT;
END;

DATA SAR1_4;
SET n;
PHI = 0.8;
U_4 = RANNOR(1);
U_3 = RANNOR(1);
U_2 = RANNOR(1);
U_1 = RANNOR(1);
U = PHI*U_4 + RANNOR(1);
U_4 = U_3;
U_3 = U_2;
U_2 = U_1;
U_1 = U;
DO t = –100 TO n BY 1;
U = PHI*U_4 + RANNOR(1); /* translation of equation (6.14) in
  SAS */
U_4 = U_3;
U_3 = U_2;
U_2 = U_1;
U_1 = U;
IF (t > 0) THEN OUTPUT;
END;
```

A6.2 Construction of orthogonal polynomials with Gram-Schmidt procedure

The use of orthogonal polynomials in multiple regression analysis in general and in temporal and spatial heterogeneity analysis in particular is motivated by the statistical independence of the estimated slopes. This independence implies that the selection of $P_{O3}(t)$ is not influenced by the presence of $P_{O1}(t)$ or $P_{O2}(t)$ in the model. It may seem surprising that monom t is "related" (mathematically) to monoms t^2 and t^3, whereas $P_{O1}(t) = t - 8$ is not "related" (mathematically) to $P_{O2}(t) = t^2 - 16t + 45.33$ and $P_{O3}(t) = t^3 - 24t^2 - 158.6t - 244.72$ (Subsection 6.3.1). To get some insight into this, consider the parabola $y = t^2$ and the straight line $x = t$ ($t = 1, \ldots, n$). The slope of the simple linear regression of the former on the latter is not zero but a positive number! (If the parabola had been $y = -t^2$, the slope would have been negative.) $P_{O1}(t)$, $P_{O2}(t)$ and $P_{O3}(t)$ are orthogonal to each other by construction. After centering t to $\bar{t} = \frac{n+1}{2}$, $P_{O2}(t)$ can be seen as the residual of the regression of t^2 on $P_{O1}(t)$, and $P_{O3}(t)$ as the residual of the regression of t^3 on $P_{O1}(t)$ and $P_{O2}(t)$. The construction algorithm is given below. Note that the orthogonality of $P_{O1}(t)$, $P_{O2}(t)$ and $P_{O3}(t)$ for the classical Euclidean scalar product used in OLS fitting does not imply their orthogonality for other scalar products used in EGLS fitting.

With:

(i) $< \boldsymbol{u}, \boldsymbol{v} >$, the Euclidean scalar product of two vectors with n components $\boldsymbol{u} = (u_1, \ldots, u_n)$ and $\boldsymbol{v} = (v_1, \ldots, v_n)$, that is, $< \boldsymbol{u}, \boldsymbol{v} > = \sum_{t=1}^{n} u_t v_t$;
(ii) $\|.\|$, the Euclidean norm (Section 2.3);
(iii) $\boldsymbol{1}$, the vector of n ones $(1, \ldots, 1)$;
(iv) $\boldsymbol{t}$, $\boldsymbol{t}^2$, and $\boldsymbol{t}^3$, the vectors with n components $(1, 2, 3, \ldots, n)$, $(1, 4, 9, \ldots, n^2)$, and $(1, 8, 27, \ldots, n^3)$,

the vectors of n values for the orthogonal polynomials of degree 1, 2, and 3 are obtained by

$$\boldsymbol{t}_O = \boldsymbol{t} - < \boldsymbol{1}, \boldsymbol{t} > \frac{\boldsymbol{1}}{\|\boldsymbol{1}\|^2};$$

$$\boldsymbol{t}_O^2 = \boldsymbol{t}^2 - < \boldsymbol{1}, \boldsymbol{t}^2 > \frac{\boldsymbol{1}}{\|\boldsymbol{1}\|^2} - < \boldsymbol{t}_O, \boldsymbol{t}^2 > \frac{\boldsymbol{t}_O}{\|\boldsymbol{t}_O\|^2}; \text{ and}$$

$$\boldsymbol{t}_O^3 = \boldsymbol{t}^3 - < \boldsymbol{1}, \boldsymbol{t}^3 > \frac{\boldsymbol{1}}{\|\boldsymbol{1}\|^2} - < \boldsymbol{t}_O, \boldsymbol{t}^3 > \frac{\boldsymbol{t}_O}{\|\boldsymbol{t}_O\|^2} - < \boldsymbol{t}_O^2, \boldsymbol{t}^3 > \frac{\boldsymbol{t}_O^2}{\|\boldsymbol{t}_O^2\|^2}.$$

Orthogonal polynomials can be normalized by dividing them by their respective Euclidean norm.

A6.3 Least-squares fitting procedures for linear models and nonlinear models whose parameters separate

The matrix algebra operations involved in the OLS and GLS fittings of linear models and the computation of multi-frequential periodograms are described below. Knowing these operations allows one to derive interesting properties and characteristics of the parameter estimators and the statistics, in relation to the heterogeneity of time series.

Linear models (e.g., multiple regression models) can be written in matrix notation as follows:

$$\boldsymbol{u} = \mathbf{X}\boldsymbol{\beta} + \boldsymbol{\varepsilon}, \tag{A6.1}$$

where $\boldsymbol{u}$ is the vector of observations $(u(1), \ldots, u(n))'$ (i.e., the time series); $\boldsymbol{\varepsilon}$ is the vector of random errors $(\varepsilon(1), \ldots, \varepsilon(n))'$; $\boldsymbol{\beta}$ is the vector of p parameters to estimate; and $\mathbf{X}$ is the corresponding $n \times p$ matrix of known quantities. Note that the first column of $\mathbf{X}$ is usually a column of ones and the corresponding parameter to estimate is the intercept, while the other $p - 1$ parameters are slopes of explanatory variables in multiple linear regression. (The prime symbol $'$ here denotes the transpose operator in matrix algebra.)

The vector of OLS estimators of $\boldsymbol{\beta}$, which minimizes the squared Euclidean norm of the difference between $\boldsymbol{u}$ and $\mathbf{X}\boldsymbol{\beta}$, can be written as

$$\hat{\boldsymbol{\beta}}_{\text{OLS}} = (\mathbf{X}'\mathbf{X})^{-1}\mathbf{X}'\boldsymbol{u}, \tag{A6.2}$$

while the GLS estimator of $\boldsymbol{\beta}$, assuming the variance–covariance matrix Σ of $\boldsymbol{\varepsilon}$ is known and positive definite, is

$$\hat{\boldsymbol{\beta}}_{\text{GLS}} = (\mathbf{X}'\boldsymbol{\Sigma}^{-1}\mathbf{X})^{-1}\mathbf{X}'\boldsymbol{\Sigma}^{-1}\boldsymbol{u}, \tag{A6.3}$$

and minimizes the square of a different norm of the difference between $\boldsymbol{u}$ and $\mathbf{X}\boldsymbol{\beta}$, because of the presence of the inverse of $\boldsymbol{\Sigma}$, $\boldsymbol{\Sigma}^{-1}$, in the equation.

It can be shown that $\hat{\boldsymbol{\beta}}_{\text{GLS}}$ is the OLS estimator of $\boldsymbol{\beta}$ in the linear model defined below for $\boldsymbol{u}$ after its prewhitening (including variance stabilization in the presence of heteroscedasticity)

$$\boldsymbol{\Sigma}^{-1/2}\boldsymbol{u} = \boldsymbol{\Sigma}^{-1/2}\mathbf{X}\boldsymbol{\beta} + \boldsymbol{\Sigma}^{-1/2}\boldsymbol{\varepsilon}, \tag{A6.4}$$

where $\boldsymbol{\Sigma}^{-1/2}$ denotes the inverse of the square root of $\boldsymbol{\Sigma}$, that is, $\mathbf{A}^2 = \boldsymbol{\Sigma}^{-1}$ with $\mathbf{A} = \boldsymbol{\Sigma}^{-1/2}$.

It follows from (A6.2) and (A6.3) that

$$\mathrm{E}[\hat{\boldsymbol{\beta}}_{\mathrm{OLS}}] = \boldsymbol{\beta} \text{ and } \mathrm{Var}[\hat{\boldsymbol{\beta}}_{\mathrm{OLS}}] = (\mathbf{X}'\mathbf{X})^{-1}\mathbf{X}'\boldsymbol{\Sigma}\mathbf{X}(\mathbf{X}'\mathbf{X})^{-1}, \quad \text{(A6.5)}$$

$$\mathrm{E}[\hat{\boldsymbol{\beta}}_{\mathrm{GLS}}] = \boldsymbol{\beta} \text{ and } \mathrm{Var}[\hat{\boldsymbol{\beta}}_{\mathrm{GLS}}] = (\mathbf{X}'\boldsymbol{\Sigma}^{-1}\mathbf{X})^{-1}. \quad \text{(A6.6)}$$

While $\hat{\boldsymbol{\beta}}_{\mathrm{OLS}}$ and $\hat{\boldsymbol{\beta}}_{\mathrm{GLS}}$ are both unbiased estimators of $\boldsymbol{\beta}$, the variance of OLS estimators is biased by autocorrelation and heteroscedasticity, but not that of GLS estimators. In applications, GLS is replaced by EGLS, because $\boldsymbol{\Sigma}$ is then unknown and substituted by a consistent estimator $\hat{\boldsymbol{\Sigma}}$. The diagonal elements of $\mathrm{Var}(\hat{\boldsymbol{\beta}}_{\mathrm{EGLS}})$ (which is a variance–covariance matrix) are used to obtain standard errors for the EGLS slope estimators. Searle (1971) and Alpargu and Dutilleul (2001), among others, provide complements of information on this.

A nonlinear model whose parameters separate (Golub and Pereyra, 1973) has linear parameters such as $\boldsymbol{\beta}$ in equation (A6.1) and nonlinear parameters contained in a pre-multiplied matrix. Pure trigonometric models with unknown frequencies and joint polynomial–trigonometric models with a pre-specified trend model and unknown frequencies can be written in this form (Dutilleul, 1990, 2001). In the case of pure trigonometric models,

$$\boldsymbol{u} = \mathbf{X}(\boldsymbol{\omega})\boldsymbol{\beta} + \boldsymbol{\varepsilon}, \quad \text{(A6.7)}$$

where $\mathbf{X}(\boldsymbol{\omega})$ is composed of a column of ones and as many pairs of columns with cosine and sine values for $t = 1, \ldots, n$ as there are frequencies $\boldsymbol{\omega}$. (The inclusion of a column of ones obviates any preliminary centering to the sample mean.) In the case of joint polynomial–trigonometric models, a number of columns with values of $P_{\mathrm{O}1}(t)$, $P_{\mathrm{O}2}(t)$ or $P_{\mathrm{O}3}(t)$ are inserted after the first column of ones, providing a matrix $\mathbf{X}^*(\boldsymbol{\omega})$ that remains a function of frequencies $\boldsymbol{\omega}$ only

$$\boldsymbol{u} = \mathbf{X}^*(\boldsymbol{\omega})\boldsymbol{\beta} + \boldsymbol{\varepsilon}. \quad \text{(A6.8)}$$

In this framework, the frequencies are estimated by maximizing, respectively,

$$I^M(\boldsymbol{\omega}) = \boldsymbol{u}'\mathbf{X}(\boldsymbol{\omega})\{\mathbf{X}(\boldsymbol{\omega})'\mathbf{X}(\boldsymbol{\omega})\}^{-1}\mathbf{X}(\boldsymbol{\omega})'\boldsymbol{u} \text{ and} \quad \text{(A6.9)}$$

$$I^{M*}(\boldsymbol{\omega}) = \boldsymbol{u}'\mathbf{X}^*(\boldsymbol{\omega})\{\mathbf{X}^*(\boldsymbol{\omega})'\mathbf{X}^*(\boldsymbol{\omega})\}^{-1}\mathbf{X}^*(\boldsymbol{\omega})'\boldsymbol{u}. \quad \text{(A6.10)}$$

It has been demonstrated theoretically and empirically that $I^M(\boldsymbol{\omega})$ and $I^{M*}(\boldsymbol{\omega})$ are maximized at the true frequencies on average, in the absence or in the presence of temporal autocorrelation, provided the number of frequencies is correctly specified.

To be complete,

$$I_{jj'}^{M*}(\omega) = \boldsymbol{u}'_j \mathbf{X}_j^*(\omega)\{\mathbf{X}_j^*(\omega)'\mathbf{X}_j^*(\omega)\}^{-1}\{\mathbf{X}_{j'}^*(\omega)'\mathbf{X}_{j'}^*(\omega)\}^{-1}\mathbf{X}_{j'}^*(\omega)'\boldsymbol{u}_{j'}. \tag{A6.11}$$

That is, the scalar product of the vectors of OLS projections of $\boldsymbol{u}_j$ on $\mathbf{X}_j^*(\omega)$ and $\boldsymbol{u}_{j'}$ on $\mathbf{X}_{j'}^*(\omega)$.

In the absence of trends, $I_{jj'}^{M}(\omega) = \boldsymbol{u}'_j \mathbf{X}(\omega)\{\mathbf{X}(\omega)'\mathbf{X}(\omega)\}^{-1}\mathbf{X}(\omega)'\boldsymbol{u}_{j'}$.

7 · *Heterogeneity analysis of spatial surface patterns*

The purpose of this chapter is to present appropriate models, methods, and procedures for analyzing the heterogeneity of spatial surface patterns statistically. To avoid redundancy with Chapter 6, the spectral or frequency-domain approach is not repeated here. Instead, it is replaced by the geostatistical approach, which is at the core of Chapter 7, but can also be followed to analyze time series. The geostatistical approach is based on the analysis of the semivariance function [equation (2.6)], its estimation by an experimental variogram, and the fitting of a model to the latter. That approach is introduced gradually in Section 7.1, where spatial heterogeneity due to autocorrelation is studied under the assumption that the underlying spatial stochastic process is weakly stationary, instead of intrinsically stationary. In doing this, we will see the importance of defining distance classes in the heterogeneity analysis of spatial surface patterns, because sampling locations are more susceptible to be unequally spaced than sampling times, and distances calculated in 2-D space are, themselves, unequally spaced by construction, even when the sampling grid is perfectly regular. In Section 7.2, the analyses of spatial heterogeneity of the mean and spatial heterogeneity due to autocorrelation are performed jointly in a statistical method of multi-scale analysis linked to the geostatistical approach. This method has two phases: one univariate and one multivariate, the latter allowing the analysis of correlations at different scales of variability. Section 7.3 deals with the questions of spatial heterogeneity of the variance and how it can be incorporated in the analysis of a mixed model, together with the two other types of spatial heterogeneity. A distance-based method of correlation analysis with spatial surface patterns is presented in Section 7.4. With the exception of Section 7.1 in which simulated data are mainly used to help establish guidelines, real data are analyzed in the frame of examples in other sections. Complementary readings on the topic of spatial statistics, or statistics for spatial data, are proposed in Section 7.5. Details about simulation procedures and geostatistical modeling are grouped in Appendix A7. MATLAB® and SAS codes implementing different analytical procedures are available on the CD-ROM accompanying the book.

7.1 Spatial autocorrelation analysis

Like time series, spatial surface patterns are partial realizations of stochastic processes, and are composed of a finite number of continuous quantitative data collected in accordance with a given sampling scheme. Unlike time series, the stochastic processes underlying spatial surface patterns may be one-, two-, or three-dimensional, and the space index associated with both the pattern and the process is continuous instead of discrete. In this chapter, most of the spatial stochastic processes studied will be two-dimensional, and so will be the observed spatial surface patterns. Formally, a spatial surface pattern can be denoted $\{u(s_i) \mid i = 1, \ldots, n \text{ and } s_i \in \mathbb{R}^d\}$, with s_i the coordinates of sampling locations, $u(s_i) \in \mathbb{R}$ and $d = 1, 2$, or 3; the underlying spatial stochastic process is then denoted $\{U(s) \mid s \in \mathbb{R}^d\}$, with $U(s) \in \mathbb{R}$ for any given s, or $\{U(x, y) \mid (x, y) \in \mathbb{R}^2\}$ if $d = 2$.

Replications are rare with spatial surface patterns. Only in the framework of the ANOVA with spatial repeated measures are replications frequent, but spatial repeated measures are not actual spatial surface patterns (see Chapter 9). Therefore, the fundamental equations used here for heterogeneity analysis are the general ones: $U(s) = \mu(s) + \varepsilon(s)$ in the univariate case and $\boldsymbol{u}(s) = \boldsymbol{\mu}(s) + \boldsymbol{\varepsilon}(s)$ in the multivariate case (Chapter 2), or $U(x, y) = \mu(x, y) + \varepsilon(x, y)$ and $\boldsymbol{u}(x, y) = \boldsymbol{\mu}(x, y) + \boldsymbol{\varepsilon}(x, y)$ if $d = 2$, with $(x, y) \in \mathbb{R}^2$. Under these conditions, restrictions on the mean function [i.e., $\mu(s)$, $\boldsymbol{\mu}(s)$, $\mu(x, y)$, $\boldsymbol{\mu}(x, y)$] or the autocovariance function of the random component [i.e., $\varepsilon(s)$, $\boldsymbol{\varepsilon}(s)$, $\varepsilon(x, y)$, $\boldsymbol{\varepsilon}(x, y)$], or both, through assumptions on the underlying spatial stochastic process, are inevitable, because otherwise there would be as many mean parameters as observations (i.e., n) and an even larger number of variance and autocorrelation parameters [i.e., $\frac{n(n+1)}{2}$].

Focusing on variance and autocorrelation parameters, functions with a very small number of parameters can be used to model the theoretical autocovariance function $C(\cdot)$ of a spatial stochastic process. Examples are presented in Fig. 7.1(a). Each of the three curves plotted there is a function of the distance between any two locations in the spatial domain, which implies that the corresponding theoretical autocovariance functions are isotropic (Section 2.3), for $d = 1$, 2, or 3. The main

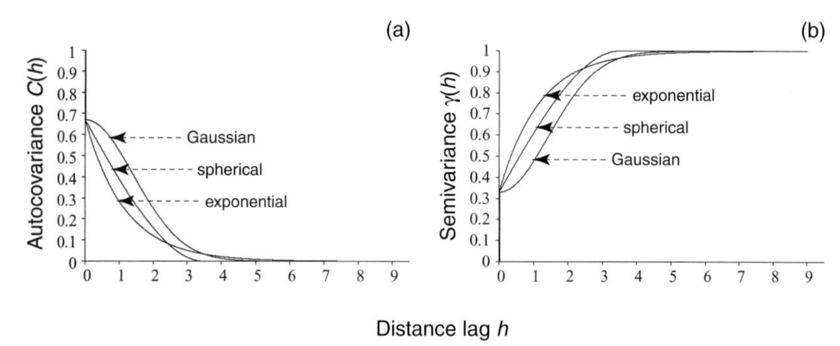

Fig. 7.1. Theoretical autocovariance and semivariance functions for exponential, spherical and Gaussian variogram models with a population variance of 1 and a nugget effect of $\frac{1}{3}$; see the ordinate near the zero distance lag in (b). By subtraction, the sill (i.e., variance component) associated with the autocorrelated random component of the stochastic process is $\frac{2}{3}$. In general terms, the two functions exist and are related, under stationarity at second order, by the equality $C(h) = \sigma^2 - \gamma(h)$, where σ^2 is the constant population variance. In particular, $C(h) = \exp(-\frac{3h}{a})$ for $h > 0$ in the exponential model with practical range a; $C(h) = 1 - 1.5\frac{h}{a} + 0.5\frac{h^3}{a^3}$ for $0 < h \leq a$, and 0 otherwise, in the spherical model with range a; and $C(h) = \exp(-\frac{3h^2}{a^2})$ for $h > 0$ in the Gaussian model with practical range a. Here, $a = 3.5$ and the right-hand side of each equation must be multiplied by $\frac{2}{3}$. The use of distance lags to define the functions implies that these are isotropic in applications in 2-D or 3-D space, and the descent of autocovariance functions to 0 with increasing distance lag and the corresponding plateau reached by semivariance functions at 1, after a smaller or larger number of distance units depending on the model, show that the stochastic processes are stationary at second order (Section 2.3).

part of the curves, which start at $\frac{2}{3}$ near the origin of the horizontal axis (i.e., near the zero distance lag) and descend to 0 at different speeds with increasing distance lag h, follows the equations of the models below:

$$C(h) = \frac{2}{3}\exp\left(-\frac{3h}{a}\right) \quad \text{for } h > 0 \text{ in the exponential model;} \tag{7.1}$$

$$C(h) = \frac{2}{3}\left(1 - 1.5\frac{h}{a} + 0.5\frac{h^3}{a^3}\right) \quad \text{for } 0 < h < a$$

$$\text{and 0 for } h \geq a \text{ in the spherical model;} \tag{7.2}$$

$$C(h) = \frac{2}{3}\exp\left(-\frac{3h^2}{a^2}\right) \quad \text{for } h > 0 \text{ in the Gaussian model.} \tag{7.3}$$

The parameter a in the equations above represents the range of spatial autocorrelation; strictly speaking, it is the practical range in equations (7.1) and (7.3), and simply the range in equation (7.2). The reason for this distinction is that there is no spatial autocorrelation beyond the range – that is, two observations separated by a distance $h \geq a$ are expected to be uncorrelated, which is the case by definition for the spherical model; in the exponential and Gaussian models, where spatial autocorrelation equals zero at infinitely large distances only, the practical range is defined as the distance at which $C(h)$ is equal to 5% of the sill (see below). In Fig. 7.1(a), $a = 3.5$ for all three models. Concerning the maximum value of $C(h)$, it is reached at the zero distance lag, and is the population variance. It is 1 here, instead of $\frac{2}{3}$, because of the choice made to include in the random component of the spatial stochastic process, in addition to the autocorrelated part, a part that is not autocorrelated. This part, called the "nugget effect" in the geostatistical literature, is the equivalent of the "white noise" in time series analysis. While the variance component (i.e., sill) associated with the autocorrelated part of the process is $\frac{2}{3}$ in Fig. 7.1(a), the nugget effect is characterized by a variance component (i.e., sill) of $\frac{1}{3}$, leading to the population variance of 1.

The presence of a nugget effect is not easy to see in Fig. 7.1(a). It is just reflected by the discontinuity at the zero distance lag, between $C(0) = 1$ (i.e., σ^2) and the beginning of the curves described by equations (7.1)–(7.3) for $h > 0$. The constant character of the population variance and the definition of theoretical autocovariance as a function of the distance between *any* two locations in the spatial domain imply that the three spatial stochastic processes are stationary at second order (Section 2.3). The presence of a nugget effect is easier to see in Fig. 7.1(b), where an alternative theoretical function, equivalent to autocovariance under second-order stationarity, is used to model spatial autocorrelation, as explained below.

The theoretical semivariance function $\gamma(\cdot)$ exists under intrinsic stationarity [equation (2.6)]. Despite its name, weak stationarity, which includes stationarity at second order, is stronger than intrinsic stationarity, and allows one to relate $C(h)$ and $\gamma(h)$ through the equality

$$C(h) = \sigma^2 - \gamma(h) \text{ for any } h \geq 0. \quad (7.4)$$

The equations for $\gamma(\cdot)$ corresponding to equations (7.1)–(7.3) for $C(\cdot)$ in Fig. 7.1 follow:

$$\gamma(h) = 1 - \frac{2}{3}\exp\left(-\frac{3h}{a}\right) \text{ for } h > 0 \text{ in the exponential model;} \tag{7.5}$$

$$\gamma(h) = \frac{1}{3} + \frac{2}{3}\left(1.5\frac{h}{a} - 0.5\frac{h^3}{a^3}\right) \text{ for } 0 < h < a$$
$$\text{and 1 for } h \geq a \text{ in the spherical model;} \tag{7.6}$$

$$\gamma(h) = 1 - \frac{2}{3}\exp\left(-\frac{3h^2}{a^2}\right) \text{ for } h > 0 \text{ in the Gaussian model.} \tag{7.7}$$

Thus, the descending behavior of theoretical autocovariance functions with increasing distance lag is replaced by an ascending behavior in theoretical semivariance functions, because of the negative sign before $\gamma(h)$ in equation (7.4). The nugget effect is also more apparent, both in Fig. 7.1(b) and in equations (7.5)–(7.7), in particular for the spherical model; see the first term (i.e., $\frac{1}{3}$) on the right-hand side of equation (7.6). Note that by definition [equation (2.6)], $\gamma(0) = 0$, since $\gamma(0) = \frac{1}{2}\text{Var}[U(s) - U(s)] = \text{Var}(0)$.

In the end, each curve is completely described with only three parameters, which are the same for both theoretical functions under weak stationarity: the range of spatial autocorrelation a (e.g., 3.5), for the autocorrelated part of the random component of the spatial stochastic process, and two variance components (i.e., sills), the one for the nugget effect (e.g., $\frac{1}{3}$) and the other for the autocorrelated part (e.g., $\frac{2}{3}$). In these conditions, the constant population variance σ^2 (e.g., 1) is equal to the sum of the two variance components.

Before considering the question of estimation of $C(h)$ and $\gamma(h)$ in the next subsections, it is important to familiarize our eyes with spatial surface patterns characterized by heterogeneity due to autocorrelation of various strengths under different models. From Fig. 7.1(a), it appears that spatial autocorrelation (i.e., spatial autocovariance divided by σ^2 here) is positive in all three models, but descends more rapidly to zero in the exponential model than in the spherical and Gaussian models in this order. Accordingly, patches of high or low values can be expected to be of smaller size in partial realizations of the exponential model relative to the two other models, for the same value of the range of spatial autocorrelation. This is, in fact, the case in the maps (left panels) of Fig. 7.2 vs. Figs. 7.3 (spherical model) and 7.4 (Gaussian model). All these partial realizations of spatial stochastic processes were produced by simulation

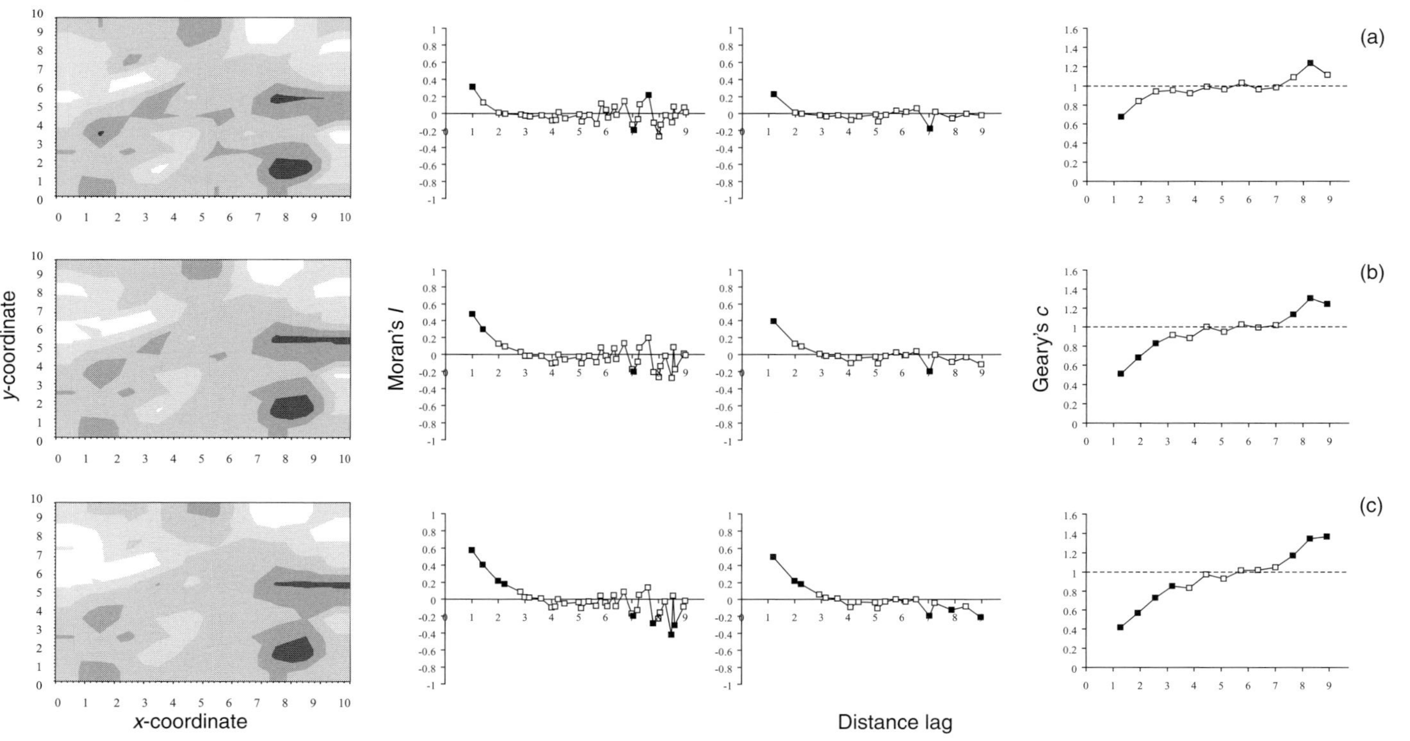

Fig. 7.2. Exponential variogram models in 2-D space, with practical ranges of (a) 2, (b) 3.5, and (c) 5 distance units. From left to right, for a given practical range: the contour map drawn from a partial realization simulated on a 10 × 10 regular grid located inside the [0, 10] × [0, 10] square – light and dark gray tones indicate, respectively, low and high values (with contours from −2 to 2 by increments of 1 here and in Figs. 7.3 and 7.4), and (0.5, 0.5) and (9.5, 9.5) are the bottom left and top right nodes of the 10 × 10 regular grid; Moran's *I* spatial correlogram computed at true vs. equal-frequency distances; and Geary's *c* spatial correlogram computed at equally spaced distance classes – filled symbols indicate that the corresponding spatial correlogram statistic value is significantly ($P < 0.05$) different from 0 (Moran's *I*) or 1 (Geary's *c*).

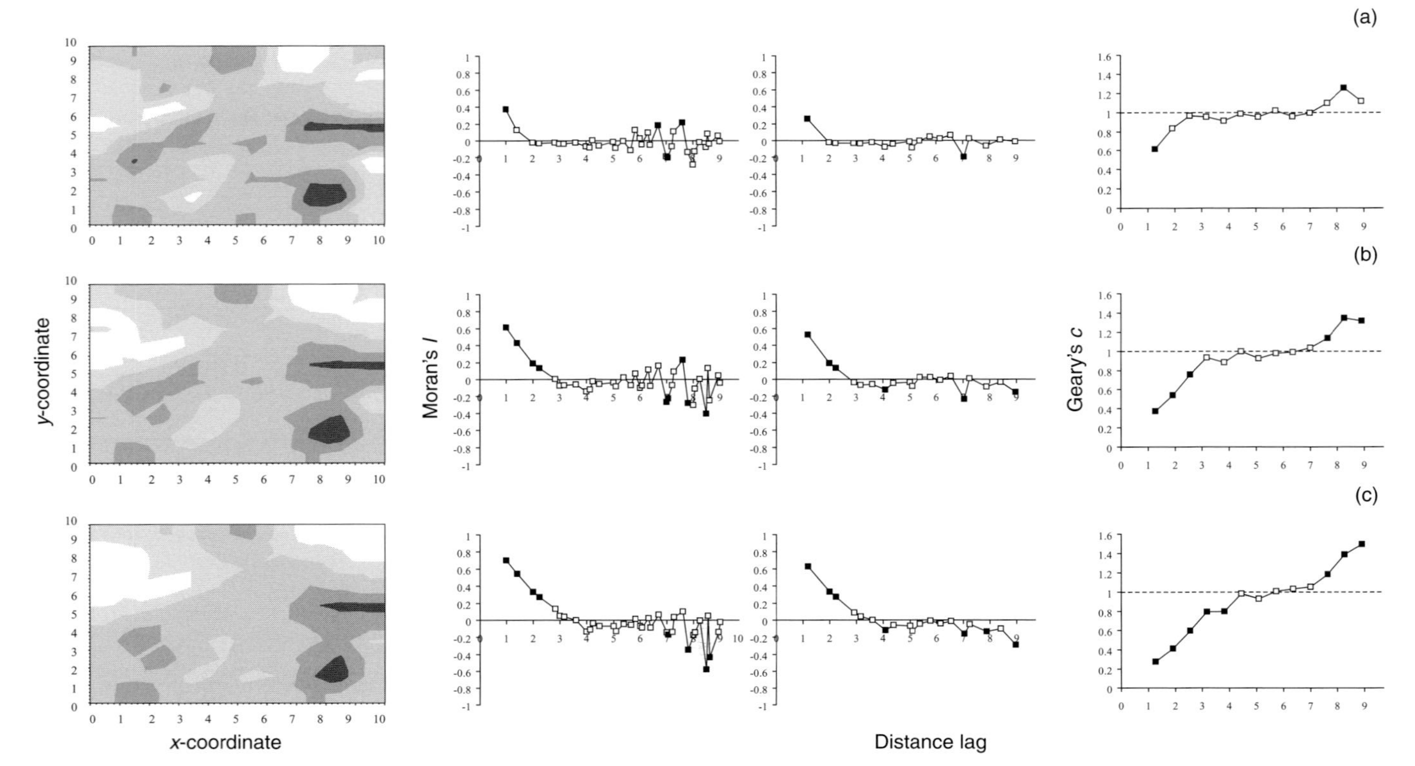

Fig. 7.3. Spherical variogram models in 2-D space, with ranges of (a) 2, (b) 3.5, and (c) 5 distance units. The definition of panels and the meaning of symbols are the same as in Fig. 7.2.

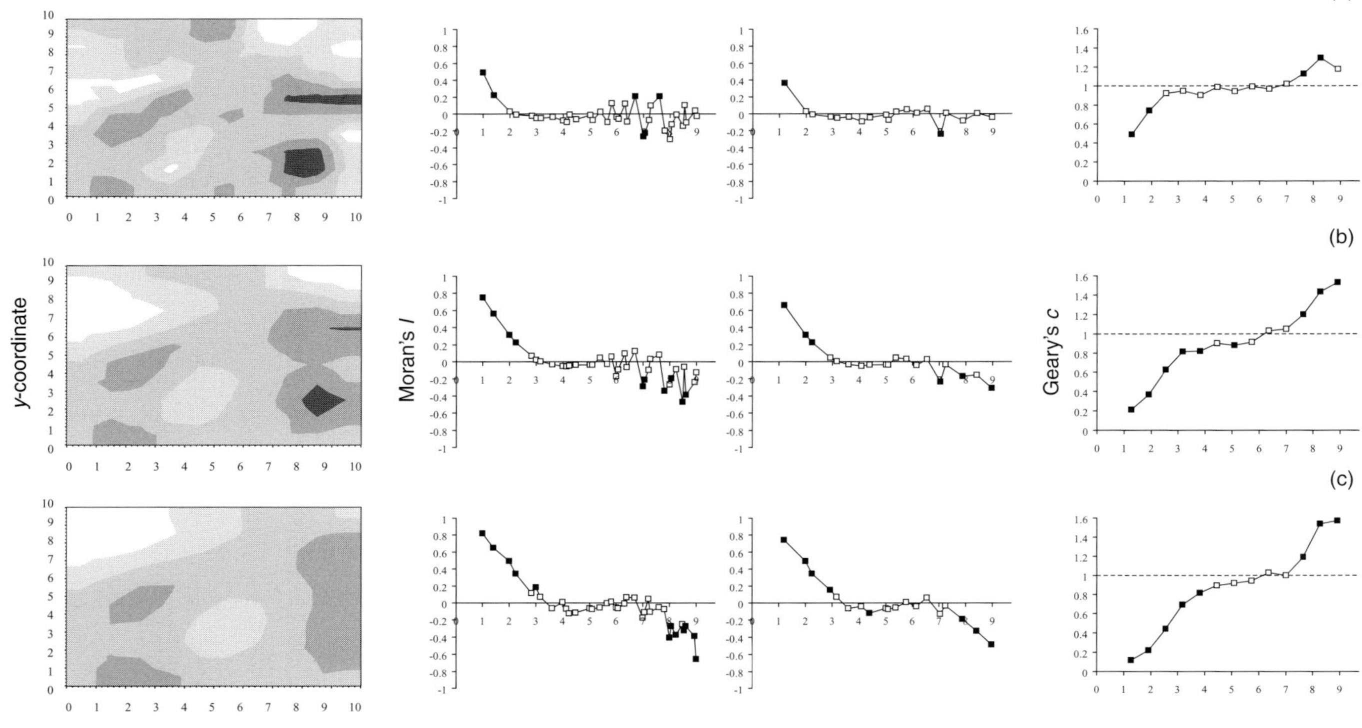

Fig. 7.4. Gaussian variogram models in 2-D space, with practical ranges of (a) 2, (b) 3.5, and (c) 5 distance units. The definition of panels and the meaning of symbols are the same as in Fig. 7.2.

and the procedures used for this are described in Section A7.1. (*Note*: The same seed was used in simulations, that is, initial pseudo-random numbers used to generate the spatial surface patterns before introduction of spatial autocorrelation were the same for all nine maps, which can be seen in the figures.) Looking at the maps again, this time for a given model, one can see that patches are being flattened and become larger, as patches that were smaller when the range was shorter are merging while the range of spatial autocorrelation increases from (a) 2 to (b) 3.5, and (c) 5. Thus, the size of patches in a spatial surface pattern characterized by heterogeneity due to autocorrelation is dependent on the type of model and the value of the range, because these influence the descent of $C(h)$ to zero and the ascent of $\gamma(h)$ to σ^2. Quantitative analyses, however, are necessary because it is difficult to distinguish graphically partial realizations arising from an exponential model with range 3.5 and a spherical model with range 2, and it is the same thing for partial realizations from a spherical model with range 5 and a Gaussian model with range 3.5.

7.1.1 Spatial correlograms

A first natural quantification of the autocorrelation contained in a spatial surface pattern is through an extension of the sample simple autocorrelation coefficients $r(k)$ used in time series analysis (Chapter 6). This extension based on Moran's (1950) I statistic provides the first of two spatial (auto)correlograms defined and discussed here. The extension is twofold: in terms of dimension, 1 in time vs. 1, 2, or 3 in space, and the use of lags in time vs. distance lags in space, where distance classes must be defined when the sampling grid is irregular and true distances *may* be used in the case of a regular sampling grid. The definition of distance classes may seem to be a technicality, but it has implications for the analysis of spatial autocorrelation and its statistical assessment, as we will see below.

Given the 2-D spatial surface pattern, $\{u(x_i, y_i) \mid i = 1, \ldots, n \text{ and } (x_i, y_i) \in \mathbb{R}^2\}$, observed at the sampling locations (x_i, y_i) $(i = 1, \ldots, n)$, Moran's I statistic value at distance lag h is given by

$$I(h) = \frac{\dfrac{1}{A(h)} \displaystyle\sum_{(i,i')} \{u(x_i, y_i) - \bar{u}\}\{u(x_{i'}, y_{i'}) - \bar{u}\}}{\hat{\sigma}^2}, \qquad (7.8)$$

where $\bar{u} = \frac{1}{n}\sum_{i=1}^{n} u(x_i, y_i)$, $\hat{\sigma}^2 = \frac{1}{n}\sum_{i=1}^{n} \{u(x_i, y_i) - \bar{u}\}^2$, and the summation $\sum_{(i,i')}$ is performed over the $A(h)$ different pairs of sampling locations (x_i, y_i) and $(x_{i'}, y_{i'})$, separated by a distance that falls in the class corresponding to lag h; this distance class can be the distance lag h itself if the sampling grid is regular and true distances are used as lags; otherwise, it is an interval with a lower bound and an upper bound. Distance classes in the form of intervals can be defined in two ways: equal-frequency, with as much as possible the same number A of different pairs of sampling locations in each class (the value of h is then fixed to the mean value of the distances falling in the interval); or equally spaced, when the upper bounds of distance classes are multiples of some predetermined distance value – in this case, upper bounds are generally used in graphs.

To fully appraise the implications of the definition of distance classes on the results of spatial autocorrelation analysis based on Moran's I or any other statistic, it is important to take into account the outcome of tests of significance at different lags. Asymptotic z-tests are available under two distinct assumptions. For a given h, the standard error associated with $I(h)$ can be calculated, either by assuming that the continuous quantitative data of the spatial surface pattern are normally distributed (normality assumption), or by randomizing the numbering $i = 1, \ldots, n$ of sampling locations with equally likely random permutations (randomization assumption). The latter assumption was used in Figs. 7.2–7.4. I refer to Cliff and Ord (1981, Section 2.3) for the expression of the standard error of $I(h)$ or its square, Var[$I(h)$]; see also Legendre and Legendre (1998, pp. 719–720).

We can now look at Moran's I spatial correlograms presented in Figs. 7.2–7.4. They could be computed at true distances (second column), because of the 10×10 regular sampling grid, and were also computed at "equal-frequency" distance classes (third column), the equally spaced distance classes being left for Geary's (1954) c spatial correlograms (fourth column; see below). Maybe paradoxically, the definition of perfect equal-frequency distance classes (i.e., with exactly the same number of pairs of sampling locations per distance lag) is not possible with a regular sampling grid, as shown in Table 7.1(b). All the correlograms were computed up to the distance lag of 9 (i.e., distance between the extremities of a row or a column of the sampling grid). This may be too large for modeling purposes, but there is no modeling involved at this stage and it will be interesting to discuss the behavior of correlograms (or variograms) at larger distances.

Table 7.1. *Distribution of the number of different pairs of sampling locations per distance lag in a spatial correlogram when (a) true distances, (b) equal-frequency distance classes, and (c) equally spaced distance classes are used with a 10 × 10 regular sampling grid*

(a)				(b)		(c)	
Distance class	Number of pairs	Distance class	Number of pairs	Distance class	Number of pairs	Distance class	Number of pairs
1.0000	180	6.0828	144	1.4142	342	0.6364	0
1.4142	162	6.3246	128	2.0000	160	1.2728	180
2.0000	160	6.4031	120	2.2361	288	1.9092	162
2.2361	288	6.7082	112	3.0000	268	2.5456	448
2.8284	128	7.0000	60	3.1623	252	3.1820	520
3.0000	140	7.0711	158	3.6056	224	3.8184	224
3.1623	252	7.2111	96	4.1231	336	4.4548	434
3.6056	224	7.2801	96	4.4721	290	5.0912	460
4.0000	120	7.6158	84	5.0000	268	5.7276	412
4.1231	216	7.8102	80	5.0990	180	6.3640	492
4.2426	98	8.0000	40	5.3852	160	7.0004	292
4.4721	192	8.0623	144	5.8310	212	7.6368	434
5.0000	268	8.2462	64	6.3246	352	8.2732	328
5.0990	180	8.4853	32	6.7082	232	8.9096	148
5.3852	160	8.5440	56	7.0711	218		
5.6569	72	8.6023	60	7.2801	192		
5.8310	140	8.9443	48	8.0623	348		
6.0000	80	9.0000	20	8.5440	152		
				9.2195	244		

Distance class = true distances in (a); in (b) and (c), distance class actually is the upper bound of the class.
Except by distributing pairs of observations separated by the same distance in different classes, perfect equal-frequency distance classes are not possible with a regular sampling grid; see (b).

Starting with the simulated partial realizations from exponential models with practical ranges of 2, 3.5, and 5 in Fig. 7.2(a)–(c), Moran's I spatial correlograms show positive autocorrelation at small distances, which is in agreement with the theoretical function in Fig. 7.1(a). The lengthening of the practical range in Fig. 7.2 is reflected by an increase of the number of statistically significant ($P < 0.05$) statistic values at first distance lags, from 1 to 2 and 4 when true distances are used (second column) and from 1 to 3 when equal-frequency distance classes are used (third column). By comparison, more positive autocorrelation at small distances is observed for the spherical model in Fig. 7.3 (second and third columns); for the range value of 3.5, for example, there are four statistically significant statistic values at first distance lags with true distances and three with equal-frequency distances. Actually, Moran's I spatial correlograms tend to start higher and descend to zero more slowly for the spherical model. This behavior is further accentuated in the case of the Gaussian model (Fig. 7.4), in agreement with the theoretical functions in Fig. 7.1(a). At larger distances, Moran's I spatial correlograms show important fluctuations when computed at true distances whatever the model, because of the smaller numbers of pairs of observations involved in the computation of the statistic at those distances [Table 7.1(a)]. These fluctuations contrast with the more steady behavior of the statistic, generally near or around zero at larger distances, when equal-frequency distance classes are used. For all three models also, the increased number of statistically significant *negative* values of Moran's I at very large distances when the range or practical range is 5 follows from patches with high and low values on opposite sides of the sampling area. This is particularly true for the Gaussian model; see the bottom contour map in Fig. 7.4(c).

The second spatial (auto)correlogram defined and discussed here is based on Geary's (1954) c statistic, which can be seen as a standardized semivariance estimate. In fact, the variance of the difference between two random variables X and Y with equal population variances $\sigma_X^2 = \sigma_Y^2 = \sigma^2$ is given by $\mathrm{Var}[X - Y] = \sigma^2 - 2\,\mathrm{Cov}[X, Y]$, so that $\frac{1}{2}\frac{\mathrm{Var}[X-Y]}{\sigma^2} = 1 - \mathrm{Corr}[X, Y]$. Geary's c statistic estimates the right-hand side of the last equality, with $X = U(s)$ and $Y = U(s')$ in general notations, under weak stationarity of the underlying spatial stochastic process. Accordingly, the sample variance of increments $u(s_i) - u(s_{i'})$ for a given distance lag is at the core of the statistic's definition below.

Given the 2-D spatial surface pattern, $\{u(x_i, y_i) \mid i = 1, \ldots, n$ and $(x_i, y_i) \in \mathbb{R}^2\}$, Geary's c statistic value at distance lag h is given by

$$c(h) = \frac{1}{2A(h)} \frac{\sum_{(i,i')} \{u(x_i, y_i) - u(x_{i'}, y_{i'})\}^2}{S^2}, \tag{7.9}$$

where $S^2 = \frac{1}{n-1} \sum_{i=1}^{n} \{u(x_i, y_i) - \bar{u}\}^2$ is the classical sample variance, with $\bar{u}$ the classical sample mean, and the summation $\sum_{(i,i')}$ is performed again over the $A(h)$ different pairs of sampling locations (x_i, y_i) and $(x_{i'}, y_{i'})$, separated by a distance that falls in the class corresponding to lag h. Note that under weak stationarity, which includes stationarity at first order (i.e., homogeneity of the mean), there is no need to center $u(s_i)$ and $u(s_{i'})$ with respect to $\bar{u}$ under the square in the numerator of the right-hand side of equation (7.9), or if they were centered, the double centering would simplify. Equally spaced distance classes are used to evaluate $c(h)$ in Figs. 7.2–7.4 (fourth column). Lower and upper bounds of those classes are multiples of 0.6364 (i.e., the maximum Euclidean distance between two locations on the 10×10 regular sampling grid, $\sqrt{162}$, divided by 20), resulting in 13 non-empty distance classes with relatively even numbers of pairs of observations between 0 and 9 [Table 7.1(c)]. This choice of distance classes for Geary's c was made to avoid redundancy with respect to Moran's I, because the two types of spatial correlograms show similar behaviors when computed at true distances and equal-frequency distance classes, in terms of statistically significant values at first distance classes and fluctuations at larger distances.

In Figs. 7.2–7.4, the smoothest behavior is shown by Geary's c spatial correlograms (fourth column). This is essentially due to the use of a smaller number of distance classes (i.e., 13), compared to the true distances (i.e., 36) and the equal-frequency distance classes (i.e., 19) used for Moran's I spatial correlograms; if the latter had been computed at 13 equally spaced distance classes, they would have looked smoother. Here, "smoothness" is related to the number of pairs of observations involved in the evaluation of statistics at a given distance lag (Table 7.1); the larger this number, the lower the variance of $c(h)$ and $I(h)$ for a given h (Cliff and Ord, 1981, Section 2.3). That being said, Geary's c spatial correlograms in Figs. 7.2–7.4(a)–(c) show very clearly (i) the differences existing between the exponential, spherical, and Gaussian variogram models in terms of spatial autocorrelation at small distances, up to the (practical) range – in agreement with the theoretical functions in Fig. 7.1(b), and (ii) differences related to the value of the (practical) range for a given model.

Key note: With irregular sampling grids, equal-frequency distance classes are recommended over equally spaced distance classes, because the former then provide a more even precision and power in the estimation and testing of all autocorrelation coefficients (Dutilleul and Legendre, 1993).

Summary: Moran's I *and Geary's* c *statistics provide two types of spatial correlogram, with different interpretations concerning the presence or absence of autocorrelation. In particular, closeness to zero in the former vs. closeness to one in the latter refers to absence of spatial autocorrelation. The definition of distance classes is very important in the analysis of spatial correlograms, where tests of significance are usually performed at the different distance lags, without modeling. Modeling is part of the variogram analysis presented below, as a follow-up to Geary's* c *spatial correlograms.*

7.1.2 Direct experimental variograms

The population semivariance function $\gamma(\cdot)$ [equation (2.6)] is the "target" in the analysis of direct experimental variograms, which are to be seen as sample statistics or estimators. Unlike $\gamma(h)$, which is a continuous function of the distance lag h under appropriate assumptions [Fig. 7.1(b)], direct experimental variograms are computed at a finite number of distance classes and plotted against discrete distance lags. In the geostatistical approach, the goal is to fit a model (e.g., exponential, spherical, Gaussian) to the direct experimental variogram "as well as possible." The resulting model parameter estimates provide information: about the spatial variability of the random variable U, through the variance components (called "sills" here) associated with the nugget effect and the autocorrelated part of the stochastic process of which the spatial surface pattern is a partial realization, and about spatial autocorrelation, through the (practical) range for the extent and the relative values of the two sills for the strength. Some of the semivariance estimators (e.g., Cressie and Hawkins, 1980) are more robust to the presence of atypical observations in the spatial surface pattern than others. Alternatively, the Box and Cox (1964) transformation can be used in a preliminary step to improve the normality of the distribution of sample data and, at the same time, to reduce the effect of "outliers." The classical semivariance estimator, due to Matheron (1962), is nothing but the non-standardized version of Geary's c spatial correlograms:

$$\hat{\gamma}(h) = \frac{1}{2A(h)} \sum_{(i,i')} \{u(x_i, y_i) - u(x_{i'}, y_{i'})\}^2, \tag{7.10}$$

with the same meaning for $A(h)$ and the summation $\sum_{(i,i')}$ as in equation (7.9).

There exist different procedures for fitting a variogram model (e.g., maximum likelihood and restricted maximum likelihood with simulated annealing: Pardo-Igúzquiza, 1997; Marchant and Lark, 2007), and a discussion of the pros and cons of each procedure falls beyond the scope of this book. In the following subsections, preference will be given to the EGLS fitting procedure developed by Pelletier *et al.* (2004); see also Pelletier *et al.* (2009a). Compared with the simpler OLS and WLS fitting procedures, EGLS here takes into account the inequality of variances and the correlation between ordinates of the experimental variogram, especially at first distance lags, and thus provides model parameter estimates with suitable properties (i.e., absence of bias and smaller standard errors). It must be noted that in the OLS fitting procedure the same weight is given to all distance lags, and the squared Euclidean distance between the experimental variogram and the corresponding ordinates in the fitted model is used to evaluate discrepancies; in trying to fit a model to an experimental variogram from small to large distances visually, our eyes "are" OLS, and this procedure is definitively not recommended.

Traditionally in geostatistical applications, the fitted variogram model is used for interpolation – that is, to predict the random variable U where it was not observed, with some version of the method called "kriging" (Isaaks and Srivastava, 1989; Goovaerts, 1997). As we will see in the following subsections, the variogram models fitted in "coregionalization analysis with a drift" (Pelletier *et al.*, 2009a, 2009b) are used for different purposes. These include the estimation of model parameters to assess the relative importance of variance components that are non-spatial vs. spatial at small scale for each variable, and the analysis of the corresponding correlations between variables, by working with data collected at the sampling locations.

Key note: The use of a main regular sampling grid, with smaller regular sampling grids centered on some nodes of the main grid, might help improve the variogram model fitting at very small distances and hence the evaluation of the non-spatial variance component (see Chapter 9).

In Geary's c spatial correlograms, the presence of autocorrelation is tested at different distance lags, but the largest distance lag at which the absence of spatial autocorrelation is rejected without disruption starting with the very first one tends to be smaller than the "true" value of the range (Figs. 7.2–7.4). This is the case especially for the exponential model, in which spatial autocorrelation decreases rapidly with increasing distance. By comparison, direct experimental variograms offer the possibility of fitting a model, and a reliable estimate of the (practical) range of spatial autocorrelation will figure among the model parameter estimates if an appropriate estimation procedure (e.g., EGLS; Pelletier *et al.*, 2009a) is used.

7.2 Joint analysis of spatial heterogeneity of the mean and spatial autocorrelation

The joint presence of heterogeneity of the mean and autocorrelation in a spatial surface pattern prevents the use of classical multiple regression on spatial coordinates to capture the former and the analysis of direct experimental variograms on raw data to model the latter. Some information is required on one source of spatial heterogeneity to handle the other properly. This information is about spatial autocorrelation when the objective is to capture spatial heterogeneity of the mean by EGLS fitting of a "trend surface," globally or locally. It is about a mean function of the spatial coordinates, in order to apply variogram analysis on the differences between the data and that mean function instead of the raw data. Phase I in the method of "coregionalization analysis with a drift" (CRAD; Pelletier *et al.*, 2009a, 2009b), which will be presented in Subsection 7.2.1, provides an appropriate framework for this "interactive" inferential process; Phase II of CRAD (Subsection 7.2.2) is multivariate instead of univariate, and concerns the analysis of correlations at different scales.

Later in this section and in Section 7.4, the following spatial versions of the fundamental equation (2.1) will be applied:

$$U(x, y) = \mu(x, y) + \varepsilon(x, y) \text{ in 2-D space and} \tag{7.11}$$

$$U(x) = \mu(x) + \varepsilon(x) \text{ in 1-D space,} \tag{7.12}$$

where $\mu(x, y)$ and $\mu(x)$ are deterministic mean functions, and $\varepsilon(x, y)$ and $\varepsilon(x)$ are weakly stationary random deviations from the mean that are possibly autocorrelated.

Classically, trend surface models in 2-D space are polynomial models of the mean function $\mu(x, y)$, which thus extend the polynomial models used for the trend $\mu(t)$ in heterogeneity analysis of time series (Subsection 6.3.1):

- degree 1, linear or planar: $\mu(x, y) = a + bx + cy$;
- degree 2, quadratic: $\mu(x, y) = a + bx + cy + dx^2 + ey^2 + fxy$;

- degree 3, cubic: $\mu(x, y) = a + bx + cy + dx^2 + ey^2 + fxy + gx^3 + hy^3 + ix^2y + jxy^2$.

Hereafter, the joint analysis of spatial heterogeneity of the mean and spatial autocorrelation is presented in an approach with strong geostatistical influence, in the framework of a statistical method (i.e., CRAD) designed for multi-scale analysis of multivariate 2-D and 1-D spatial data. It must be noted that the spectral approach of Chapter 6 can almost readily be applied to the analysis of 1-D spatial data, except that there are two nearest neighbors for one interior sampling location on a regular transect; it follows that $U(x) = \phi\frac{U(x-1)+U(x+1)}{2} + \varepsilon(x)$ for the 1-D spatial AR(1) process, which corresponds to the exponential model because the theoretical autocorrelation function, $\rho(h) = \phi^h$ is equivalent to $\exp(-\frac{3h}{a})$, with $a = -\frac{3}{\log(\phi)}$. In addition, in the same way that equation (7.4) relates the theoretical autocovariance function $C(h)$ and the theoretical semivariance function $\gamma(h)$ under weak stationarity and isotropy, other equations can be written to obtain a theoretical spectral density function $f(\omega_1, \omega_2)$ with two arguments ω_1 and ω_2 in 2-D space under weak stationarity, by applying the Fourier transform of the theoretical autocovariance and semivariance functions $C(\mathbf{h})$ and $\gamma(\mathbf{h})$ – the possible anisotropy justifies the use of a vector lag $\mathbf{h}$ instead of a distance lag h (see Recommended readings in Section 7.5).

7.2.1 The univariate case

With a single partial realization of a spatial stochastic process, and the mean function (i.e., the drift) and the random component both unknown, the decomposition $U(x, y) = \mu(x, y) + \varepsilon(x, y)$ is inevitably characterized by some arbitrariness in the choice of models for the estimation of the two components (Cressie, 1993, p. 112; Chilès and Delfiner, 1999, p. 233). In Phase I of CRAD (Pelletier *et al.*, 2009a), the two main criteria applied in performing the decomposition are (i) modeling spatial heterogeneity of the mean by a deterministic trend surface, globally or locally, while incorporating spatial autocorrelation of the random component through an EGLS fitting of the trend surface model, and (ii) modeling spatial autocorrelation from the direct experimental variogram of residuals when these can be considered a partial realization of a weakly stationary process, that is, when the direct experimental variogram of residuals allows the fitting of a model with a (practical) range. Note that the decomposition at sampling locations $U(x_i, y_i) = \hat{\mu}(x_i, y_i) + \hat{\varepsilon}(x_i, y_i)$ $(i = 1, \ldots, n)$, where hats denote estimates, and the use of an EGLS estimator for the drift resemble the approach followed in "universal kriging" or "kriging with a trend model" (Cressie, 1993, p. 151; Goovaerts, 1997, p. 139).

The estimation procedure is univariate and iterative in Phase I of CRAD (Pelletier *et al.*, 2009a). As a starting point, a trend surface model is fitted by OLS over the whole set of sampling locations (global drift estimation), or on subsets within a moving circular window centered at each sampling location (local drift estimation), which provides initial drift estimates $\hat{\mu}(x_i, y_i)$ $(i = 1, \ldots, n)$. Thereafter, the model fitted by WLS (Zhang *et al.*, 1995) to the direct experimental variogram of residuals $\hat{\varepsilon}(x_i, y_i) = U(x_i, y_i) - \hat{\mu}(x_i, y_i)$ $(i = 1, \ldots, n)$ is used to update the drift estimates, through EGLS fitting of the trend surface model. The mutual update of drift estimates $\hat{\mu}(x_i, y_i)$ $(i = 1, \ldots, n)$ and residuals $\hat{\varepsilon}(x_i, y_i)$ $(i = 1, \ldots, n)$ is continued until the change in variogram model parameter estimates for the latter is minimal.

Local drift estimation is more flexible and general than its global counterpart, allowing the capture of spatial heterogeneity of the mean at intermediate and large scales in a greater variety of surface patterns (Pelletier *et al.*, 2009a). The size of the circular moving window (i.e., its radius) is an important parameter in a local drift estimation procedure. In CRAD, increasingly larger window sizes are tried and the selected one provides the direct experimental variogram of residuals for which the objective function in the EGLS model fitting is the lowest. Theoretically, the smallest window would contain just one sampling location, and the largest all n sampling locations. Such choices are not practical because the former would provide drift estimates equal to the observations, $\hat{\mu}(x_i, y_i) = U(x_i, y_i)$ $(i = 1, \ldots, n)$, so that $\hat{\varepsilon}(x_i, y_i) = 0$ $(i = 1, \ldots, n)$, and the latter would correspond to a global instead of local drift estimation procedure. Accordingly, the search must be made within an interval of radius values between these two extremes. Pelletier *et al.* (2009a) found that the use of a planar model within the moving window in procedure L_1, as local approximation of the mean function, is appropriate and that the optimal window size is dependent on the trend surface model.

Key note: The selection of an optimal size for the moving window in local drift estimation, combined with the use of EGLS for model fitting, allows a minimization of the bias inherent to the direct experimental variogram of residuals (Cressie, 1993, p. 166; Pelletier et al., *2009a).*

As for variogram model fitting, the (practical) range of spatial autocorrelation is first estimated by EGLS in Phase I of CRAD; then, given

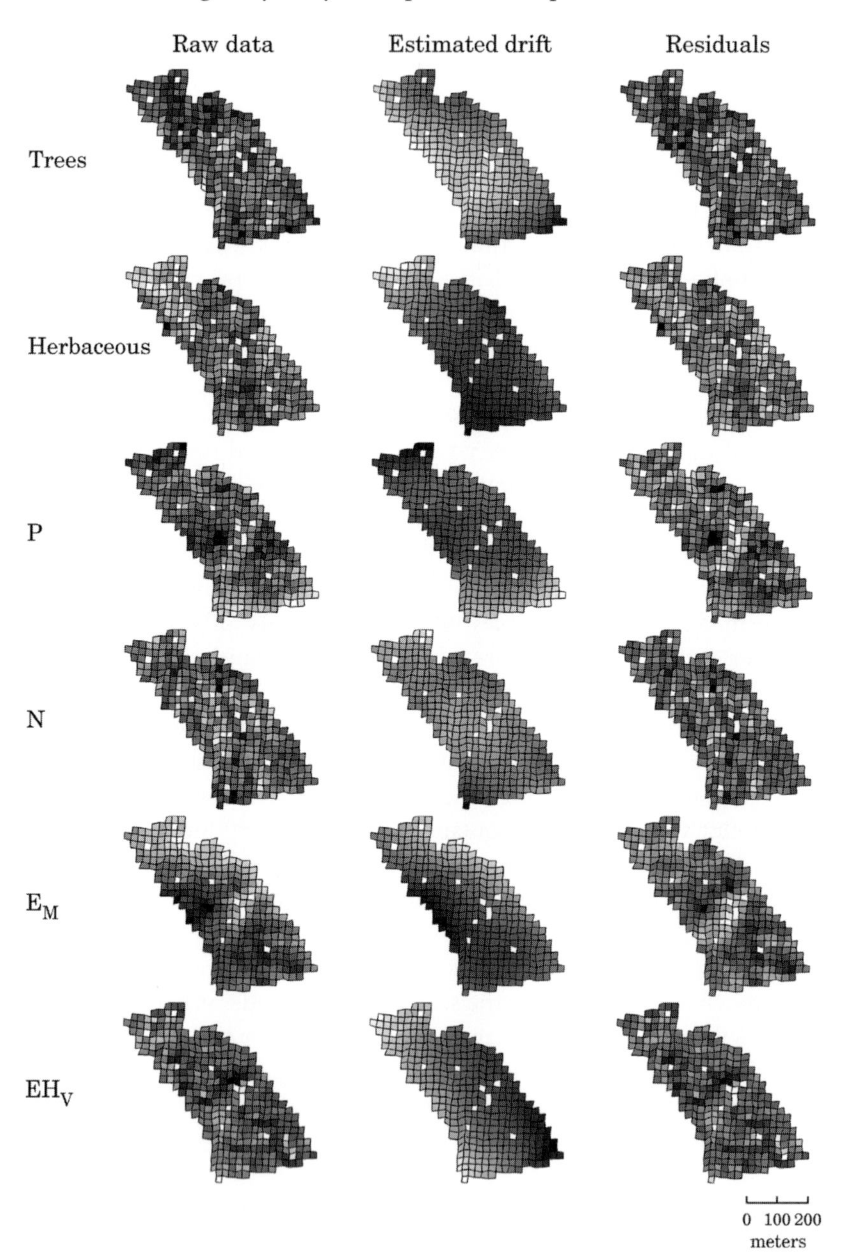

Fig. 7.5. Example of heterogeneity analysis of spatial surface patterns for plant diversity in relation to soil properties and topography at the Molson Ecological Reserve of McGill University in Île-Perrot and Terrasse-Vaudreuil (Québec, Canada). Maps of the raw data (left), estimated drift (middle) and residuals (right)

the estimated range, the sills for the nugget effect and the autocorrelated random component are estimated by EGLS (Pelletier *et al.*, 2004, 2009a). As discussed in Section 7.1, the spherical model (which has been found appropriate in the examples of this and the next subsection) falls between the exponential and Gaussian models in terms of descent of spatial autocorrelation with increasing distance lag (Fig. 7.1); it is also a model characterized by a range instead of a practical range of spatial autocorrelation.

The following example with real data (to be continued in the next subsection) illustrates the type of results that may be obtained in Phase I of CRAD, and shows how to interpret them.

Plant diversity study example

The general purpose of the ecological study described below (see Pelletier *et al.*, 2009a, 2009b, for complements) was to assess the relationships of plant species diversity with soil properties and topographical indices derived from elevation models in a forested area of southern Québec. The study field is the 22-ha forested area of the Molson Reserve of McGill University in Île-Perrot and Terrasse-Vaudreuil, near Montréal (Québec, Canada). Although the variogram had been used to analyze spatial patterns of diversity in former studies, Pelletier *et al.*'s study appears to be the first in which spatial patterns were decomposed into scales of variability (this example) and scale-dependent relationships between diversity and environmental variables were examined (continued example in Subsection 7.2.2), by application of Phases I and II of the CRAD method.

The same 339 quadrats as in Pelletier *et al.* (2009a, 2009b), with an area of about 625 m^2 (i.e., 25 m × 25 m) each, provided the support for data collection and analysis (Fig. 7.5). The spatial location of the four corners of each quadrat was georeferenced with a Global Positioning

---

Fig. 7.5. (*cont.*) for two plant diversity indices (Trees, Herbaceous), two soil properties (extractable P, total N), and two topographical variables (mean elevation E_M, elevation variability index EH_V), following the decomposition of observed spatial surface patterns (i.e., raw data = estimated drift + residuals) in Phase I of the method of coregionalization analysis with a drift (CRAD). Drift estimation was performed locally within a moving window of a size selected so that weak stationarity could be assumed on residuals; see the direct experimental variograms and fitted models in Fig. 7.6(b) and their cross-counterpart in Fig. 7.6(c). Note that "raw data" above means the data were standardized to a mean of 0 and a variance of 1 after Box–Cox transformation, and that the white spots indicate sampling locations with missing observations.

System and the geographical center of quadrats was used in spatial analyses. Vascular plants were divided into five physiognomic groups: Trees, Shrubs, Carex, Seedless (ferns and lycopodia), and Herbaceous. For each group, the value of the diversity index in quadrat i was calculated as $D_i = \exp(H_i)$ $(i = 1, \ldots, 339)$, with H_i the classical Shannon–Wiener entropy index for quadrat i (Shannon, 1948). For the purpose of this example, only results for Trees and Herbaceous diversity indices will be presented.

Composite soil samples were also collected for each quadrat. Among the variables measured on these samples, six soil variables were retained for data analyses: pH (in water), extractable Mg, K, and P (Mehlich III), soil organic matter (loss on ignition), and total N (acid digestion). These soil properties are thus "averages" over a quadrat. In addition, three topographical variables were evaluated at the quadrat level from a digital elevation model (Pelletier *et al.*, 2009b). These three topographical variables are: mean elevation (E_M); fractal dimension, calculated following Burrough (1981); and an elevation variability index (EH_V, called "elevation variance heterogeneity" by Pelletier *et al.*, 2009b) – low values of EH_V indicate that topography is either flat or very rough over the quadrat, whereas high values indicate a mixture of zones with low and high variability of elevation inside the quadrat.

For the purpose of this example, results for two soil variables (i.e., extractable P and total N) and two topographical variables (i.e., mean elevation and the elevation variability index) will be presented.

Key note: Although all variables here (i.e., diversity, soil, topography) were sampled at the quadrat level, the observed patterns are surface patterns, as partial realizations on the study field of spatial stochastic processes with continuous quantitative values.

The six variables mapped in Fig. 7.5 show different patterns of spatial heterogeneity of the mean (middle maps with estimated drifts), and only small patches of very high or very low values can be seen in residuals (right maps). In particular, residuals for mean elevation and the elevation variability index are less noisy than for other variables, suggesting a stronger spatially autocorrelated component relative to the nugget effect for the two topographical variables. Such information would hardly be obtained from raw data (left maps).

Various effects of the presence of spatial heterogeneity of the mean on the behavior of the six direct experimental variograms computed from raw data, denoted $\hat{\gamma}_j(h)$ $(j = 1, \ldots, 6)$, can be observed in Fig. 7.6(a).

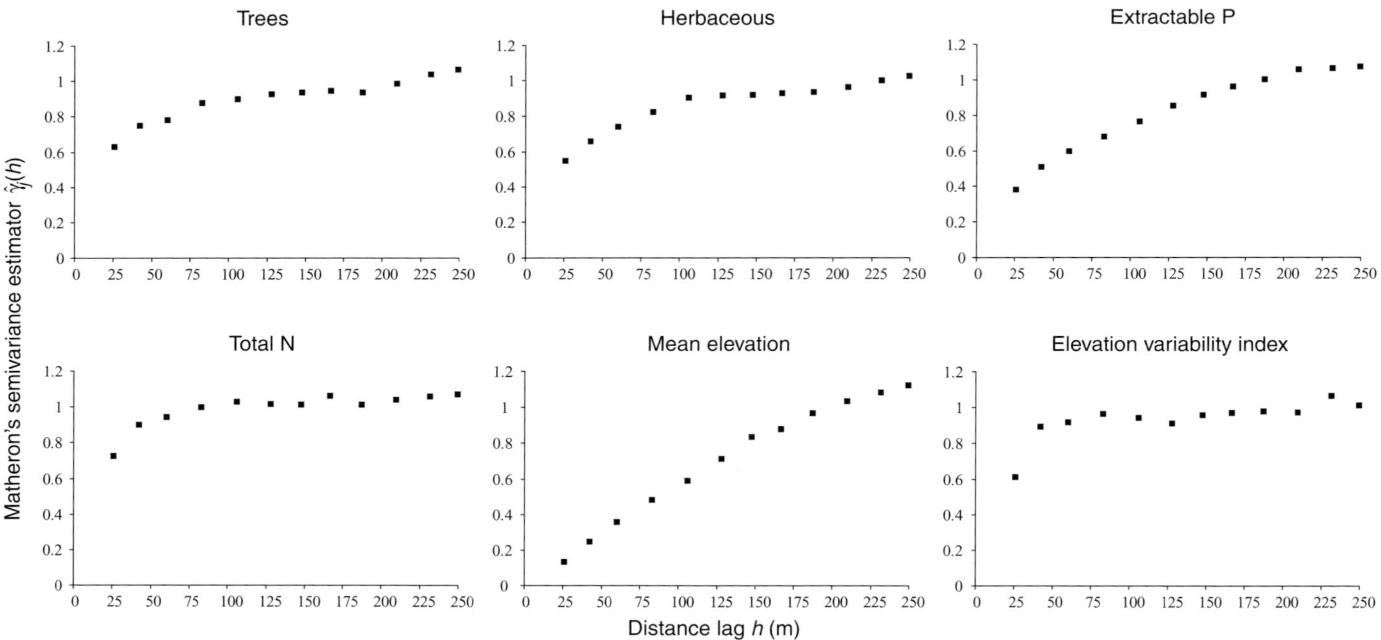

Fig. 7.6. Continued example of heterogeneity analysis of spatial surface patterns for plant diversity in relation to soil properties and topography at the Molson Ecological Reserve of McGill University in Île-Perrot and Terrasse-Vaudreuil (Québec, Canada). (a) Direct experimental variograms computed from raw data for the six variables of Fig. 7.5; (b) direct experimental variograms for the same six variables, but computed from residuals; continuous curves represent variogram models based on a spherical component plus a nugget effect when justified, fitted separately by estimated generalized least squares (EGLS) in Phase I of CRAD (see Table 7.2 for complementary numerical results); (c) cross-experimental variograms computed from residuals for some of the pairs of variables including the Tree and Herbaceous diversity indices (variables to explain) in Phase II of CRAD – in this phase, compared to Phase I, a common range is estimated first (i.e., 82 m) and then used in the EGLS fitting of a linear model of coregionalization to all experimental variograms (direct and cross) of residuals.

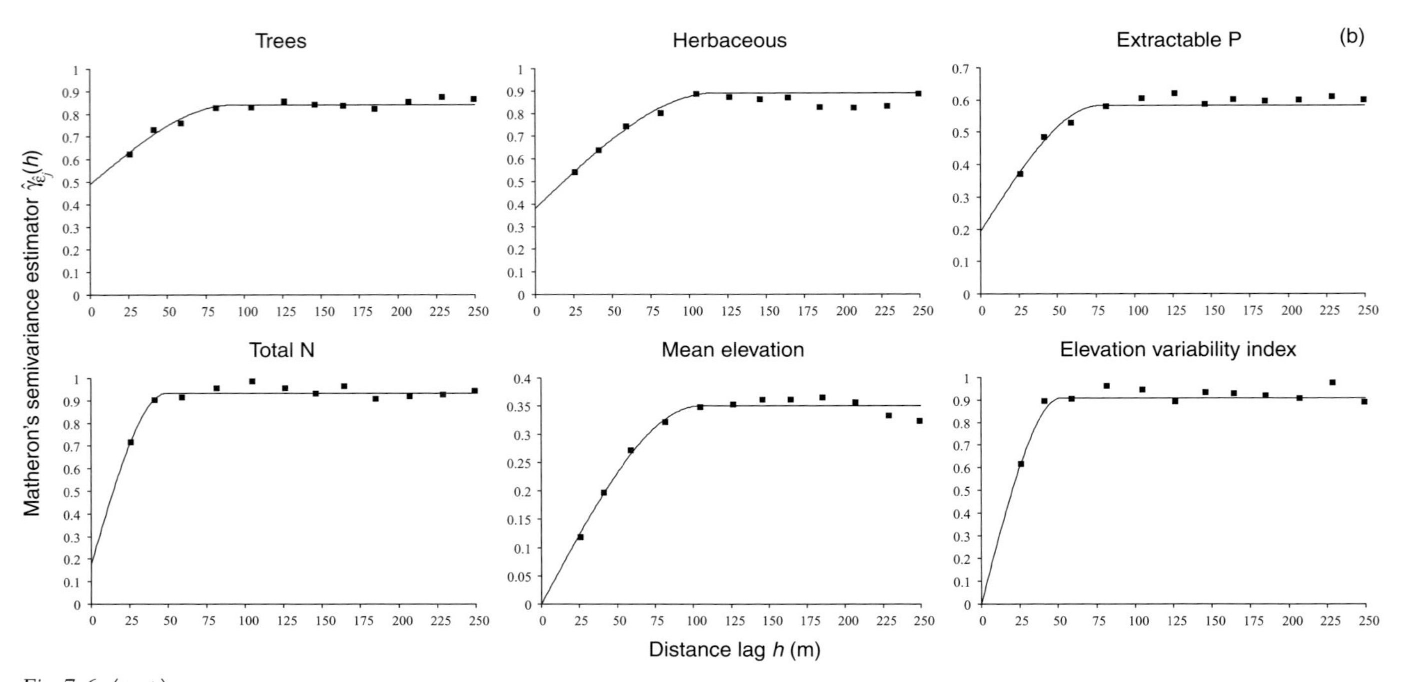

(b)
Trees
Herbaceous
Extractable P
Total N
Mean elevation
Elevation variability index
Matheron's semivariance estimator $\hat{\gamma}_{\varepsilon_j}(h)$
Distance lag h (m)

Fig. 7.6. (*cont.*)

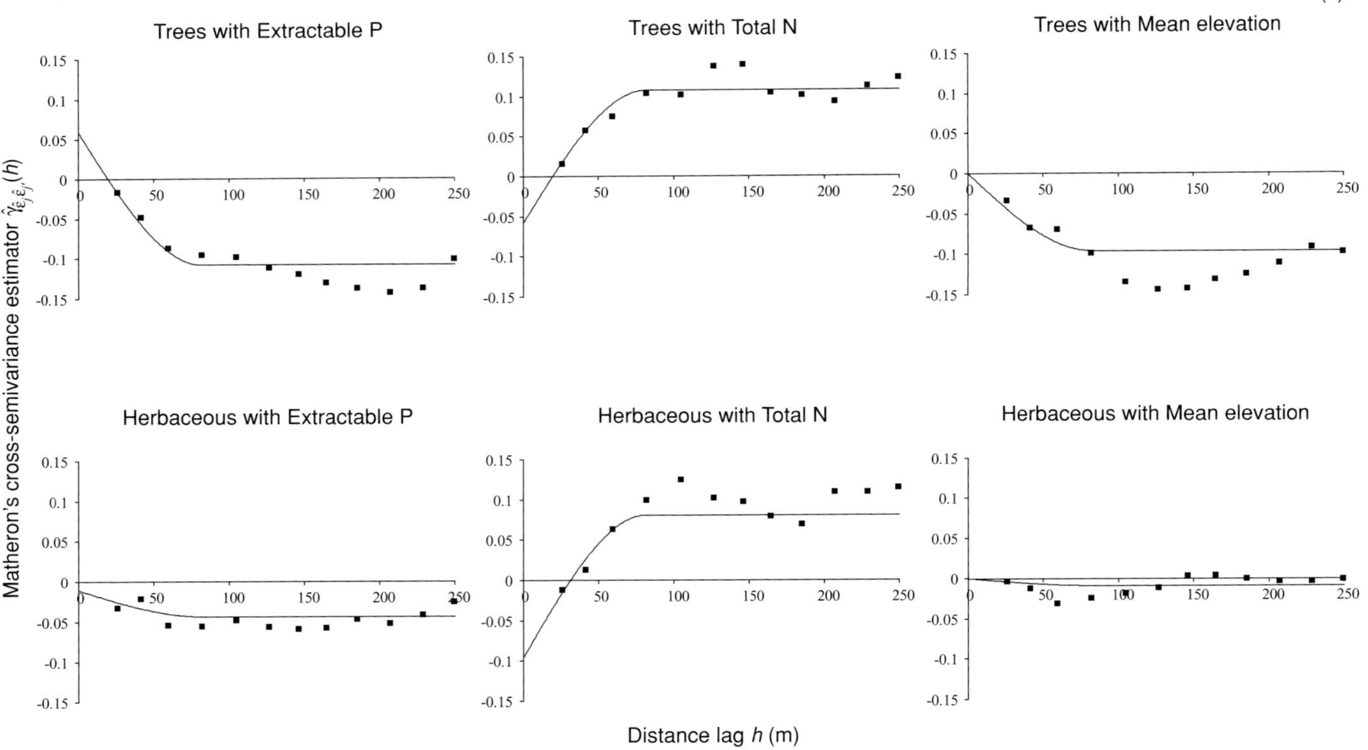
(c)
Trees with Extractable P
Trees with Total N
Trees with Mean elevation
Herbaceous with Extractable P
Herbaceous with Total N
Herbaceous with Mean elevation
Matheron's cross-semivariance estimator $\hat{\gamma}_{\hat{\varepsilon}_j \hat{\varepsilon}_{j'}}(h)$
Distance lag h (m)
0.15
0.1
0.05
0
-0.05
-0.1
-0.15
0
50
100
150
200
250

Fig. 7.6. (*cont.*)

The strongest effects are for mean elevation and extractable P, for which the semivariance estimator seems to continue to increase after the 250-m distance lag or stops increasing just before 250 m. (The distance of 250 m, which was chosen as maximum distance lag in Fig. 7.6(a)–(c), corresponds to half the length of the side of a square with same area as the study field; this distance compares to 5 in Figs. 7.2–7.4.) To a lesser degree, the Trees and Herbaceous diversity indices show some signs of spatial heterogeneity of the mean in Fig. 7.6(a), where their direct experimental variograms $\hat{\gamma}_1(h)$ and $\hat{\gamma}_2(h)$ start to increase again after 175 m after seemingly reaching a plateau from 100 or 125 m. Lastly, total N and the elevation variability index appear to be weakly stationary, at least from their direct experimental variograms, and little difference with variograms computed from residuals can be expected for these two variables (see below).

The behavior of the six direct experimental variograms computed from residuals, denoted $\hat{\gamma}_{\hat{\varepsilon}_j}(h)$ $(j = 1, \ldots, 6)$ in Fig. 7.6(b), is consistent with the visual observations made on the maps of residuals in Fig. 7.5 and with previous material discussed in Section 7.1, including Figs. 7.1–7.4. Compared with raw data in Fig. 7.6(a), all direct experimental variograms of residuals clearly reach a plateau after some distance lag in Fig. 7.6(b). As expected, total N and the elevation variability index are the variables with the most similar variograms between the two figures. A spherical model with a nugget effect [cf. equation (7.6)] is found to represent appropriately the spatial autocorrelation and non-spatial variability for four of the six variables, and a nugget effect is not required for the two topographical variables [see the continuous curves in Fig. 7.6(b)]. The goodness-of-fit is very good at small to intermediate distance lags because the EGLS fitting procedure (Pelletier *et al.*, 2004, 2009a) gives more weight, by definition, to these distance lags. It is also where "sills" of the nugget effect and the autocorrelated component of the model are estimated, together with the range of spatial autocorrelation.

In Table 7.2, the non-spatial variability of each variable is measured by the estimated sill of the nugget effect, and the spatial variability at small scale by the estimated sill of the spherical component which represents the autocorrelated spatial structure, while spatial heterogeneity of the mean is quantified by the pseudo-variance of the estimated drift. Non-spatial variability is the most important in the two plant diversity indices, and spatial heterogeneity of the mean, in extractable P and mean elevation, while the six estimated sills of spherical components are greater than 30%, with a maximum of 90% for the elevation variability index. All the

Table 7.2. *Parameter estimates in Phase I of the method of coregionalization analysis with a drift (CRAD) for each of the six variables in the first example with plant diversity studied in relation to soil properties and topography at the Molson Ecological Reserve of McGill University in Île-Perrot and Terrasse-Vaudreuil (Québec, Canada)*

Variable	Estimated sill of nugget effect	Estimated sill of spherical component	Estimated range of spherical component	Pseudo-variance of estimated drift
Trees diversity	0.4899	0.3494	91.9	0.0907
Herbaceous diversity	0.3810	0.5101	118.3	0.1490
Extractable P	0.1937	0.3893	78.7	0.4437
Total N	0.1777	0.7559	49.0	0.0735
Mean elevation E_M	0	0.3107	101.8	0.6199
Elevation variability EH_V	0	0.9094	52.3	0.1112

Estimated ranges are in meters.
The two estimated sills and the pseudo-variance do not add up to 1 because the decomposition of the total variance is based on the method of estimated generalized least squares, which is not orthogonal for the classical scalar product. (The data were standardized to a mean of 0 and a variance of 1 after the Box–Cox transformation; see Fig. 7.5.)

estimated ranges of spatial autocorrelation are much shorter than 250 m, ranging from about 50 to 120 m (Table 7.2). This low variation in range estimates shows potential for a common estimated range in Phase II of CRAD (multivariate case).

A copy of the software used to perform CRAD in this example and the following ones can be retrieved at http://environmetricslab.mcgill.ca.

7.2.2 The multivariate case

For assessing the heterogeneity of the mean with multiple spatial surface patterns, separate univariate analyses are acceptable or even recommended, since there is generally no contribution of the other variables to the estimation of the drift of a given variable (Helterbrand and Cressie, 1994). This allows one to concentrate on the estimation of variance components and auto- and cross-correlation parameters, which is performed on the direct and experimental variograms (direct and cross) of EGLS residuals in Phase II of CRAD (Pelletier *et al.*, 2009b).

Matheron's classical semivariance estimator [equation (7.10)] is based on a sum of squared differences, which is the squared Euclidean norm of the vector of increments at a given distance lag for one partial realization of a stochastic process assumed to be at least intrinsically stationary. Accordingly, the classical *cross*-semivariance estimator due to the same author is based on a sum of *cross*-products that is the scalar product between the vectors of increments at a given distance lag for two stochastic processes sampled at the same locations (x_i, y_i) $(i = 1, \ldots, n)$ in 2-D space and assumed to be at least intrinsically stationary:

$$\hat{\gamma}_{jj'}(h) = \frac{1}{2A(h)} \sum_{(i,i')} \{u_j(x_i, y_i) - u_j(x_{i'}, y_{i'})\}\{u_{j'}(x_i, y_i) - u_{j'}(x_{i'}, y_{i'})\} \tag{7.13}$$

under isotropy. Note that in Phase II of CRAD, the observations $u_j(\cdot)$ and $u_{j'}(\cdot)$ are replaced by EGLS residuals $\hat{\varepsilon}_j(\cdot)$ and $\hat{\varepsilon}_{j'}(\cdot)$, and the experimental cross-variogram is denoted by $\hat{\gamma}_{\hat{\varepsilon}_j\hat{\varepsilon}_{j'}}(h)$.

The experimental cross-variogram (7.13) is a measure of spatial association between variables. By definition, it may take positive or negative values, but a lag-based approach is not really recommended for its interpretation (Goovaerts, 1992). For parsimony reasons and a correct interpretation, it is better to follow an approach based on "structures," in which a model is fitted to all $\frac{p(p+1)}{2}$ experimental variograms (direct and cross) together, if p variables are included in the analysis. (Factorial cokriging, which is a multivariate kriging technique, produces oversmoothed predictions for spatially autocorrelated components, resulting in inflated correlations; see Larocque *et al.*, 2006.)

In simple terms, the linear model of coregionalization (Journel and Huijbregts, 1978, p. 172) is a multivariate multiple regression model for experimental variograms (direct and cross), which are the variables to explain, while the explanatory variables are basic semivariance functions (e.g., nugget effect, exponential, spherical, Gaussian), which correspond to structures. In this model, the same semivariance functions are fitted to all experimental variograms. Only the slopes (i.e., the sills) differ, and there is one matrix of sills, or "coregionalization matrix," per structure. More technical details about the model and its fitting are given in Section A7.2 and in Pelletier *et al.* (2004, 2009b). With a nugget effect and one spatially autocorrelated component (e.g., spherical), one common range of spatial autocorrelation for all variables is estimated first and then, given the estimated range, the two sill matrices are estimated iteratively by EGLS, ensuring that each sill matrix estimate is definite positive so that it can be treated as a variance–covariance matrix (i.e., its diagonal entries as variances and its off-diagonal entries as covariances). Eventually, structural coefficients of correlation are calculated in the usual way: for a given structure *str*, if $\hat{\beta}_{j,str}$ and $\hat{\beta}_{j',str}$ denote the sills estimated for the direct variograms of variables U_j and $U_{j'}$, and $\hat{\beta}_{jj',str}$ for their cross-variogram, the estimate of their correlation at structure *str* is given by

$$r_{jj',str} = \frac{\hat{\beta}_{jj',str}}{\sqrt{\hat{\beta}_{j,str}\hat{\beta}_{j',str}}}. \tag{7.14}$$

Similarly, structural coefficients of determination (i.e., structural R^2s) can be calculated from sill matrix estimates (see Section A7.2 and one of the examples below).

Concerning tests of significance, when simple and multiple correlations are to be assessed on spatial data under weak stationarity or on residuals after estimation and removal of spatial heterogeneity of the mean (e.g., in Phase II of CRAD), modified *t*- and *F*-tests taking spatial autocorrelation into account are recommended (Dutilleul, 1993b; Dutilleul *et al.*, 2008; software available at http://environmetricslab.mcgill.ca). For structural correlations, at the nugget effect (non-spatial) and at the spatially autocorrelated component (small-scale spatial), modified *t*-tests per structure that take spatial autocorrelation into account *and* incorporate the uncertainty associated with the estimation of sills exist (Dutilleul and Pelletier, in press). As for large-scale spatial structures, their number of repetitions over a study field being small, they are considered deterministic in general (e.g., in CRAD), and treated as spatial heterogeneity of the mean. Accordingly, the classical *t*-test with $n - 2$ degrees of freedom based on Spearman's rank-based statistic (Sokal and Rohlf, 2003) can be used to analyze the correlation between estimated drifts.

Key note: The statement "The larger the number of structures, the better" is not true in a coregionalization analysis. Actually, it is the contrary, and some parsimony is required when defining the linear model of coregionalization, because of the uncertainty associated with the estimation of sills and functions of sills, such as structural coefficients of correlation and determination (Larocque et al., *2007). This parsimony must be seen as an advantage as usual, instead of a constraint;*

by comparison, any spatially autocorrelated structure covers the whole spectrum of frequencies in the spectral approach (Larocque, 2008, Chapter 2). In applications, it is very important to include in the same model and analysis variables with similar scales of variability, non-spatial and spatial, as shown by their direct experimental variograms and the models fitted to them. It means that variables with short and long ranges of spatial autocorrelation should not be included together in a coregionalization analysis, as this would contradict the definition of the linear model of coregionalization. Variables with and without a nugget effect can be analyzed jointly if the appropriate adjustment is made for the estimation of sill matrices, as is the case in CRAD (Pelletier et al., *2009b).*

Four examples with real data, in plant ecology, entomology, temperature monitoring and limnology, are presented below to illustrate the various results that may be obtained in Phase II of CRAD, depending on the study conditions and objectives, and to show how to discuss them. Sample sizes in the last two examples are not large enough for a coregionalization analysis of residuals, so that correlation analyses are then limited to estimated drifts and residuals, without evaluating structural correlations separately for the nugget effect and the spatially autocorrelated structure from estimated sill matrices.

Continued plant diversity study example

This example is a follow-up of the one in Subsection 7.2.1, where the complete dataset was described. Eleven variables are considered here: two plant diversity indices (i.e., Trees and Herbaceous), as the variables to explain, and six soil properties (i.e., pH, extractable Mg, K, and P, soil organic matter, total N) and three topographical characteristics (i.e., E_M, fractal dimension, EH_V), as the explanatory variables. Focus will be on coefficients of determination (i.e., squared multiple correlations between the variables to explain and the explanatory variables), which will be computed on raw data and estimated drifts and for the nugget-effect and spherical structures. In fact, the conditions are satisfied for the analysis of structural coefficients of determination under a linear model of coregionalization for the EGLS residuals: the sample size is large ($n = 339$), and the ranges of spatial autocorrelation are short relative to the extent of the study area (Larocque *et al.*, 2007).

Ten variables were included in the analysis for the nugget effect, because the sill estimate at this structure for the direct variogram of E_M did not meet the threshold of 5%. For a similar reason, a different topographical variable (i.e., fractal dimension) was not retained in the analysis for the spherical structure; the common range of spatial

Table 7.3. *Coefficients of determination in Phase II of CRAD in the second example with plant diversity studied in relation to soil properties and topography at the Molson Ecological Reserve of McGill University in Île-Perrot and Terrasse-Vaudreuil (Québec, Canada)*

Type of data or structure	Soil and topography together	Soil alone	Topography alone	Soil minus topography	Topography minus soil
Raw data	0.1771	0.1181	0.0875	0.0896	0.0590
Nugget effect	0.5890	0.1322	0.0858	0.5032	0.4568
Spherical component	0.3799	0.3246	0.0546	0.3253	0.0553
Estimated drift	0.8308	0.6973	0.5511	0.2797	0.1335

"Minus" means that the coefficient of determination is evaluated for the variable above left, after partialling out the effect of the variable right below.

autocorrelation estimated for the other variables was 82 m. The use of this range in model fitting resulted in the sill estimate at the nugget-effect structure for the direct variogram of EH_V meeting the threshold of 5% in Phase II of CRAD; this result illustrates the kind of change that may happen from Phase I to Phase II in applications. All 11 variables were used in the analyses performed on raw data and estimated drifts.

The calculation of determination coefficients from estimated variance–covariance matrices and sill matrix estimates considered as such is detailed in Section A7.2. In simple terms, each coefficient of determination here is an "average" of two squared multiple correlations (i.e., there are two variables to explain), computed between a plant diversity index and 8 or 9 explanatory variables; such averaging is usually performed in redundancy analysis (van den Wollenberg, 1977). Concerning the interpretation of observed values, the higher the coefficient (i.e., the closer to 1), the better the "average" explanation is.

In Table 7.3, the determination coefficient values obtained with raw data are low to very low when the soil and topographical variables are used to explain plant diversity together and separately (first three columns) as well as when the explanation of the other set of variables is partialled out (last two columns). Before partialling out, the coefficient values are highest on estimated drifts and intermediate for the nugget-effect and spherical structures, and soil alone explains plant diversity better than topography alone, especially at the nugget-effect and spherical structures. Partialling has little effect on results for raw data and the spherical structure. It has important effects characterized by higher coefficient values for the nugget-effect structure, due to negative correlations between

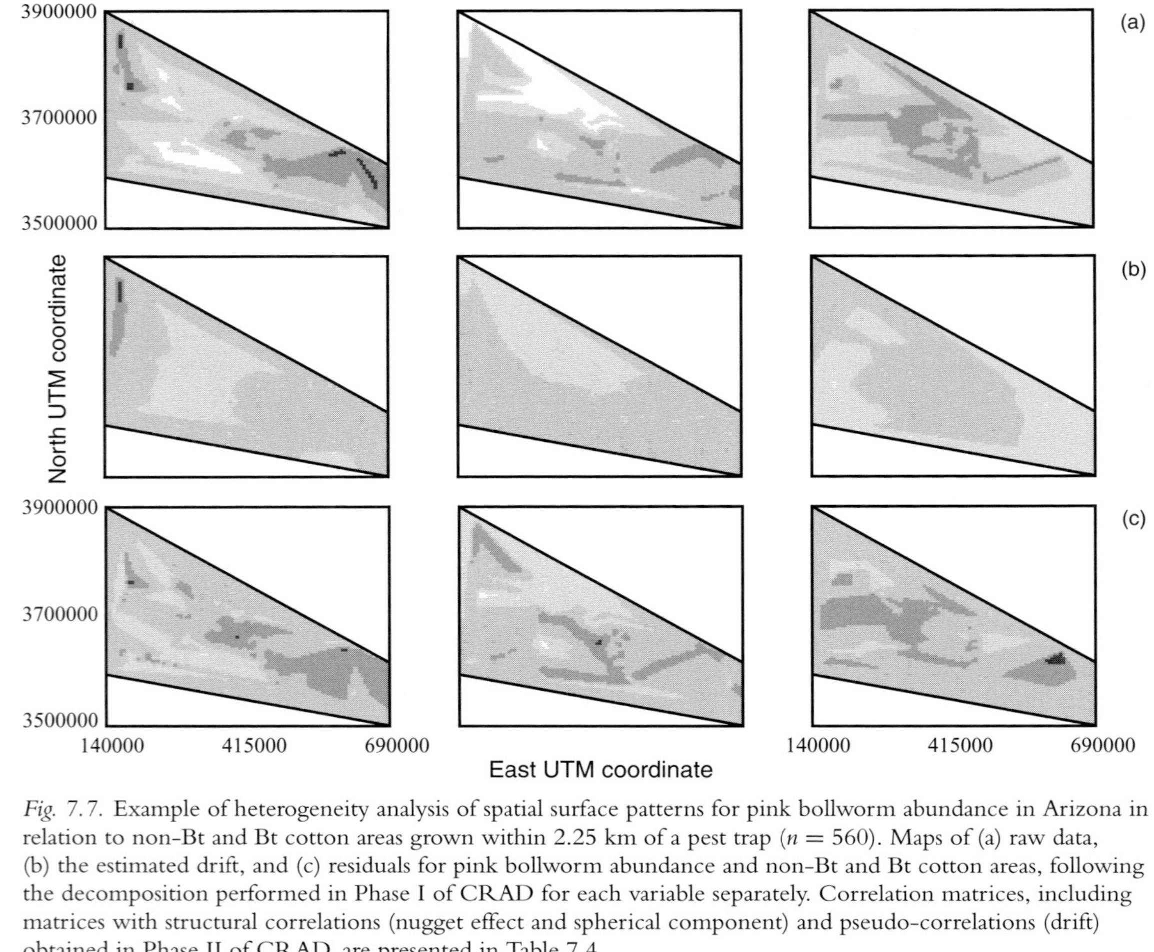

Fig. 7.7. Example of heterogeneity analysis of spatial surface patterns for pink bollworm abundance in Arizona in relation to non-Bt and Bt cotton areas grown within 2.25 km of a pest trap ($n = 560$). Maps of (a) raw data, (b) the estimated drift, and (c) residuals for pink bollworm abundance and non-Bt and Bt cotton areas, following the decomposition performed in Phase I of CRAD for each variable separately. Correlation matrices, including matrices with structural correlations (nugget effect and spherical component) and pseudo-correlations (drift) obtained in Phase II of CRAD, are presented in Table 7.4.

soil and topographical variables at this structure (numerical results not reported). It also has important effects for estimated drifts, characterized by lower coefficients due to a positive association between soil and topography at large scale (numerical results not reported). Such partialling-out effects are classical in multiple regression analysis (Whittaker, 1984).

In conclusion, the multi-scale analysis approach followed in Phase II of CRAD allows the detection of relationships that, otherwise, can run undetected in a classical redundancy analysis performed on raw data; see the low to very low determination coefficient values for raw data vs. the scale-specific coefficient values obtained in this example. Spatial heterogeneity of the mean in particular can be associated with strong relationships between variables at large scale, like plant diversity and the soil and topography variables here.

Example of pink bollworm and transgenic cotton in Arizona

Progeny of susceptible insects do not survive on transgenic crops that produce *Bacillus thuringiensis* (Bt) toxins. To delay insect resistance to Bt crops effectively, the spatial configuration of refuges (i.e., non-Bt fields) must allow extensive mating between resistant adults from transgenic fields and susceptible adults from refuges. In this context, Carrière *et al.* (2004) studied the zone of influence of refuges, with pink bollworm as pest and cotton as plant species.

The present example, in which a part of the data of Carrière *et al.* (2004) is revisited, differs from the original study by its objectives. Here, the data analysis is driven for a good part by spatial heterogeneity (i.e., mean and autocorrelation) and the assessment of correlations between pink bollworm abundance, as measured from insect capture data collected with pheromone traps, and the surrounding areas of non-Bt and Bt cotton at three scales of variability: non-spatial, small-scale spatial (spatially autocorrelated), and large-scale spatial (associated with spatial heterogeneity of the mean), compared with raw data. Accordingly, no ring analysis with rings of 0.75-km diameter as in Carrière *et al.* (2004) is performed. Instead, the areas grown with non-Bt and Bt cotton within 2.25 km of each pheromone trap are submitted to CRAD, together with pink bollworm abundance. The 2000 data ($n = 560$) are analyzed. The areas of cotton fields were measured using GPS and GIS technologies (Carrière *et al.*, 2004).

The "raw data" mapped in Fig. 7.7(a) are standardized to a mean of 0 and a variance of 1, after Box–Cox transformation. In this panel and the other two in Fig. 7.7, gray contours are comparable (i.e., from very light gray, −2, to very dark gray, +2). Maps are non-rectangular because of the shape of the sampling domain – the sampling grid was irregular, focusing on regions of Arizona where cotton was grown; regions outside the (convex) interior frame were not sampled at all.

From Phase I of CRAD, pink bollworm abundance had the strongest nugget effect (estimated sill: 0.50); Bt cotton, the strongest spatially autocorrelated component (estimated sill: 0.67); and non-Bt cotton, the strongest drift (pseudo-variance: 0.43). The estimated drifts [Fig. 7.7(b)] obtained in the same phase of data analysis are very smooth, especially for non-Bt and Bt cotton, so that the details of the spatial distribution of raw data are recovered in the maps of residuals [Fig. 7.7(a)–(c)]. Though smooth, the mapped drifts suggest a positive correlation between pink bollworm abundance and non-Bt cotton and negative correlations with Bt cotton at large scale, because of the respective spatial heterogeneity of the mean of the three variables: pink bollworm abundance tends to be higher (lower) when there is more (less) non-Bt cotton and less (more) Bt cotton in the same part (i.e., west, middle, east) of Arizona. Such correlations were hypothesized and observed on raw data to a certain degree by Carrière *et al.* (2003, 2004). They are difficult to visualize from the maps of raw data and residuals here, and the multi-scale analysis approach in Phase II of CRAD will help confirm, or not, the hypothesized correlations at other scales or structures.

From Phase II of CRAD, the common range of spatial autocorrelation estimated by using a spherical structure plus a nugget effect for the three variables is 85.15 km. Structural correlations estimated under the corresponding linear model of coregionalization are reported in Table 7.4, with correlations evaluated on raw data and (pseudo-)correlations between estimated drifts. Correlations on raw data are moderately strong, but with the expected sign [Table 7.4(a)]. At the nugget effect [Table 7.4(b)], structural correlations match the expectations even better, as two of them are declared statistically significant ($P < 0.05$) by a valid test and the third one is very close. Leaving the spherical structure last, the negative correlations between estimated drifts [Table 7.4(d)] are the strongest, which confirms the preliminary observations made in view of Fig. 7.7(b). Last but not least, correlations between pink bollworm abundance and cotton areas of the two types at the spherical structure [Table 7.4(c)] are positive! The explanation of the entomologist (Yves Carrière, personal communication) is that at the corresponding spatial scale, pink bollworm adults, which are flying and cannot distinguish between non-Bt and Bt cotton but "just" need cotton to feed and mate, tend to be

Table 7.4. *Correlation matrices in the example of pink bollworm abundance in Arizona analyzed in relation to the areas of non-Bt and Bt cotton grown within 2.25 km of a pest trap (n = 560): (a) correlations evaluated on raw data; structural correlations resulting from the EGLS fitting of a linear model of coregionalization with a nugget effect (b) and a spherical component (c) to the experimental variograms (direct and cross) in Phase II of CRAD; and (d) pseudo-correlations between estimated drifts*

	(a)			(b)		
	Pink bollworm	Non-Bt cotton	Bt cotton	Pink bollworm	Non-Bt cotton	Bt cotton
Pink bollworm	1	**0.3897** (<0.0001)	−0.0618 (0.1443)	0.4990	**0.2012** (0.0354)	**−0.1746** (0.0329)
Non-Bt cotton		1	**−0.3449** (<0.0001)		0.2945	−0.2041 (0.0560)
Bt cotton			1			0.2185
	(c)			(d)		
Pink bollworm	0.3212	0.3667 (0.2291)	0.4951 (0.1072)	0.2006	**0.4620** (<0.0001)	**−0.1895** (<0.0001)
Non-Bt cotton		0.6674	0.0100 (0.9739)		0.2038	**−0.7399** (<0.0001)
Bt cotton			0.2435			0.4293

Diagonal entries are sample variances standardized to 1 after Box–Cox transformation in (a), "sills" of variogram models fitted to direct experimental variograms in (b) and (c), and pseudo-variances in (d).
Probabilities of significance in (b) and (c) were obtained with modified t-tests per structure (Dutilleul and Pelletier, in press). The classical t-test with $n-2$ degrees of freedom was used in (a), with Pearson's statistic, and in (d), with Spearman's statistic. It must be noted that the presence of spatial heterogeneity of the mean has an effect on both the correlations and the probabilities of significance in (a); accordingly, these are reported by completeness or as a point of reference for correlations and probabilities of significance in (b)–(d).

more abundant in or near cotton fields, regardless of whether these are transgenic or not.

In conclusion, the structural correlations that are part of the output of Phase II of CRAD can be scale-dependent, by showing association between variables of different sign depending on the structure. Correlations evaluated on raw data appear to be "average" correlations, the averaging this time being over structural correlations of the nugget effect

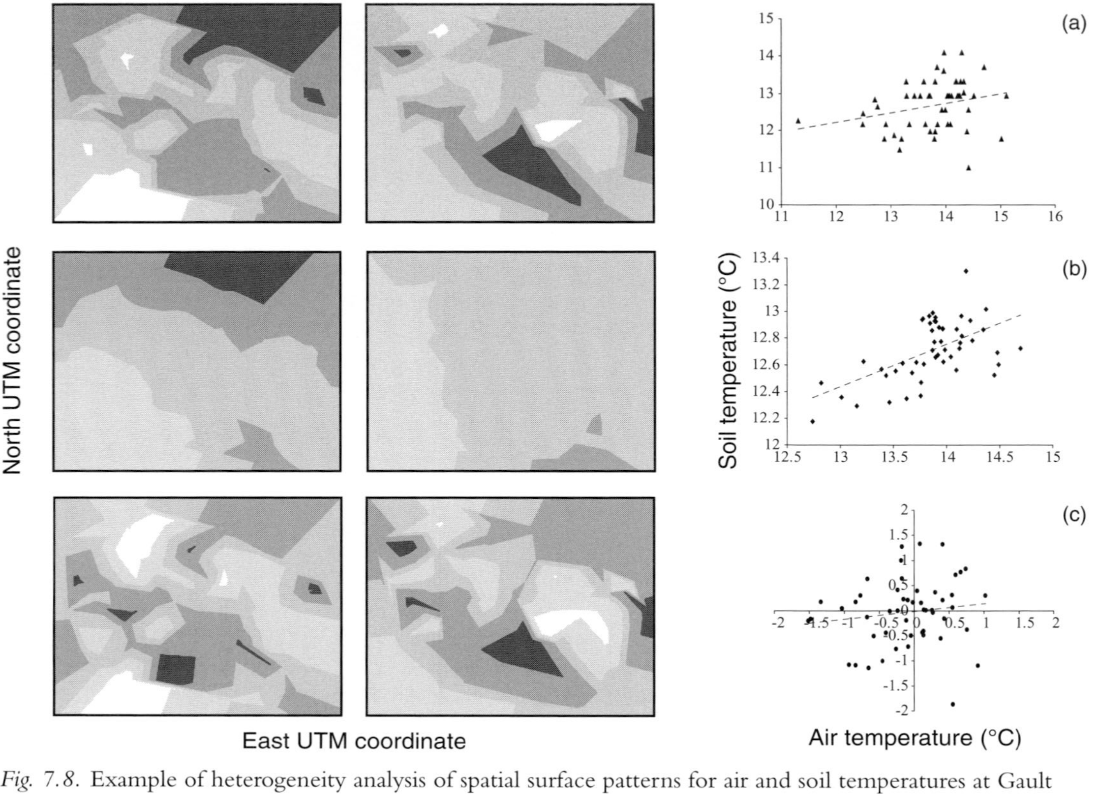

Fig. 7.8. Example of heterogeneity analysis of spatial surface patterns for air and soil temperatures at Gault Nature Reserve in Mont-Saint-Hilaire (Québec, Canada) on June 21, 2004 at midnight. Maps of (a) raw data, (b) estimated drifts, and (c) residuals (air temperature left, soil temperature middle), and the corresponding scatter plots (right). The dimensions of the sampling domain are 3.5 km (west–east) × 2.6 km (south–north). The dashed line drawn in each scatter plot is the simple linear regression line fitted by OLS; the sign, positive or negative, of the slope of this line is indicative of the sign of the correlation between variables.

and the spatially autocorrelated component and the pseudo-correlations between drifts.

MSH air–soil temperature example

This is a follow-up of the MSH air–soil temperature examples in Chapter 6. The basic difference is that instead of analyzing two time series of air and soil temperatures recorded at the same location (Chapter 6), air and soil temperature data collected at a given time at $n = 53$ sampling locations are submitted to a heterogeneity analysis of spatial surface patterns here. Two times were chosen: June 21, 2004 midnight and June 21, 2004 noon, simply to show that correlations between variables inferred from spatial data can change from time to time – they may be time-dependent. The heterogeneity analysis of spatio-temporal surface patterns will be the topic of Chapter 8. The objectives here are: (i) to identify the presence of spatial heterogeneity of the mean in each variable at a given time, through the decomposition $U_j(x, y) = \mu_j(x, y) + \varepsilon_j(x, y)$ $(j = 1, 2)$, performed in Phase I of CRAD; (ii) to evaluate, from the estimated drifts and the potentially spatially autocorrelated residuals, the correlations between variables at large scale vs. at smaller scales (non-spatial and spatial) in a simplified version of Phase II of CRAD because of the moderate sample size; and (iii) to compare these correlations with those obtained on raw data and assess whether they are time-dependent.

From the maps in Figs. 7.8(a) and 7.9(a), it appears that the warmest (coldest) zones in the reserve at midnight are not necessarily the warmest (coldest) zones at noon, especially in the air. The smooth drift estimated for soil temperature at midnight [Fig. 7.8(b)] provides residuals with a spatial surface pattern similar to raw data [Fig. 7.8(c)]. The maps in Fig. 7.9(b) suggests a correlation other than positive between air and soil temperatures at large scale at noon (confirmed by the scatter plot), as north–south seems to be the main axis of variation for the estimated drift for air temperature whereas the estimated drift for soil temperature is not linear and more in diagonal.

Quantitatively and strictly speaking, only one correlation between air and soil temperatures is statistically significant in Figs. 7.8 and 7.9. It is that between estimated drifts at midnight (Spearman's statistic: 0.4741, $P = 0.0003$) [Fig. 7.8(b)], which indicates a relationship at large spatial scale between the temperatures in the absence of solar radiation. Other correlations are either moderately positive (on raw data at midnight and noon, and on residuals at noon), almost zero (on residuals at midnight), or moderately negative (between estimated drifts at noon).

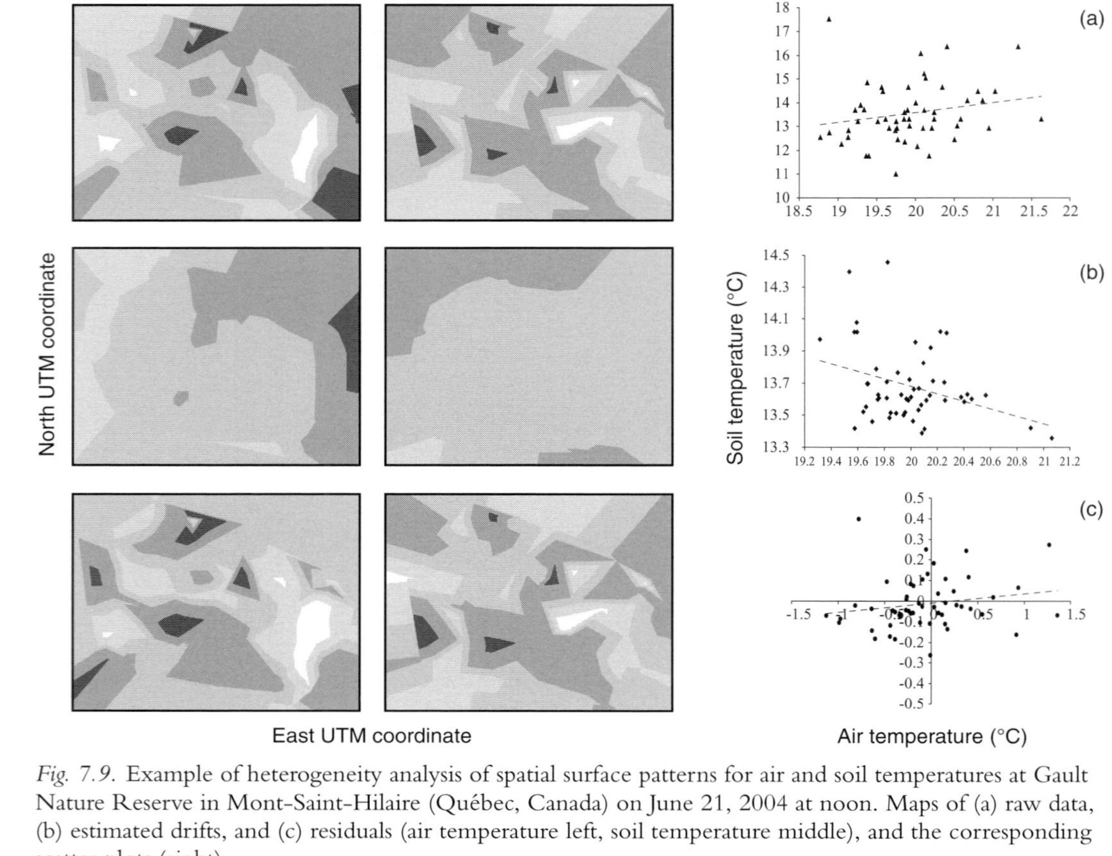

Fig. 7.9. Example of heterogeneity analysis of spatial surface patterns for air and soil temperatures at Gault Nature Reserve in Mont-Saint-Hilaire (Québec, Canada) on June 21, 2004 at noon. Maps of (a) raw data, (b) estimated drifts, and (c) residuals (air temperature left, soil temperature middle), and the corresponding scatter plots (right).

In conclusion, correlations inferred from spatial data can be time-dependent, in addition to being scale-dependent in the spatial framework. With small datasets, the fitting of a linear model of coregionalization to all the experimental variograms (direct and cross) of residuals in Phase II of CRAD is not recommended (Larocque *et al.*, 2007), but it is possible to obtain interesting information about relationships between variables after separating spatial heterogeneity of the mean from heterogeneity due to spatial autocorrelation in Phase I of CRAD. This was the case here with the relationship at large spatial scale between air and soil temperatures at midnight.

Example of macrozooplankton and abiotic and biotic factors in Lac Léman
In Pinel-Alloul *et al.* (1999), the spatial distribution of macrozooplankton in Lac Léman was studied in the spring, in relation to abiotic factors (e.g., water temperature) and biotic factors (e.g., chlorophyll *a* biomass). Mapping and spatial correlogram analysis based on Geary's *c* statistic revealed that crustacean abundances and abiotic factors were structured along a west–east gradient in the great lake basin; see Fig. 7.10, where the sampling grid ($n = 28$) is plotted. Chlorophyll *a* biomass and other biotic factors were found to have more patchy or inshore–offshore spatial distributions. Using canonical (correlation) analyses, in which space was modeled globally with a second-degree polynomial in the spatial coordinates (Subsection 7.2.1), the authors determined the relative contribution of spatial and environmental factors to the distribution of macrozooplankton species, and found that a large portion of the macrozooplankton total variation remained unexplained.

The data analysis is revisited in this example, based on the drift estimation procedure L_1 in Phase I of CRAD. Such a local procedure is more flexible than any global one, and was shown to better separate a "true" patchy drift (which contains spatial heterogeneity of the mean) from spatially autocorrelated residuals (Pelletier *et al.*, 2009a). Thus, macrozooplankton abundances will be correlated with abiotic and biotic factors, using the estimated drifts and the residuals provided by L_1, and these correlations will be compared to those obtained with raw data.

All but one correlation in Table 7.5(a)–(c) are positive. Several of them are statistically significant ($P < 0.05$) despite the small sample size, and several are not scale-dependent. Focusing on estimated drifts [Table 7.5(b)], chlorophyll *a* biomass has statistically significant positive correlations with the abundances of two macrozooplankton species,

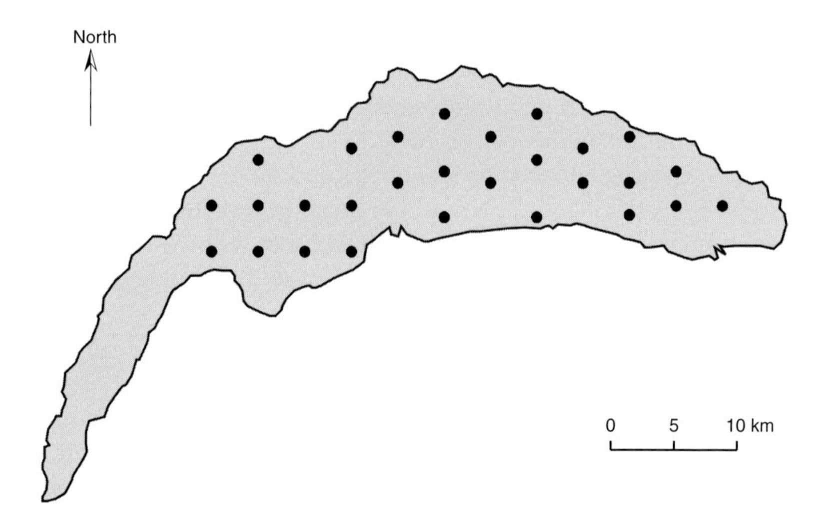

Fig. 7.10. Sampling grid in the Lac Léman example. The lake is located in Switzerland (60%) and France (40%). It is 72 km long and 13–14 km wide, and the distance between neighboring sampling locations ranges from 3 to 5 km, resulting in an almost regular grid of 28 points covering the large basin. The small basin (left tail) is known to have distinct thermal and nutrient regimes and different plankton distributions and seasonal dynamics, so the data available for it are not part of the analysis here.

suggesting that the spatial heterogeneities of the mean of the three variables, as captured by the local drift estimation procedure L_1, are similar. By comparison, the two correlations are weaker on raw data. Overall, the strongest correlations are observed between estimated drifts, but several correlations on residuals are statistically significant as well.

In conclusion, an improved estimation of the drifts supports an improved correlation analysis at large spatial scale. While spatial heterogeneity of the mean induces a bias in the spatial correlograms and variograms of raw data, very interesting relationships may be found between estimated drifts. The finding of several correlations that are not scale-dependent is likely the result of a strong gradient in the field, such as the west–east gradient in Lac Léman. As an introduction to the sampling questions discussed in Chapter 9, the use of a relatively regular sampling grid in this study (Fig. 7.10) allowed a spatial analysis with interpretable results, despite the small sample size.

Table 7.5. *Correlation matrices in the Lac Léman example, where the distribution of macrozooplankton is studied at different scales in relation to abiotic and biotic factors ($n = 28$): correlations evaluated on (a) raw data (Pearson's statistic) and on (b) drift estimates (Spearman's statistic) and (c) residuals (Pearson's statistic), following Phase I of CRAD; below each correlation is reported (in parentheses) by the corresponding probability of significance, and correlations in bold are statistically significant ($P < 0.05$); diagonal entries of 1 are replaced by sample variances*

	Water temperature	Ammonium	Chlorophyll *a*	Crustacean zooplankton 1	Crustacean zooplankton 2	Crustacean zooplankton 3	Nauplii
(a)							
Water temperature	$1.39\ 10^{-7}$	**0.5518**	**0.7691**	**0.5997**	**0.5136**	0.2667	**0.5558**
		(0.0023)	(<0.0001)	(0.0007)	(0.0052)	(0.1701)	(0.0021)
Ammonium		0.6817	0.1966	**0.5784**	**0.4861**	0.2867	**0.6351**
			(0.3160)	(0.0013)	(0.0087)	(0.1391)	(0.0003)
Chlorophyll *a*			0.1005	0.2980	0.2131	0.0885	0.2968
				(0.1235)	(0.2762)	(0.6542)	(0.1250)
Crustacean zooplankton 1				0.0183	**0.6261**	**0.7065**	**0.8099**
					(0.0004)	(<0.0001)	(<0.0001)
Crustacean zooplankton 2					10.68	**0.5770**	**0.7560**
						(0.0013)	(<0.0001)
Crustacean zooplankton 3						195321	**0.6814**
							(<0.0001)
Nauplii							6548
(b)							
Water temperature	0.5412	**0.8035**	**0.4877**	**0.6563**	**0.6519**	**0.0389**	**0.4160**
		(<0.0001)	(0.0085)	(0.0001)	(0.0002)	(0.8443)	(0.0277)
Ammonium		0.9208	0.2359	**0.7274**	**0.8785**	**0.2146**	**0.6606**
			(0.2268)	(<0.0001)	(<0.0001)	(0.2729)	(0.0001)
Chlorophyll *a*			0.5491	**0.5046**	0.0005	0.0525	**0.4132**
			(0.0062)	(0.9978)	(0.7906)	(0.0288)	(0.0062)

(cont.)

Table 7.5. *(cont.)*

	Water temperature	Ammonium	Chlorophyll *a*	Crustacean zooplankton 1	Crustacean zooplankton 2	Crustacean zooplankton 3	Nauplii
Crustacean zooplankton 1				0.7338	**0.5813** (0.0012)	**0.5884** (0.0010)	**0.8664** (<0.0001)
Crustacean zooplankton 2					0.3090	0.1850 (0.3459)	**0.5331** (0.0035)
Crustacean zooplankton 3						0.5665	**0.5610** (0.0019)
Nauplii							0.6437
(c)							
Water temperature	0.4013	0.0386 (0.8454)	**0.6223** (0.0004)	0.2778 (0.1523)	**0.4578** (0.0184)	0.2616 (0.1787)	0.2582 (0.1846)
Ammonium		0.1646	−0.1726 (0.3797)	**0.4239** (0.0204)	0.2296 (0.2398)	0.3429 (0.0740)	**0.5846** (0.0010)
Chlorophyll *a*			0.2830	0.0306 (0.8772)	0.3767 (0.0910)	0.1703 (0.3863)	0.0494 (0.8029)
Crustacean zooplankton 1				0.3411	**0.4144** (0.0243)	**0.7007** (<0.0001)	**0.8005** (<0.0001)
Crustacean zooplankton 2					0.7019	**0.4690** (0.0107)	**0.5025** (0.0067)
Crustacean zooplankton 3						0.3594	**0.6868** (<0.0001)
Nauplii							0.4288

In (c), probabilities of significance were adjusted for the presence of spatial autocorrelation, using Dutilleul's (1993b) modified t-test. The classical t-test with $n - 2$ degrees of freedom was used in (a) and (b). It must be noted that the presence of spatial heterogeneity of the mean has an effect on both the correlations and the probabilities of significance in (a); accordingly, these are reported by completeness or as a point of reference for correlations and probabilities of significance in (b) and (c).

Summary: As in heterogeneity analysis of time series (Chapter 6), the analyses of spatial heterogeneity of the mean and heterogeneity due to spatial autocorrelation must be performed jointly to be effective; one type of heterogeneity needs to be studied by taking the other into account. More specifically, drifts should be estimated by EGLS, for example, instead of OLS (because the latter assumes the absence of spatial autocorrelation, in addition to spatial homogeneity of the variance), and autocorrelation or semivariance statistics should be evaluated after estimated drifts are removed from the data (e.g., on EGLS residuals). Thus, correlations can be inferred from the raw data, EGLS drift estimates and EGLS residuals. They can also be evaluated indirectly from "sills," which are variance and covariance parameters in direct and cross-variogram models; the result is structural correlations. Beyond the mere computation of experimental variograms, the model-based approach that includes the fitting of a model – univariate or multivariate – is recommended, but requires larger datasets (because of the uncertainty inherent to the estimation of the range of spatial autocorrelation and the sill matrices). The second dimension of space in many applications makes spatial heterogeneity of the mean potentially more complex, which stresses the importance of using a flexible local procedure of drift estimation within a moving window, like L_1 *in CRAD, to capture it. In the examples, the strongest correlations were observed between estimated drifts, but these are, in fact, pseudo-correlations (because they originate from spatial heterogeneity of the mean). Eventually, the analyses of spatial heterogeneity of the mean and heterogeneity due to spatial autocorrelation in a parsimonious approach to multi-scale analysis like that followed in CRAD can lead to the detection of interesting relationships hidden in the raw data.*

7.3 Analysis of spatial heterogeneity of the variance

In the spatial framework, heterogeneity of the variance, or heteroscedasticity, is characterized by different levels of variation of the random variable U in different regions or zones. For example, it is observed when variation is low (i.e., fluctuations around mean values are of small amplitude) in some parts of the field or transect, and high (i.e., fluctuations are wide, making mean values difficult to discern) in other parts. As we will see, the two other types of spatial heterogeneity can "interact" with spatial heterogeneity of the variance, by existing at one level of variation but not at the other.

The example at the core of this section and around which the whole section is articulated is in 1-D space, which simplifies the writing of equations and the presentation and interpretation of results. Of course,

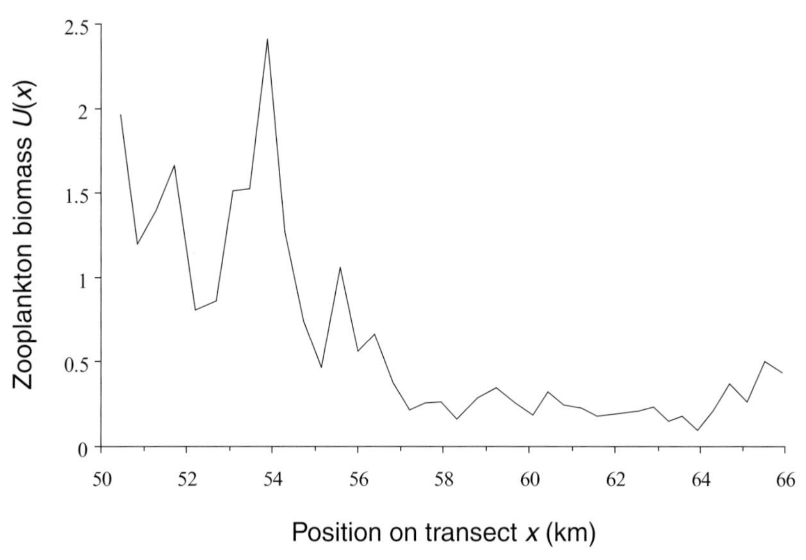

Fig. 7.11. Zooplankton biomass sample data collected along a ~16-km transect in Lake Erie ($n = 38$) depict spatial heterogeneity of the variance between the first portion (50.5–57 km) and the second portion (57–66 km) of the transect.

all the models and analytical procedures extend to 2-D space, where portions of a transect with different levels of variation are replaced by strata in a field with similar characteristics. Outside of spatial repeated measures (Chapter 9), replicated spatial data series are not frequent, and without replication, as in the Lake Erie example below, it is not possible to model spatial heterogeneity of the variance with one parameter per sampling location. From an experimental perspective, the situation is thus different from the second forestry example in Section 6.4.

Assume that zooplankton biomass data, denoted $u(x_1), \ldots, u(x_n)$, were collected at unequally spaced sampling sites along a transect in a lake, as in Lake Erie in Stockwell and Sprules (1995). Such data (in wet µg/L) are plotted in Fig. 7.11; they correspond to the dataset "Erie 5" in Dutilleul *et al.* (2000b). From about 50.5 km to 57 km on the transect, there appears to be spatial heterogeneity of the mean in zooplankton biomass and this spatial heterogeneity is accompanied by a high variation, relative to that associated with what seems to be spatial homogeneity of the mean from about 57 km to the end of the transect. Graphically, it is very difficult to say anything about the presence or strength of spatial autocorrelation, for either portion of the transect.

The apparent spatial heterogeneity of the mean in the first portion of the transect resembles that in Fig. 2.3(a), except that time is replaced by 1-D space, sampling locations are unequally spaced, and the true trend, or drift, is decreasing instead of increasing. Nevertheless, a similar model of simple linear regression can be applied to the first portion of the transect, as well as to the second one in order to assess the spatial homogeneity of the mean in that case:

$$U(x) = a + bx + \varepsilon(x), \tag{7.15}$$

where x denotes the position of a sampling location on the transect and $a + bx$ stands for $\mu(x)$. Clearly, from Fig. 7.11, different regressions apply to the two portions of the transect, and the model below reflects these differences:

$$U(x_{ij}) = a + \Delta a_i + (b + \Delta b_i)x_{ij} + \varepsilon(x_{ij}), \tag{7.16}$$

where x_{ij} ($j = 1, \ldots, n_i$; $i = 1, 2$) denote the positions of sampling locations on the two portions of the transect, and Δa_i and Δb_i represent the deviations of values of the intercept and slope specific to portion i from an average intercept a and an average slope b. As for spatial heterogeneity of the variance and spatial autocorrelation, these can be modeled through the variance–covariance matrices Σ_1 and Σ_2 of the random vectors $\boldsymbol{\varepsilon}_1 = \{\boldsymbol{\varepsilon}(x_{11}), \ldots, \varepsilon(x_{1n_1})\}$ and $\boldsymbol{\varepsilon}_2 = \{\varepsilon(x_{21}), \ldots, \varepsilon(x_{2n_2})\}$ ($n_1 = 16$ and $n_2 = 22$):

$$\text{Var}(\boldsymbol{\varepsilon}_1) = \Sigma_1 = \sigma_1^2 \Sigma_{\rho 1} \text{ and } \text{Var}(\boldsymbol{\varepsilon}_2) = \Sigma_2 = \sigma_2^2 \Sigma_{\rho 2}, \tag{7.17}$$

where $\Sigma_{\rho 1}$ and $\Sigma_{\rho 2}$ denote the two autocorrelation matrices, σ_1^2 is expected to be different from σ_2^2 because of the suspected heterogeneity of the variance between the two portions of the transect, and the estimates of $\Sigma_{\rho 1}$ and $\Sigma_{\rho 2}$ will be diagonal or not, depending on the absence or presence of spatial autocorrelation that will be found in the corresponding subset of data.

Two models can also be fitted separately, one for each portion of the transect:

$$U(x_{ij}) = a_i + b_i x_{ij} + \varepsilon(x_{ij}) \quad (i = 1 \text{ or } 2), \tag{7.18}$$

where a_i and b_i denote the intercept and slope of the regression line for portion i of the transect, the other notations and parameters being the same as in equation (7.16). Separate model fitting was retained here for the analysis of the data of Fig. 7.11.

Several variance–covariance structures (e.g., exponential, spherical, Gaussian – Section 7.1; compound symmetry – Section 6.4) were tried for estimation of the parameters (including the intercept and slope of the regression line as fixed effects) by restricted maximum likelihood in SAS PROC MIXED. Based on Akaike's and Schwarz's information criteria (Section 6.4), the Gaussian and exponential variance–covariance structures are the most appropriate for portion 1 and portion 2 of the transect, respectively; in other words, the values of information criteria were the smallest for these structures. Besides the difference in variance–covariance structure, which indicates that spatial autocorrelation decreases more rapidly with increasing distance along portion 2 than along portion 1 of the transect, there are two other noticeable differences. One: The estimates of σ_1^2 and σ_2^2 are 0.2278 and 0.0113, indicating a much higher variation (i.e., about 20-fold) of zooplankton biomass in the first portion. Two: The estimate of slope b_1 (value: –0.1660; standard error: 0.0607) is significantly different from zero ($P = 0.0162$), whereas the estimate of slope b_2 (value: 0.0141; standard error: 0.0139) is not ($P = 0.3214$). Accordingly, spatial heterogeneity of the variance has been quantified and spatial heterogeneity of the mean has been analyzed adequately, by taking spatial autocorrelation into account when necessary. Thereafter, the spatial heterogeneity or homogeneity in the distribution of zooplankton biomass can be related to other factors, such as depth and water temperature in the two portions of this transect in Lake Erie.

Summary: Even in the absence of replication, mixed models can be used to analyze heteroscedasticity in space, together with heterogeneity of the mean and heterogeneity due to autocorrelation. The three types of spatial heterogeneity can "interact" with each other. The models and analytical procedures presented and used here in 1-D space extend to 2-D space.

7.4 The Mantel test: use and misuse

In classical multivariate analysis (i.e., without spatial referencing), $n \times n$ matrices of distances between individuals are calculated from the data available for p variables, to measure dispersion in the cloud of n data points; when the variables are expressed in different units, the data are usually standardized to a mean of 0 and a variance of 1 prior to the calculation of distances. Such non-spatial distances must be distinguished from the spatial distances used in Moran's I and Geary's c correlograms

(Subsection 7.1.1) and in experimental variograms (Subsection 7.1.2). In the spatial framework, we will see that non-spatial distances calculated from the data can be used – with caution – in the heterogeneity analysis of two spatial surface patterns.

Let $\mathbf{D}_U = (d_{U,ii'})$ and $\mathbf{D}_{U'} = (d_{U',ii'})$ denote the $n \times n$ matrices of distances calculated between the same n sampling locations, using the data available for random variables U and U' separately. Then, the normalized Mantel statistic is defined as the product-moment coefficient of linear correlation between the two distance matrices:

$$\text{Normalized Mantel statistic} = \frac{\sum\sum (d_{U,ii'} - \bar{d}_U)(d_{U',ii'} - \bar{d}_{U'})}{\sqrt{\sum\sum (d_{U,ii'} - \bar{d}_U)^2 \sum\sum (d_{U',ii'} - \bar{d}_{U'})^2}}, \tag{7.19}$$

where double summations are over $i = 1, \ldots, n - 1$ and $i' = i + 1, \ldots, n$, and $\bar{d}_U$ and $\bar{d}_{U'}$ are, respectively, the mean of distances $(d_{U,ii'}; i = 1, \ldots, n - 1; i' = i + 1, \ldots, n)$ and the mean of distances $(d_{U',ii'}; i = 1, \ldots, n - 1; i' = i + 1, \ldots, n)$. *Note*: The original statistic due to Mantel (1967) was introduced in the spatio-temporal framework (see Chapter 8). In hypothesis testing, the absence of linear correlation between the distances $d_U(x, x')$ and $d_{U'}(x, x')$ is tested against a one-tailed alternative or the two-tailed alternative. The corresponding probability of significance is evaluated by permutations: the rows and columns of one distance matrix are renumbered randomly a large number of times and equation (7.19) is computed each time; eventually, the value of the Mantel statistic observed for the two initial distance matrices calculated from the data is located in the generated distribution of statistic values. In the continued Lake Erie example below, 10 000 permutations were used.

Before we see how some Mantel type of analysis can be part of the heterogeneity analysis of spatial surface patterns, the following warning message is important: the analysis of distances is not equivalent to the analysis of raw data! More specifically, the observed absence of correlation with raw data does not necessarily imply the absence of correlation between distances calculated from raw data, and applying to raw data relationships found in an analysis of distances may be, at the least, risky (Dutilleul *et al.*, 2000b). There may also be a change in sign, from correlations evaluated on raw data to correlations between distances calculated from raw data. To discuss this, consider two situations, one without change of sign and another with change of sign. First, let $U(x) = \mu(x) = 1 + 2x$ and $U'(x) = \mu'(x) = 3 + 4x$. In this case, there is a positive relationship between variables U and U' [i.e., not a positive *correlation*, because there is no random term in the definition of $U(x)$ and $U'(x)$] and there is a positive relationship between the derived (Euclidean) distances. To be convinced of the latter, let us calculate U-distances between $x = 1$ and $x' = 2$ and between $x = 1$ and $x'' = 10$ – they are 2 and 18, respectively – and U'-distances for the same two pairs of locations – these are 4 and 36. In fact, $d_{U'} = 2d_U$ in this situation (and in the next one), and the announced positive relationship between distances is thus demonstrated. Second, let $U(x) = \mu(x) = 1 + 2x$ and $U'(x) = \mu'(x) = 3 - 4x$. In that case, there is a negative relationship between raw data, but the positive relationship between the distances $d_U(x, x')$ and $d_{U'}(x, x')$ is unchanged! Because the replacement of $+4x$ by $-4x$ in the definition of $U'(x)$ does not affect the calculation of U'-distances. In applications, the observations of $U(x)$ and $U'(x)$ will generally be the sum of mean functions, $\mu(x)$ and $\mu'(x)$, and some random terms, $\varepsilon(x)$ and $\varepsilon'(x)$. Accordingly, there is potential for a new type of Mantel analysis arising from the separation of $\mu(x)$ from $\varepsilon(x)$ in $U(x)$ and of $\mu'(x)$ from $\varepsilon'(x)$ in $U'(x)$ in the context of spatial heterogeneity analysis. This idea is developed below, and Phase I of CRAD is finding another application in this development.

Assume Euclidean distances were calculated from the raw data and from the estimated drifts and residuals provided by Phase I of CRAD (Pelletier *et al.*, 2009a) for two variables of interest. The objective in doing this is to compare (1) the relationships found between distances calculated from estimated drifts and from residuals, with those between distances calculated from raw data, and (2) the correlations (or pseudo-correlations) between estimated drifts and between residuals, with those between the derived distances, following Dutilleul *et al.* (2000b). We already know from Subsection 7.2.2 that correlations between raw data can be different from correlations between estimated drifts and between residuals (i.e., the data). It must be noted that (i) the decomposition of space in Phase I of CRAD, which is at the basis of the new type of Mantel analysis illustrated hereafter, is different from the decomposition of space performed in the Mantel correlogram, where spatial distance lags are used (Oden and Sokal, 1986; Sokal, 1986); (ii) an example with 1-D spatial data is presented here, but 2-D spatial data or time series could have been used as well.

Example of fish and zooplankton biomass in Lake Erie
The source of the data is the same as in Section 7.3 (i.e., Stockwell and Sprules, 1995). In Dutilleul *et al.* (2000b), this dataset is called "Erie 7*" ($n = 61$, after the removal of two outliers). The raw data for fish and zooplankton biomass are plotted in Fig. 7.12(a), and their respective estimated drifts and residuals in Fig. 7.12(b) and (c).

The correlation between raw data is weak and negative (Pearson's $r = -0.0636$; $P = 0.6262$, with the classical two-tailed t-test), and corresponds to the "arrow effect" that may be observed in the scatter plot of raw data [Fig. 7.13(a), top panel]. The scatter plot of distances calculated from raw data offers a different picture [Fig. 7.13(a), bottom panel], with a stronger negative correlation coefficient (Mantel statistic = –0.1105) statistically significant at different levels depending on the test used ($P = 0.0096$, left-hand one-tailed; $P = 0.0590$, two-tailed). This result is explained by relatively small distances in fish biomass combined with relatively large distances in zooplankton biomass in the first third of the transect.

The following results for the estimated drifts and residuals are new. Results for estimated drifts are particularly interesting. Spearman's statistic evaluated between estimated drifts (i.e., the data) is negative and highly significant (value = –0.5333; $P < 0.0001$, two-tailed); Pearson's r provides similar results. The interpretation is that the estimated drift of zooplankton biomass is mainly decreasing after 10.5 km on the transect, while the increase in that of fish biomass is sustained from beginning to end [Fig. 7.12(b)]. As for distances calculated from estimated drifts, the value of Mantel statistic is highly significant, but positive (value = 0.2296; $P < 0.0001$, right-hand one-tailed and two-tailed). This change of sign could be expected; see the two situations discussed above. The explanation is that on the main part of the transect (after 10.5 km again), larger differences in fish biomass between two positions correspond to larger differences in zooplankton biomass [Fig. 7.12(b)]. Results for residuals are less interesting, and are reported for completeness: Pearson's $r = 0.0667$ ($P = 0.6097$, with the classical two-tailed t-test) and Mantel statistic = 0.0129 ($P = 0.3689$, right-hand one-tailed; $P = 0.8540$, two-tailed). (Any heteroscedasticity in the raw data should not affect the probability of significance of Spearman's statistic for estimated drifts, and when classical probabilities of significance for correlations between raw data and between residuals (i.e., the data) are far from thresholds, a modified t-test should not lead to different conclusions; see Dutilleul *et al.* (2000b) in the case of raw data.)

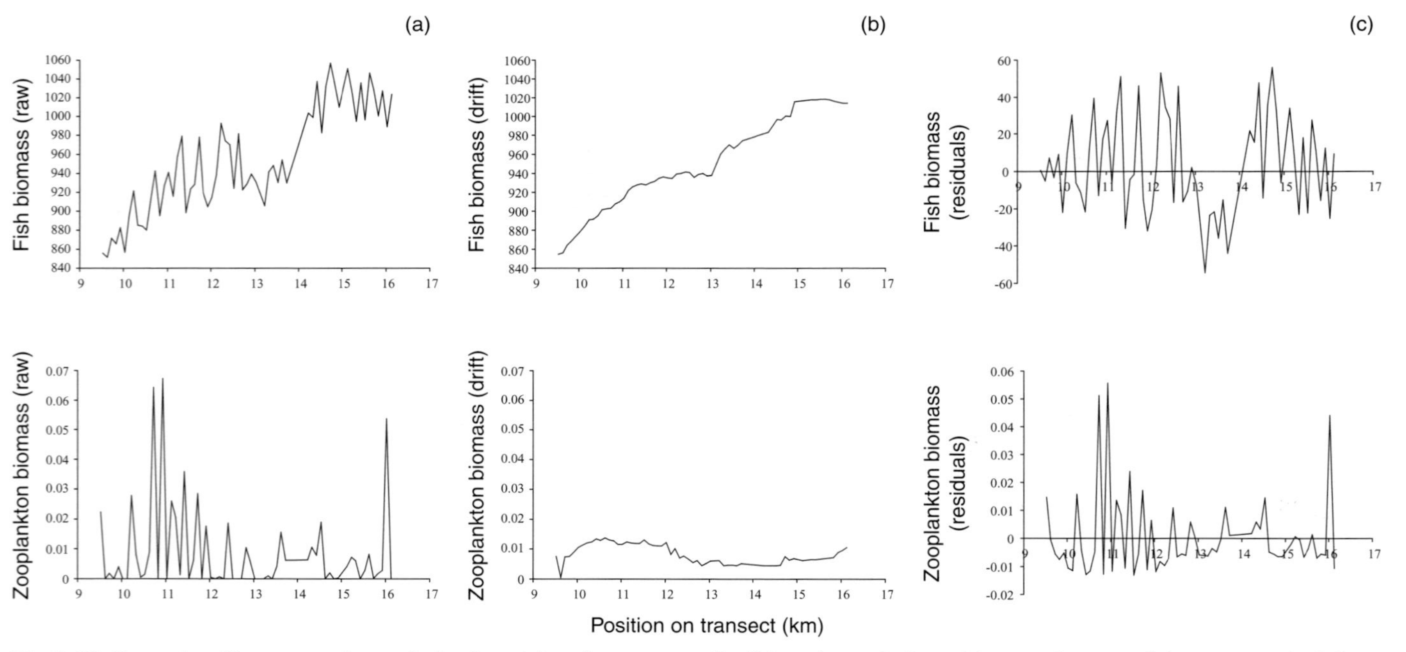

Fig. 7.12. Example of heterogeneity analysis of spatial surface patterns for fish and zooplankton biomass along a ~7-km transect in Lake Erie ($n = 61$). Plots of (a) raw data, (b) estimated drifts, and (c) residuals (Phase I of CRAD) against the position on transect for the two variables.

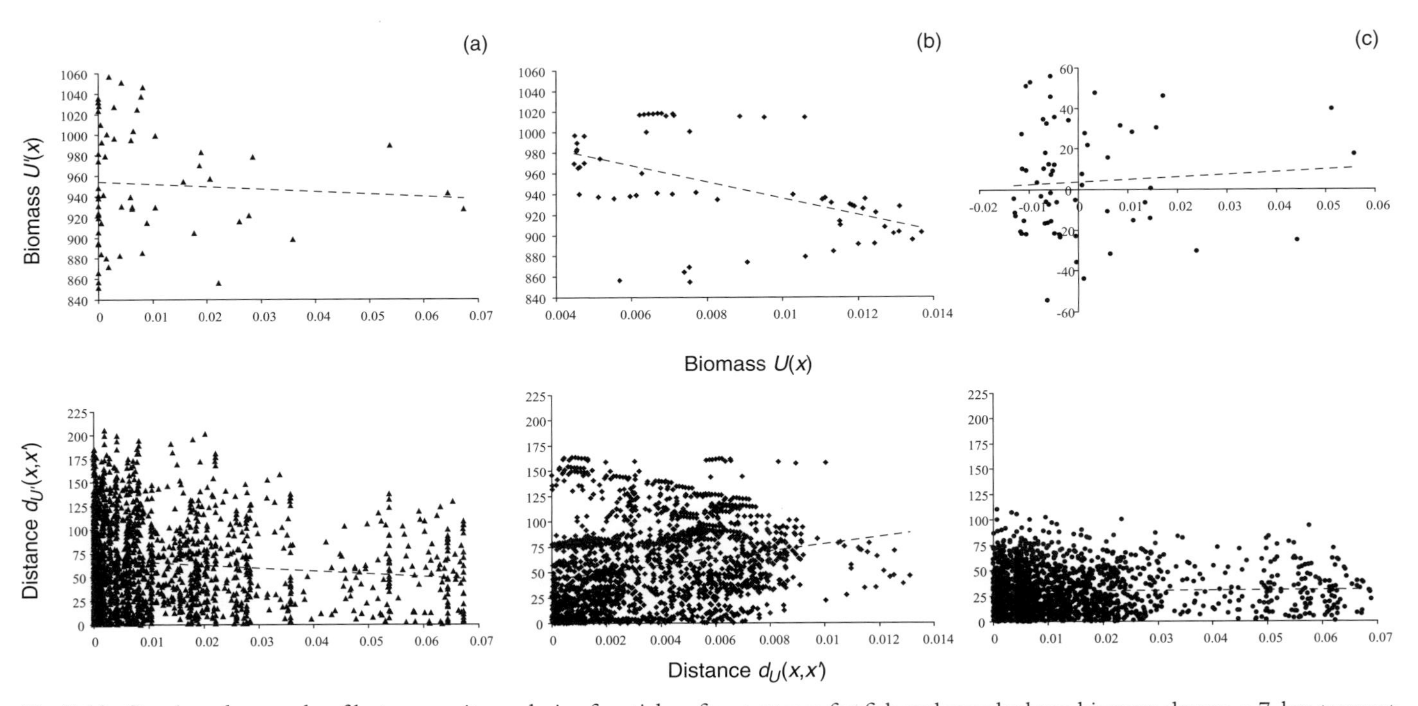

Fig. 7.13. Continued example of heterogeneity analysis of spatial surface patterns for fish and zooplankton biomass along a ~7-km transect in Lake Erie ($n = 61$). (a) Simple linear correlation analysis between raw data (top) and Mantel analysis between corresponding Euclidean distances (bottom); (b) and (c) same analyses for estimated drifts and residuals, respectively.

Summary: Extreme care should be taken when analyzing correlations between distances derived from the data, to investigate relationships between variables in a multivariate dataset. Almost anything is possible, and applying to raw data relationships found between distances may be problematic or is, at the least, hazardous. Proof of this was given here for two variables. Correlations between distances are correlations between measures of dispersion that are not sample variances based on the squared Euclidean distance between all the observations and the sample mean. A new type of Mantel analysis was presented that allows separate correlation analysis of distances associated with sources of spatial heterogeneity of the mean and analyses for distances calculated for random deviations from mean functions in spatial surface patterns.

7.5 Recommended readings

The interested reader will find in Ripley (1981, Chapter 5) and Schabenberger and Gotway (2005, Chapter 4) a number of sections on periodogram analysis and the estimation of the spectral density function from 2-D spatial data, two topics that were not covered here. Cliff and Ord (1981, Chapter 2) is a classical reference for spatial correlograms based on Moran's I and Geary's c statistics; Legendre and Legendre (1998, Chapter 13) provide an accessible presentation of the same material. Cressie (1993, Chapter 2) and Schabenberger and Gotway (2005, Chapter 4) are two strong references for variogram analysis; the latter draw an impressive list of theoretical autocovariance and semivariance functions (with the corresponding spectral density functions) for spatial stochastic processes. In *Estimating and Choosing* published in 1989 (i.e., the English version of *Estimer et Choisir*, 1978), Georges Matheron gives a fabulous vision on the fundamental aspects of geostatistics. The book of Isaaks and Srivastava (1989) represents an excellent introduction to applied geostatistics; Goovaerts (1997) goes deeper into the subject of geostatistics, especially on the aspects of kriging (Chapters 5 and 6); and Journel and Huijbregts (1978) may be the first geostatistics book that included a definition of the linear model of coregionalization. The article series of Pelletier *et al.* (2004, 2009a, 2009b) and Larocque *et al.* (2006, 2007) describe the path followed in the development of the method of coregionalization analysis with a drift (CRAD). Lastly, Dutilleul *et al.* (2000b) provides an interesting comparison between the classical correlation analysis with raw data and the Mantel analysis performed on derived distance matrices, while Dutilleul (1993b) and Dutilleul *et al.* (2008) present the details of modified t- and F-tests to be used for simple and multiple correlation analyses with spatial data.

Appendix A7: Simulation procedures and geostatistical complements

A7.1 Simulation of partial realizations of spatial stochastic processes

The simulation procedures presented for discrete-time processes in Section A6.1 cannot be used to generate partial realizations of spatial stochastic processes on a sampling transect or grid. As mentioned before, there is no order in space in general. In 1-D space, one cannot simulate "from left to right" or "from right to left" a 1-D spatial AR(1) process characterized by the equation $U(x) = 0.5\rho[U(x-1) + U(x+1)] + \varepsilon(x)$, because both $u(x-1)$ and $u(x+1)$ are required to obtain $u(x)$. Things worsen in 2-D space, because each sampling location (x_i, y_i) in the interior of a regular grid has four nearest neighbors.

Other simulation procedures are thus necessary. In the case of the 2-D spatial AR(1) process, the matrix notation provides an element of solution, with a drawback in the form of edge effects. It consists in writing the $n \times 1$ random vector of observations $\boldsymbol{u} = \{U(x_1, y_1), \ldots, U(x_n, y_n)\}'$ as a linear transformation of itself by using a matrix of weights $\mathbf{W}$, plus a random vector of errors $\boldsymbol{e} = \{\varepsilon(x_1, y_1), \ldots, \varepsilon(x_n, y_n)\}'$. The resulting equation $\boldsymbol{u} = \rho\, \mathbf{W} \boldsymbol{u} + \boldsymbol{e}$ replaces the equation $U(x, y) = 0.25\rho[U(x-1, y) + U(x+1, y) + U(x, y-1) + U(x, y+1)] + \varepsilon(x, y)$. Unfortunately, the weights contained in matrix $\mathbf{W}$ are not the same on the corners and borders of a regular sampling grid as in its interior, since the number of nearest neighbors is 2, 3, and 4, respectively. In the case of the 1-D spatial AR(1) process, the problem concerns the extremities of a regular sampling transect. In both cases, more than n data should be simulated (i.e., more in 2-D space relative to 1-D space), and to reduce edge effects, only those well inside the sampling transect or grid should be kept for later processing. This simulation procedure, however, is specific to regular sampling schemes and autoregressive spatial processes.

The SAS codes used to simulate the autocorrelated spatial surface patterns in Figs. 7.2–7.4 were written with SAS/IML language, and are on the CD-ROM. They are all based on the same basic principle – that is, first, to simulate an $n \times 1$ vector $\boldsymbol{e}$ of pseudo-random numbers (i.e., random errors) from a normal distribution with a zero mean and a variance of 1, using the function RANNOR (SAS Institute Inc., 2009), and then to pre-multiply $\boldsymbol{e}$ by the "square root" of the variance–covariance matrix Σ of the requested type (e.g., exponential, spherical, or Gaussian, with a given range or practical range):

$$\boldsymbol{u} = \mathbf{B}'\boldsymbol{e}, \tag{A7.1}$$

where the prime symbol $'$ denotes the transpose operator in matrix algebra, and $\mathbf{B}$ $(n \times n)$ is such that $\mathbf{B}'\mathbf{B} = \Sigma$. (This operation is called the Cholesky decomposition of Σ when matrix $\mathbf{B}$ is triangular superior, which is the case with function HALF in PROC IML.) In a preliminary step, the sampling grid must be defined and the (Euclidean) distances between sampling locations must be calculated. Such a simulation procedure is general, and can be easily modified for irregular grids. It suffices to import the spatial coordinates of sampling locations or to simulate them in PROC IML. The example below is for an exponential variogram model with a practical range of 2 [see Fig. 7.2(a)].

```
PROC IML;
/** definition of sampling grid **/
/* the spatial coordinates of the n = nX*nY sampling locations are
   contained in array GridXY */
nX = 10;
nY = nX;
GridXY = 0.0*J(nX*nY,2,1.0);
DO ii = 1 TO nX;
DO jj = 1 TO nY;
GridXY[jj + (ii – 1)*nX,1] = ii;
GridXY[jj + (ii – 1)*nX,2] = jj;
END;
END;
PRINT GridXY;
/** calculation of Euclidean distances between sampling locations **/
/* the calculated distances are stored in array D */
D = 0.0*I(nX*nY);
DO ii = 1 TO nX*nY – 1;
DO jj = ii + 1 TO nX*nY;
x1 = GridXY[ii,1];
x2 = GridXY[jj,1];
y1 = GridXY[ii,2];
y2 = GridXY[jj,2];
```

```
D[ii,jj] = SQRT((x1 – x2)**2 + (y1 – y2)**2);
D[jj,ii] = D[ii,jj];
END;
END;
/** simulation of n random errors **/
/* the simulated errors are stored in array e */
e = 0.0*J(nX*nY,1,1.0);
DO ii = 1 TO nX*nY;
e[ii,1] = RANNOR(1);
END;
/** definition of the variance–covariance matrix of the random vector
   of observations **/
practicalrange = 2.0;
VARCOV = I(nX*nY);
DO ii = 1 TO nX*nY – 1;
DO jj = ii+1 TO nX*nY;
h = D[ii,jj];
h_r = h/practicalrange;
VARCOV[ii,jj] = EXP(-3*h_r);
VARCOV[jj,ii] = VARCOV[ii,jj];
END;
END;
/** simulation of n observations **/
/* the simulated observations are stored in array u */
HALFV = HALF(VARCOV);
u = 0.0*J(nX*nY,1,1.0);
u = t(HALFV)*e; /* translation of equation (A7.1) in SAS */
PRINT GridXY u;
RUN;
```

A7.2 The linear model of coregionalization

In the example of plant diversity study in Subsections 7.2.1 and 7.2.2, most of the variables (diversity, soil, topography) had a non-spatial random component that was modeled by a nugget effect in variogram analysis, and all of them had a spatially autocorrelated random component (with correlation between sampling locations within a certain distance) that could be modeled with a spherical variogram model with common range in a "coregionalization analysis." In terms of variables, it corresponds to the decomposition $\boldsymbol{u}(x, y) = \boldsymbol{\mu}(x, y) + \boldsymbol{\varepsilon}_1(x, y) + \boldsymbol{\varepsilon}_2(x, y)$, with $\boldsymbol{\varepsilon}_1(x, y)$ and $\boldsymbol{\varepsilon}_2(x, y)$, the random structural components corresponding to the nugget effect and the spherical variogram model, respectively. With some constraints, the linear model of coregionalization thus allows the use of more than one scale of variability (i.e., more than one structure) in the analysis of multivariate spatial data. In the formalization below, it is assumed for convenience that $\boldsymbol{\mu}(x, y) = \boldsymbol{\mu}$ for all (x, y) – there is spatial homogeneity of the means.

The linear model of coregionalization is usually written in terms of experimental variograms and semivariance functions (Journel and Huibregts, 1978, p. 172; Goovaerts, 1997, Section 4.3). There is one experimental direct variogram $\hat{\gamma}_j(h)$ for each of the p variables and as many experimental cross-variograms $\hat{\gamma}_{jj'}(h)$ as there are pairs of variables, and the number of structures STR is equal to the number of semivariance functions to be fitted. In the equation below (where isotropy is assumed because a distance lag h is used), the left-hand side is the random matrix $\hat{\boldsymbol{\gamma}}(h)$ of the $\frac{p(p+1)}{2}$ experimental variograms at lag h, and the right-hand side is built as a linear model, with the semivariance functions $\gamma_{str}(h)$ as explanatory variables and the sill matrices $\mathbf{B}_{str} = (\beta_{jj',str})$ as slopes ($str = 1, \ldots, STR$), plus an error term:

$$\hat{\boldsymbol{\gamma}}(h) = \sum_{str=1}^{STR} \gamma_{str}(h)\mathbf{B}_{str} + \text{error}. \tag{A7.2}$$

In applications, model (A7.2) is fitted over a number of distance lags $h_1, \ldots, h_K$; one appropriate way to define these is to use half the square root of the area covered by the sampling domain as upper bound h_K, with a number of equally spaced distance lags K that ensures at least 100 pairs of observations in each distance class. The combination of a nugget effect and a spherical model was found adequate in the CRAD applications of Subsection 7.2.2; the first explanatory variable in equation (A7.2) is then a vector of ones and the corresponding "sill" estimate indicates where the fitted model starts in the corresponding direct or cross-variogram, while the second is a vector with components $1.5\frac{h}{a} - 0.5\frac{h^3}{a^3}$ for some value of the range a.

(The theoretical function corresponding to the experimental cross-variogram defined in equation (7.13) is the cross-semivariance, referred to in Chapter 2. It is defined as follows under intrinsic stationarity:

$$\gamma_{jj'}(s' - s) = 0.5\text{Var}[U_j(s') - U_{j'}(s)] \text{ for any } (s, s'), \tag{A7.3}$$

where s and s' denote two spatial indices.)

Pelletier *et al.* (2004) presented the EGLS fitting algorithm for the linear model of coregionalization, and showed that EGLS estimators of sills have better properties than other estimators, especially for cross-variograms. Larocque *et al.* (2007) scrutinized uncertainty in coregionalization analysis, and quantified it in terms of bias and variance of estimators. They found that uncertainty is increasing (i) with the number of structures (i.e., non-spatial: nugget effect; spatially autocorrelated: exponential, spherical, or Gaussian) included in the linear model of coregionalization, and (ii) from sills for direct variograms to sills for cross-variograms and functions of them, such as structural correlations [equation (7.14)] and structural coefficients of determination, which can be highly uncertain. Fortunately, a sufficiently large number of sampling locations and a small value of the range of spatial autocorrelation relative to the extent of the sampling domain can keep the uncertainty to a reasonable level with two structures in the linear model of coregionalization (Larocque *et al.*, 2007).

8 · *Heterogeneity analysis of spatio-temporal surface patterns*

In this chapter, we will see how some of the models, methods and procedures presented in Chapters 6 and 7 can be used readily or with adjustment to analyze the heterogeneity of spatio-temporal surface patterns. In particular, the analysis of heterogeneity due to spatio-temporal autocorrelation will require a certain adjustment. In the absence of a very practical spatio-temporal distance, spatio-temporal auto- and cross-correlations do exist, but their analysis cannot be based on spatio-temporal correlograms or variograms computed under the isotropy condition. This key question is discussed in Section 8.1, together with the simpler cases of heterogeneity of the mean and heterogeneity of the variance for spatio-temporal surface patterns. Following this discussion, three avenues will be explored: a spatio-temporal version of the method of coregionalization analysis with a drift (CRAD), which is presented in Section 8.2; the use of state-space models, which were originally designed for the analysis of multiple time series and can be applied for the analysis of spatio-temporal auto- and cross-correlations (Section 8.3); and the Mantel analysis with spatial and temporal distances (Section 8.4). These statistical methods address specific aspects of heterogeneity for a spatio-temporal surface pattern and their conditions of application are different; for example, we remember from Chapter 7 the implications of performing correlation analysis from distances instead of raw data. These specific aspects and differences are illustrated with a series of examples, using air and soil temperature data collected at Gault Nature Reserve in Mont-Saint-Hilaire (Québec, Canada). In some examples, the vertical dimension of space will be involved in the data analysis. Finally, the loop on the heterogeneity analysis of surface patterns is closed by a number of recommended readings in the spatio-temporal framework (Section 8.5). Copies of the Matlab and SAS codes used for the data analyses of the examples are on the CD-ROM accompanying the book.

8.1 How to approach the heterogeneity analysis of a spatio-temporal surface pattern?

As we will rapidly realize, the answer to the question above is not by merging without discernment all the models, methods and procedures available for the heterogeneity analysis of purely spatial and purely temporal surface patterns. Consider the case of heterogeneity due to spatio-temporal autocorrelation, since this is the most difficult type of heterogeneity to analyze for a spatio-temporal surface pattern, and assume air temperature is being recorded every hour by hobos located on a sampling grid in a given site, because this is a main example used for illustration in the following sections. Thus, a number of time series equal to the number of hobos is collected in one day and each time series is composed of 24 observations of air temperature, if there is no interruption in the recording. This dataset can be presented in a table with rows corresponding to hobos and columns corresponding to the hourly observations of air temperature. To make the geographical coordinates of hobo sites explicit, the same data can be presented as a spatio-temporal surface pattern $\{u(x_i, y_i, t_j) \mid i = 1, \ldots, n \text{ and } j = 1, \ldots, 24\}$, with (x_i, y_i) indexing space in 2-D horizontally and t_j indicating the sampling time. Furthermore, in order to incorporate altitude (i.e., the vertical, third spatial axis) in the indexing, the spatio-temporal surface pattern can be written as $\{u(x_i, y_i, z_i, t_j) \mid i = 1, \ldots, n \text{ and } j = 1, \ldots, 24\}$. It must be noted that (i) the total number of observations is unchanged and equal to $24n$, since the number of hobos, n, remains the same – only the index characterizing their spatial location has expanded; (ii) for a given t_j, the surface pattern is purely spatial, and for a given (x_i, y_i) or (x_i, y_i, z_i), it is purely temporal or a time series.

From (ii) above, it is clear that the heterogeneity analysis of purely spatial or purely temporal "sections" of a spatio-temporal surface pattern can be performed readily by using material from Chapters 7 and 6, but what about the analysis of the spatio-temporal surface pattern as a whole? This is much less clear and certainly not straightforward when spatio-temporal autocorrelation is concerned. For example, consider the air temperatures recorded by one hobo located in (x, y, z) at time t, $U(x, y, z, t)$, and by another in (x', y', z') at time t', $U(x', y', z', t')$. If the two sampling locations are not too remote and the same holds true for the two sampling times, the two air temperature records are likely to be correlated, so $\mathrm{Corr}[U(x, y, z, t), U(x', y', z', t')]$ is likely to be different from zero. This is the correlation of one random variable with itself at two different space-time points, so one can speak of autocorrelation, in space-time instead of only in space or in time. Under the property of stationarity at order 2 for the spatio-temporal stochastic process, the theoretical autocorrelation above is equal to $\rho(x' - x, y' - y, z' - z, t' - t)$ [equation (2.4)], and under the property of isotropy, it is given by $\rho(\|(x' - x, y' - y, z' - z)\|, |t' - t|)$ [equation (2.7)]; see Section 2.3.

The separate use of the Euclidean norm $\|\cdot\|$ to measure the spatial distance between (x, y, z) and (x', y', z') and of the absolute value $|\cdot|$ (i.e., the Euclidean norm in 1-D) to measure the proximity between times t and t' in equation (2.7) is characteristic of the absence of a spatio-temporal distance. In fact, in accordance with the fundamental difference in nature existing between space and time, a squared Euclidean distance cannot be calculated between (x, y, z, t) and (x', y', z', t') because it would imply the sum of quantities [i.e., $(x' - x)^2$, $(y' - y)^2$, $(z' - z)^2$ vs. $(t' - t)^2$] expressed in different units (e.g., squared meters vs. squared hours), which is impossible.

Key note: The absence of a spatio-temporal distance prevents the use of correlograms and variograms computed as functions of a single spatio-temporal distance lag under isotropy, to analyze spatio-temporal autocorrelation. Therefore, alternative approaches must be sought.

The graphical representation of a theoretical spatio-temporal autocorrelation function $\rho(\|(x' - x, y' - y, z' - z)\|, |t' - t|)$ against two distance lags, one spatial and one temporal, provides a "map" starting from 1.0 at (0, 0) and expected to decrease to 0 from the edges (where $\|(x' - x, y' - y, z' - z)\| = 0$ with $|t' - t| > 0$ and $|t' - t| = 0$ with $\|(x' - x, y' - y, z' - z)\| > 0$), to the opposite sides. The estimation of such a function, based on sample spatio-temporal autocorrelation coefficients, is not very practical because it requires (i) large numbers of sampling locations and sampling times, to ensure sufficient numbers of pairs of observations in the different distance classes, and (ii) a coordinated sampling in space and time that avoids scarce observations in some areas over some time intervals, to obtain representative estimates. Its modeling beyond separability [equation (2.9)] may quickly be complicated by multiple terms of space-time interaction and submitted to stationarity constraints (Gneiting, 2002).

(In classical (i.e., Euclidean) geometry and classical (i.e., non-relativistic) physics, there is no spatio-temporal distance. Such type of distance exists, however, in the Minkowski space, where Albert Einstein formulated his theory of special relativity. In that space, all vectors have four components, three for space and one for time, and the squared norm of a vector is equal to the sum of squares of its spatial coordinates minus the square of the temporal one. Throughout this book, we are *not* working in the Minkowski space.)

In the next sections, three possible avenues are presented for the analysis of spatio-temporal autocorrelation and the heterogeneity analysis of spatio-temporal surface patterns in general. The first two do not require large numbers of sampling locations *and* sampling times, and can be accommodated with a few sampling locations and a large number of sampling times, or the contrary. They are: a spatio-temporal version of coregionalization analysis with a drift (CRAD; Pelletier *et al.*, 2009a, 2009b) and the state-space models (Akaike, 1974b; Priestley, 1981, pp. 797–800). Both these avenues are based on a model in which the same response at a few sampling locations is considered as a multivariate response or several correlated responses, so spatio-temporal autocorrelations are analyzed as temporal cross-correlations. In CRAD, it does not matter whether the few sampling units are in space or in time, but in the case of state-space models the few sampling units must be in space because these models are written as if the multivariate stochastic process was first-order autoregressive in discrete time. Another difference between the two avenues is that the analysis of the heterogeneity of the mean is integrated to CRAD via its Phase I, but stationarity at order 1 figures among the conditions of application of state-space models, so any trend must be removed prior to their application. The third avenue explored thereafter is a version of Mantel's (1967) test with spatial and temporal distances. This is model-free when the objective is limited to the assessment of space-time interactions, based on distances to be seen as measures of dispersal or variation between sampling units. Some modeling is involved, however, when the Mantel test is performed with different sets of spatial and temporal distances, depending on whether these are calculated from the data before or after removing the heterogeneity of the mean.

Fortunately, the cases of spatio-temporal heterogeneity of the mean and spatio-temporal heterogeneity of the variance are simpler to deal with. In the multivariate approach followed in CRAD (Section 8.2), the temporal heterogeneity of the mean will be first analyzed at each of the sampling locations separately and then compared among sampling locations; such a procedure is supported by a similar practice in geostatistics (Helterbrand and Cressie, 1994). In the same approach, the sills (i.e., variance components) estimated in Phase II of CRAD for the non-temporal random component and the temporally autocorrelated component of the response at each of the sampling locations provide the basis for some spatio-temporal heterogeneity analysis of the variance.

This conceptual section is closed on a thought on the three dimensions of space. Conducting a sampling in which the three dimensions of space are equally explored is not easy. In many studies, for each horizontal position there are only a few altitudes if not just one. When the indices (x, y, z) used to locate a response in 3-D are expressed in the same unit (e.g., m), 3-D distances as well as 2-D and 1-D distances, horizontal and vertical, are also expressed in the same unit (e.g., m). This does not mean that a horizontal 2-D spatial distance and a vertical 1-D spatial distance of the same value represent the same remoteness or proximity in biological and environmental terms. In particular, the observation of spatial heterogeneity at some distance horizontally does not imply that it also exists at the same distance vertically. This will be illustrated and developed in the examples of Sections 8.2–8.4, and includes differences in the mean and variance of the response as well as auto- and cross-correlations.

8.2 Coregionalization analysis with a drift

This statistical method due to Pelletier *et al.* (2009a, 2009b) was presented in the spatial framework in Section 7.2. It is presented here in the spatio-temporal framework with essentially the same two phases: I, to estimate the deterministic and random components of each "variable," and II, to estimate the "correlations" between deterministic components and those between random components, at two structures for the latter if the sample size is sufficiently large to fit a linear model of coregionalization to all the experimental variograms (direct and cross) with non-excessive uncertainty (Section A7.2). An important difference between frameworks in the application of the CRAD method lies in the models used to represent the observed surface patterns or the corresponding stochastic processes, although they are all variants of the fundamental equation (2.1).

- In the 2-D spatial framework, the model used in CRAD is $\boldsymbol{u}(x, y) = \boldsymbol{\mu}(x, y) + \boldsymbol{\varepsilon}(x, y)$, with $\boldsymbol{u}(x, y) = (U_1(x, y), \ldots, U_p(x, y))'$, where $U_j(x, y)$ and $U_{j'}(x, y)$ $(j \neq j')$ denote two different variables observed at the same location indexed by (x, y), and $\boldsymbol{\mu}(x, y) = \mathrm{E}[\boldsymbol{u}(x, y)]$.
- In the spatio-temporal framework, with a few sampling locations in 2-D space and a large number of sampling times, it is $\boldsymbol{u}(t) = \boldsymbol{\mu}(t) + \boldsymbol{\varepsilon}(t)$, with $\boldsymbol{u}(t) = (U_1(t), \ldots, U_p(t))'$, where $U_j(t)$ and $U_{j'}(t)$ $(j \neq j')$ denote

the same variable observed in two different locations at the same time t, and $\boldsymbol{\mu}(t) = \mathrm{E}[\boldsymbol{u}(t)]$.

- In the spatio-temporal framework, but with a large number of sampling locations in 2-D space and a few sampling times, the model is $\boldsymbol{u}(x, y) = \boldsymbol{\mu}(x, y) + \boldsymbol{\varepsilon}(x, y)$, with $\boldsymbol{u}(x, y) = (U_1(x, y), \ldots, U_p(x, y))'$, where $U_j(x, y)$ and $U_{j'}(x, y)$ $(j \neq j')$ denote the same variable observed at two different times in the same location (x, y), and $\boldsymbol{\mu}(x, y) = \mathrm{E}[\boldsymbol{u}(x, y)]$.

The first of the two spatio-temporal frameworks above corresponds to the examples presented later in this section, with a relatively small number of hobos recording air and soil temperatures at Gault Nature Reserve in Mont-Saint-Hilaire (Québec, Canada) on an hourly basis over a 15-day period. In this framework, spatial distances between hobo sites are not taken into account explicitly, but this is not a real drawback and there are, actually, several advantages associated with that: (i) it is possible to relate the strongest correlations observed to specific characteristics at the hobo sites; (ii) the closest sites spatially do not necessarily show the strongest correlations temporally, because of altitude and orientation effects; (iii) in situations like those of the examples presented hereafter, it appears more appropriate to assume stationarity in time than stationarity and isotropy in space after the removal of estimated drifts. Above all, spatio-temporal autocorrelations are analyzed through temporal cross-correlations $\mathrm{Corr}[U_j(t), U_{j'}(t')] = \rho_{jj'}(|t' - t|)$ or temporal cross-semivariances $\gamma_{jj'}(|t' - t|)$, where the spatial distance between hobo sites in 1-D, 2-D, or 3-D is reflected by the pair of subscripts (j, j').

Before moving to the announced examples, note that the second spatio-temporal framework above would apply to a situation in which satellite photos of vegetation cover are taken of a site at a fine spatial resolution on a seasonal basis within a year. Spatio-temporal autocorrelations would then be analyzed through spatial cross-correlations $\mathrm{Corr}[U_j(x, y), U_{j'}(x', y')] = \rho_{jj'}(h)$ or spatial cross-semivariances $\gamma_{jj'}(h)$, where $h = \|(x', y') - (x, y)\|$ and temporal proximity would be reflected by the pair of subscripts (j, j').

8.2.1 MSH air temperature example (CRAD)

The central objective of this and the following example is to better understand the fluctuations of temperature at Gault Nature Reserve in Mont-Saint-Hilaire (Québec, Canada) spatially and temporally, at a critical time in the plant growing season, that is, the first half of May; the

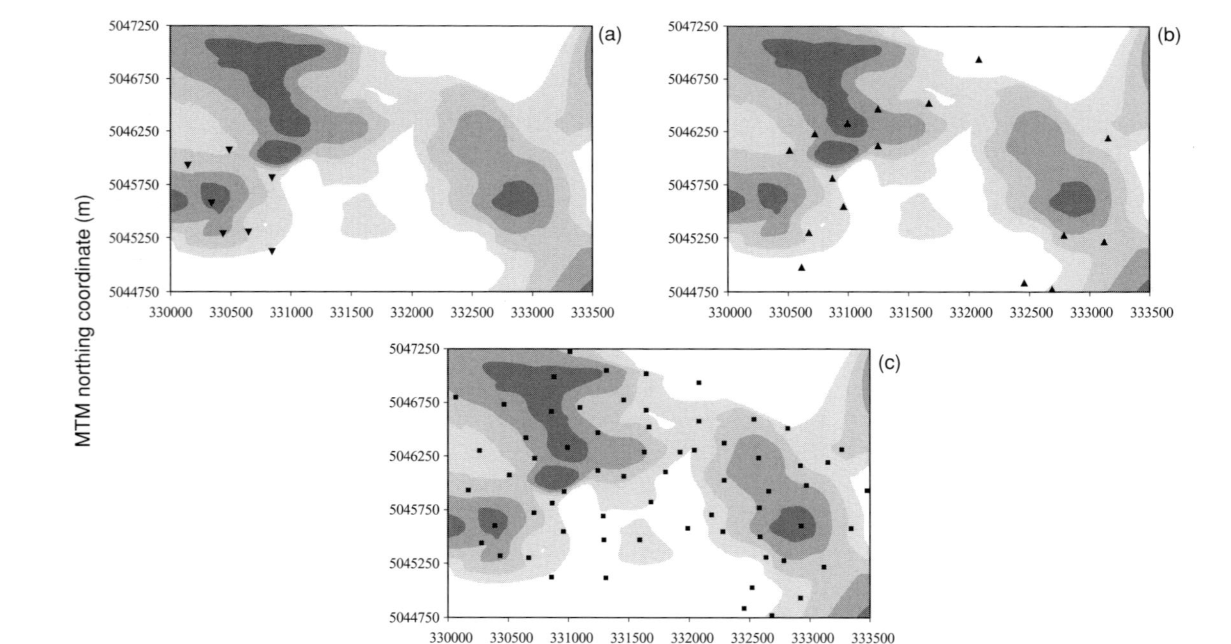

Fig. 8.1. Maps of the Gault Nature Reserve site in Mont-Saint-Hilaire (Québec, Canada), showing the sampling grids in examples of coregionalization analysis with a drift (CRAD) and Mantel analysis with spatio-temporal data, with altitude represented by gray contours superimposed on them. (a) and (c) Air temperature over the period May 1–15, 2004 (hourly means, $n = 360$) – for CRAD (seven sampling locations) and Mantel analysis with spatial distances and distances between temporal extrema (65 sampling locations); (b) soil temperature over the period May 1–15, 2004 (hourly means, $n = 360$) – for CRAD and Mantel analysis (16 sampling locations). Altitude gray contours vary from 200 m (white) to 400 m (dark gray) by increments of 50 m. MTM = Modified Transverse Mercator, useful in Canada only.

Table 8.1. *Modified Transverse Mercator (MTM) coordinates (in m) for the seven hobo sites of Fig. 8.1(a) and the 16 hobo sites of Fig. 8.1(b), with their respective number*

(a)			(b)		
Hobo number	Easting MTM	Northing MTM	Hobo number	Easting MTM	Northing MTM
Hobo06	330861.9	5045118.5	Hobo01	330958.8	5045547.1
Hobo09	330430.7	5045317.2	Hobo03	331670.0	5046521.8
Hobo10	330392.1	5045598.9	Hobo04	332456.0	5044831.9
Hobo46	330866.6	5045809.0	Hobo07	333119.6	5045214.0
Hobo48	330167.4	5045931.1	Hobo15	332688.5	5044753.7
Hobo63	330512.0	5046071.7	Hobo22	330611.4	5044978.8
Hobo64	330669.9	5045301.0	Hobo24	332784.1	5045273.9
			Hobo37	331246.3	5046464.5
			Hobo40	332082.8	5046934.8
			Hobo46	330866.6	5045809.0
			Hobo55	330720.2	5046225.7
			Hobo60	333154.2	5046192.1
			Hobo62	330991.2	5046327.9
			Hobo63	330512.0	5046071.7
			Hobo64	330669.9	5045301.0
			Hobo67	331243.4	5046115.8

Hobo46, Hobo63, and Hobo64 are common to Fig. 8.1(a) and (b).

year is the same as in the other MSH examples in Chapters 6 and 7, that is, 2004. This study of temperature space-time dynamics will include the three types of spatio-temporal hererogeneity, in the mean, in the variance, and that due to autocorrelation, as explained above.

Air temperature records made by seven hobos located around and on a small hill on the west side of the reserve [Fig. 8.1(a), Table 8.1(a)] were analyzed for heterogeneity. Temporally, these records are hourly means of air temperature over the period May 1–15, 2004 (i.e., $n = 360$). Visually, the seven time series show relatively clear peaks mostly in the first half of the afternoon every 24 h [see the continuous curves in Fig. 8.2(a1)–(g1)]. Differences between hobos in the local shape of temperature fluctuations (i.e., at small scale) can be observed, but more global fluctuations (i.e., at intermediate and large scales) are very similar from May 1 to May 15. It also froze at some but not all hobo sites on May 4 (72–96 h) and 8 (168–192 h).

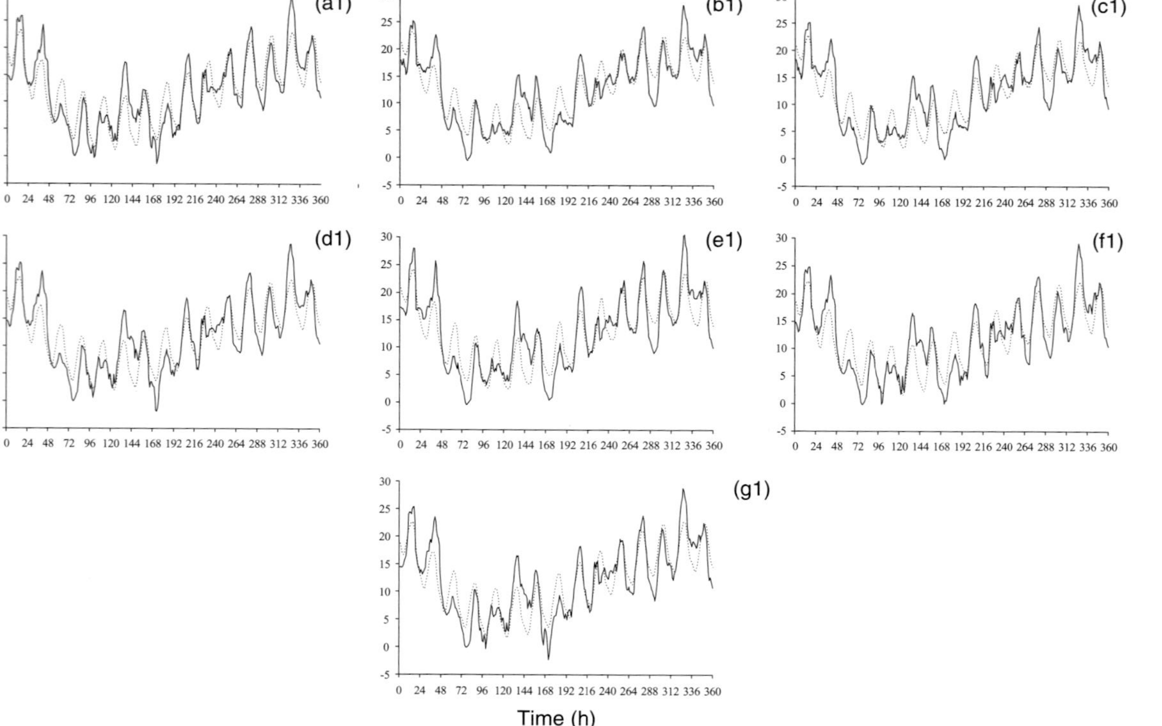

Fig. 8.2. Example of hourly mean air temperature over the period May 1–15, 2004 ($n = 360$) in seven sampling locations [Fig. 8.1(a), Table 8.1(a)] at Gault Nature Reserve in Mont-Saint-Hilaire (Québec, Canada): (a1)–(g1) observed (continuous curve) and predicted (dashed) time series – predicted time series are the result of the joint EGLS fitting of an orthogonal polynomial of degree 3 and four cosine and sine waves with periods of 24, 12, 8, and 6 h; (a2)–(g2) the corresponding EGLS residuals (i.e., observed minus predicted time series).

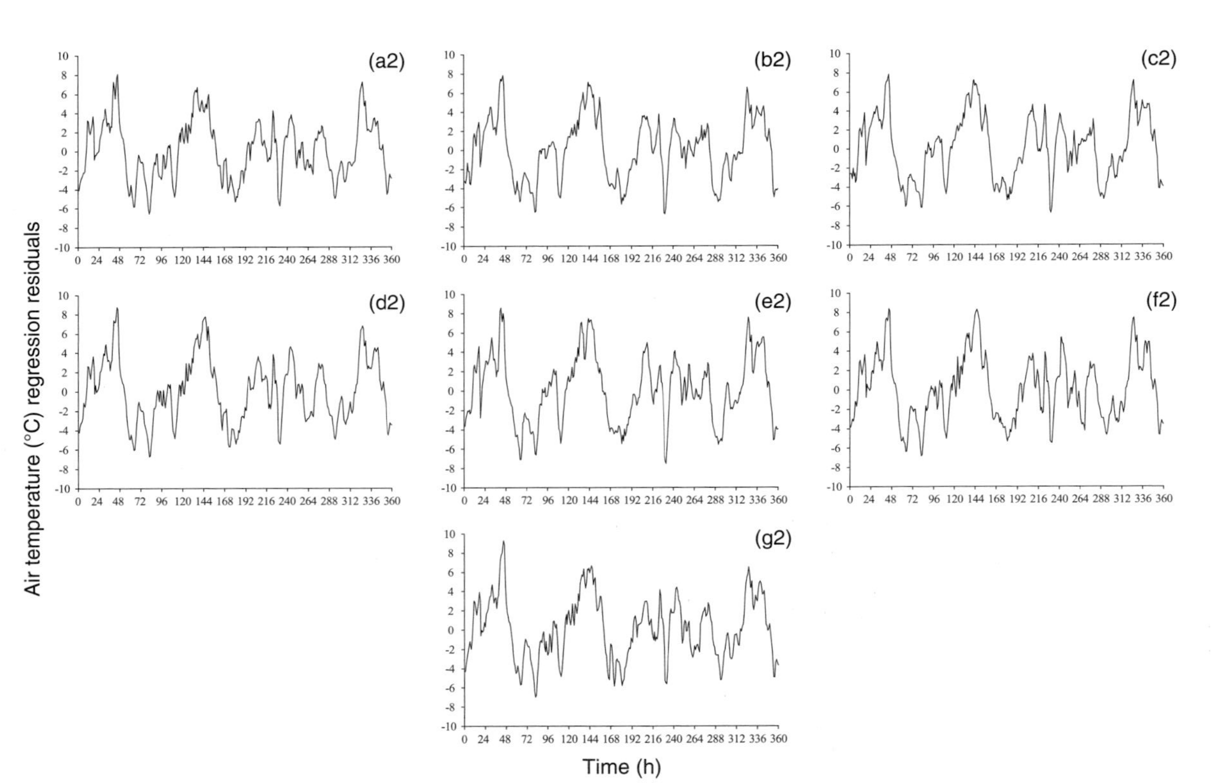
(a2)
(b2)
(c2)
(d2)
(e2)
(f2)
(g2)
Air temperature (°C) regression residuals
Time (h)

Fig. 8.2. (*cont.*)

Because the deterministic component $\boldsymbol{\mu}(t)$ of the CRAD model is restricted to large-scale heterogeneity of the mean while 24-h periodic components in time series of 360 hourly means represent intermediate-scale heterogeneity of the mean, a joint polynomial–trigonometric model [equation (6.24)] was first fitted to each time series, prior to performing Phase I of CRAD. That model was composed of an orthogonal polynomial of degree 3 plus four cosine and four sine waves corresponding to the fundamental period of 24 h and the harmonics of 12, 8, and 6 h. Its EGLS fitting was performed with SAS PROC AUTOREG, as in Subsection 6.3.2. The predicted time series and EGLS residual counterparts are plotted in Fig. 8.2(a1)–(g1) and (a2)–(g2). The R^2 value for the deterministic part of the model was 0.43–0.45 in general, and reached 0.53 and 0.47 for the hobo sites with lowest and second lowest altitudes [i.e., Hobo06 and Hobo46; see panels (a) and (d)]. The pseudo-variances of estimated temporal drifts are not heterogeneous in space and, by difference, the same holds true for the variances of estimated random components. (Strictly speaking, the equality between the total variance and the sum of the pseudo-variance of the estimated drift and the variance of the estimated random component is only approximate, because of the use of an EGLS fitting procedure to obtain the estimated drift.)

The seven series of EGLS residuals were then submitted to Phase I of CRAD. In the case of air temperature, no further significant trace of temporal heterogeneity of the mean was found in the residual series by the local drift estimation procedure based on a moving window. The estimated sills for a nugget effect in the fitted variogram models did not meet the threshold of 5%, and the range values estimated for the spherical structure varied from 26.5 to 28.0 h in Phase I of CRAD. In view of the partial realizations of discrete-time AR(2) and ARMA(1,1) stochastic processes plotted in Fig. 6.2(a) and (c), it is reasonable to assume that the fluctuations in the residual series of Fig. 8.2(a2)–(g2) are autocorrelated random. In CRAD, the geostatistical approach is followed, so a purely spherical variogram model was fitted to each direct experimental variogram separately first (Phase I) and then to all experimental variograms (direct and cross) jointly (Phase II). The estimated common range is 26.5 h. It does not mean that the sample temporal autocorrelation coefficients are or should be significantly different from zero between the zero lag and 26.5 h; actually, as the lag increases and approaches 26.5 h, the value of the sample autocorrelation coefficient is expected to be closer to zero and hence non-significant.

Table 8.2. *Spatio-temporal correlations between sampling locations (i.e., hobo sites) at the time scale of 26.5 h (i.e., the common range of temporal autocorrelation estimated in Phase II of CRAD) in the example of hourly mean air temperature at Gault Nature Reserve in Mont-Saint-Hilaire (Québec, Canada); these are structural correlations calculated from sills estimated for a spherical variogram model with range 26.5, fitted to the 28 experimental variograms (direct and cross) computed from the EGLS residual time series [see panels (a2)–(g2) in Fig. 8.2 and Table 8.1(a);* n = *360]*

	Hobo06	Hobo09	Hobo10	Hobo46	Hobo48	Hobo63	Hobo64
Hobo06	1	0.739	0.755	0.884	0.741	0.844	0.913
Hobo09		1	0.909	0.737	0.812	0.697	0.745
Hobo10			1	0.743	0.861	0.721	0.764
Hobo46				1	0.745	0.864	0.872
Hobo48					1	0.740	0.741
Hobo63						1	0.833
Hobo64							1

Probabilities of significance are not reported because they are all below 0.05, so all the structural correlations are statistically significant at that level. The modified *t*-test per structure (Dutilleul and Pelletier, in press) was used to evaluate the probabilities of significance.

Pseudo-correlations between estimated drifts are all very close to 1.0, confirming the visual observations made about the heterogeneity of the mean at intermediate and large scales from the observed and predicted time series of air temperature [Fig. 8.2(a1)–(g1)]. In other words, spatio-temporal heterogeneity of the mean is essentially temporal! For their part, the estimated structural correlations between random components, using a spherical variogram model with a range of 26.5 h, vary from 0.697 to 0.913 (Table 8.2). The two highest ones are for the pairs of hobo sites (06, 64) – with the lowest altitude, Hobo06, and its nearest neighbor – and (09, 10) – with the highest altitude, Hobo10, and its nearest neighbor [Fig. 8.1(a), Table 8.1(a)].

8.2.2 MSH soil temperature example (CRAD)

In terms of period covered, a similar material (i.e., hourly means over May 1–15, 2004; $n = 360$) was used for soil temperature. Spatially, records from 16 hobos distributed over a good part of the reserve, essentially on the northwest main hill and around the east hill, were analyzed for heterogeneity [Fig. 8.1(b), Table 8.1(b)]; this sampling grid has three hobo sites in common with that of the MSH air temperature example (CRAD).

Unlike air temperature, soil temperature was always positive during the studied period in the area sampled, and fluctuations in soil temperature are smoother and narrower than in air temperature [Figs. 8.2 and 8.3(a1)–(p1)]. This has to do with the gaseous nature of air compared to soil which is solid, and confirms previous observations made in view of Fig. 6.10(a) and (e). Like air temperature, the 16 time series of hourly means of soil temperature show relatively clear peaks every 24 h, but around the middle of the afternoon (i.e., before or after 18:00) [see the continuous curves in Fig. 8.3(a1)–(p1)]. For soil temperature, there appears to be a greater spatial heterogeneity among hobo sites; see the narrow fluctuations in Fig. 8.3(a1) (i.e., Hobo01) vs. the wider fluctuations in Fig. 8.3(g1) (i.e., Hobo24).

The same models and procedures as for air temperature were used to analyze the heterogeneity of the 16 time series of hourly means of soil temperature, with the following results and differences. The EGLS fitting of the joint polynomial–trigonometric model provided R^2 values varying from 0.54–0.55 (Hobo07 and Hobo22) to 0.68–0.69 (Hobo62 and Hobo15). These values are higher by 11%–15% and vary slightly more than for air temperature. It must be noted that such R^2 values are relative measures of temporal heterogeneity of the mean, using the total variation of each time series as a basis.

In Phase I of CRAD, a significant trace of temporal heterogeneity of the mean was found by the local drift estimation procedure L_1 in 13 of the series of EGLS residuals plotted in Fig. 8.3(a2)–(p2), but relative to the total variation of the time series this represented an increase of a few percents in the R^2 value. The three series where no further temporal heterogeneity of the mean was found are those of Hobo07 (which already had the lowest R^2), Hobo24, and Hobo46. The main implication is that Hobo55 finally ranked with the hobo sites with highest R^2 values for the deterministic components of soil temperature at intermediate and large temporal scales.

How can one discuss the results above in environmental terms? On the one hand, lower R^2 values for deterministic components indicate that temporal heterogeneity of the mean is weaker or less well regulated. This is the case for soil temperature at Hobo07 and Hobo22 [see panels (d1) and (f1) in Fig. 8.3], which are both characterized by an average altitude (about 300 m), a southeast orientation (relative to a different hill), and a

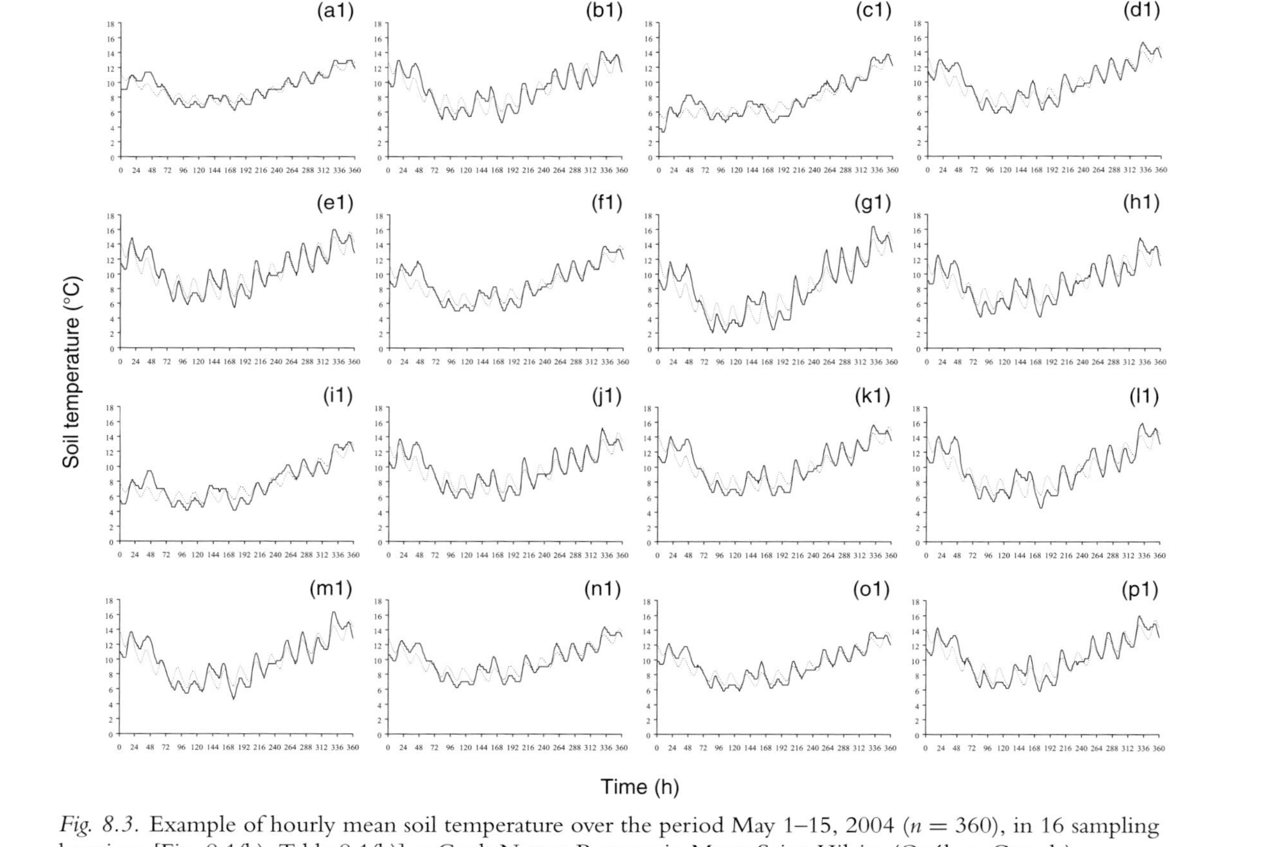

Fig. 8.3. Example of hourly mean soil temperature over the period May 1–15, 2004 ($n = 360$), in 16 sampling locations [Fig. 8.1(b), Table 8.1(b)] at Gault Nature Reserve in Mont-Saint-Hilaire (Québec, Canada): (a1)–(p1) observed (continuous curve) and predicted (dashed) time series – predicted time series are the result of the joint EGLS fitting of an orthogonal polynomial of degree 3 and four cosine and sine waves with periods of 24, 12, 8, and 6 h; (a2)–(p2) the corresponding EGLS residuals (i.e., observed minus predicted time series).

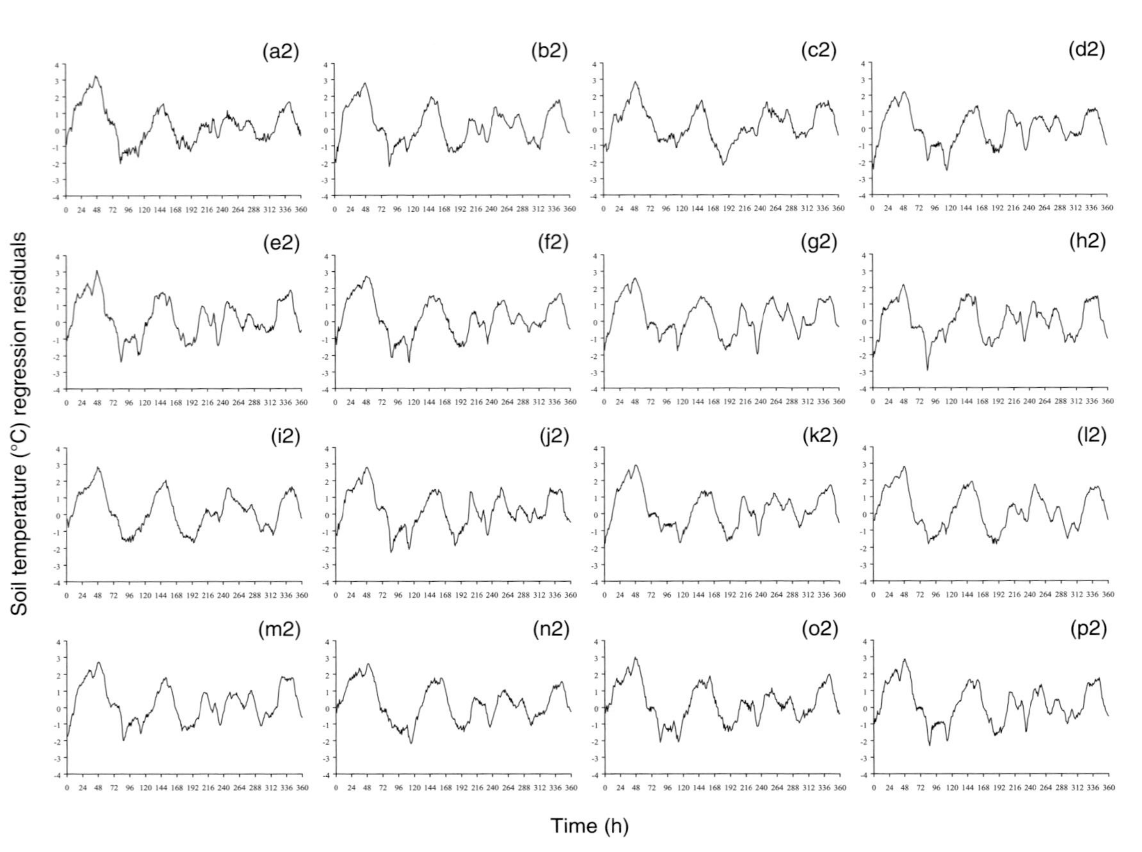
(a2)
(b2)
(c2)
(d2)
(e2)
(f2)
(g2)
(h2)
(i2)
(j2)
(k2)
(l2)
(m2)
(n2)
(o2)
(p2)
Soil temperature (°C) regression residuals
Time (h)
0 24 48 72 96 120 144 168 192 216 240 264 288 312 336 360
4 3 2 1 0 -1 -2 -3 -4

Fig. 8.3. (cont.)

high solar radiation. On the other hand, higher R^2 values correspond to soil temperatures that are more easily predictable without autocorrelated random component. That is the case for Hobo15, Hobo55, and Hobo62, which are characterized by different environmental factors compared to Hobo07 and Hobo22. Solar radiation is low at Hobo15 and Hobo55, which are oriented west of a (different) hill at low (Hobo15) vs. average (Hobo55) altitudes. Hobo62 has an even higher altitude (i.e., the highest altitude of all hobo sites in this example) and an average solar radiation, but is very close to Hobo55 spatially, without being its nearest neighbor [Fig. 8.1(b), Table 8.1(b)].

As it was the case for air temperature, the estimated sills for a nugget effect in the models fitted to direct experimental variograms in Phase I of CRAD did not meet the threshold of 5% for soil temperature. Compared to air temperature, the estimated range of temporal autocorrelation in fitted variogram models is longer, which is in accordance with the greater smoothness observed for hourly mean soil temperatures (Fig. 8.3). The range values estimated for each hobo site separately vary from 44.5 to 55.0 h. The common range estimated in Phase II of CRAD is 50.5 h.

Pseudo-correlations between estimated drifts remain positive and statistically significant, but are weaker than for air temperature, as they vary from 0.541 to 0.997; this confirms the greater spatial heterogeneity among hobo sites anticipated in view of Fig. 8.3(a1)–(p1). Under a spherical variogram model with a range of 50.5 h, the estimated structural correlations between random components are of moderate to intermediate strength, varying from 0.167 to 0.620 (Table 8.3). The two strongest ones are observed for the pairs (Hobo15, Hobo37) and (Hobo24, Hobo62). Since the hobo sites of each pair are distant spatially and differ in altitude, orientation, and solar radiation, there must be different environmental factors involved, such as the type of soil and its properties at the hobo sites and the tree canopy openness around early May. In other words, such results obtained in a spatio-temporal heterogeneity analysis give researchers insight into correlations that they might not expect, at small temporal scale between sampling locations, and offer them avenues as to how they could collect complementary data to test new hypotheses.

Summary: Despite some common points, the spatio-temporal heterogeneity of air temperature appeared to be different from that of soil temperature at Gault Nature Reserve in the middle of spring 2004. The fundamental difference in medium nature (i.e., air vs. soil) is reflected by the respective fluctuations of the two types of temperature time series. Pseudo-correlations of 1.0 between estimated

Table 8.3. *Spatio-temporal correlations between sampling locations (i.e., hobo sites) at the time scale of 50.5 h (i.e., the common range of temporal autocorrelation estimated in Phase II of CRAD) in the example of hourly mean soil temperature at Gault Nature Reserve in Mont-Saint-Hilaire (Québec, Canada); these are structural correlations calculated from sills estimated for a spherical variogram model with range 50.5, fitted to the 136 experimental variograms (direct and cross) computed from the EGLS residual time series [see panels (a2)–(p2) in Figure 8.3 and Table 8.1(b);* n = *360]*

	Hobo01	Hobo03	Hobo04	Hobo07	Hobo15	Hobo22	Hobo24	Hobo37	Hobo40	Hobo46	Hobo55	Hobo60	Hobo62	Hobo63	Hobo64	Hobo67
Hobo01	1	0.380	0.239	0.239	0.354	0.214	0.188	0.393	0.207	0.232	0.196	0.295	0.275	0.209	0.294	0.372
Hobo03		1	0.317	0.478	0.521	0.355	0.413	0.569	0.382	0.473	0.201	0.535	0.537	0.191	0.398	0.454
Hobo04			1	0.367	0.354	0.238	0.404	0.346	0.327	0.326	0.332	0.379	0.413	0.190	0.302	0.295
Hobo07				1	0.412	0.433	0.478	0.351	0.316	0.472	0.457	0.538	0.538	0.325	0.383	0.454
Hobo15					1	0.447	0.456	0.608	0.301	0.527	0.265	0.463	0.566	0.207	0.480	0.554
Hobo22						1	0.459	0.321	0.302	0.458	0.349	0.446	0.502	0.335	0.330	0.454
Hobo24							1	0.332	0.356	0.543	0.541	0.499	0.620	0.408	0.437	0.508
Hobo37								1	0.286	0.473	0.196	0.454	0.466	0.167	0.429	0.480
Hobo40									1	0.289	0.306	0.388	0.420	0.188	0.331	0.345
Hobo46										1	0.282	0.513	0.571	0.263	0.514	0.591
Hobo55											1	0.427	0.441	0.496	0.349	0.413
Hobo60												1	0.506	0.333	0.409	0.456
Hobo62													1	0.303	0.484	0.559
Hobo63														1	0.316	0.387
Hobo64															1	0.439
Hobo67																1

Probabilities of significance are not reported because they are all below 0.05 with one exception (i.e., Hobo37 with Hobo63), so all the structural correlations but one are statistically significant at that level. The Dutilleul–Pelletier modified *t*-test per structure (Dutilleul and Pelletier, in press) was used to evaluate the probabilities of significance.

drifts for air temperature mean that peaks and troughs in predicted series are perfectly synchronized between hobo sites. For soil temperature, however, pseudo-correlations can be as low as 0.5, indicating a greater variability in temporal heterogeneity at intermediate-to-large scales between hobo sites; this point will be followed up in the next section. Similarly, the positive structural correlations between estimated random components are stronger for air temperature, suggesting a greater coordination of small-scale temporal fluctuations because of the turbulence of the air and its fast circulation in the corresponding area of the reserve. A drawback of air turbulence and the more sudden changes for air temperature, due in part to wind effects unrelated to the daily cycle, is that R^2 *values of joint polynomial–trigonometric models are lower than for soil temperature.*

A common point for all the temperature series analyzed here, whether for air or soil, is the absence of a nugget effect in the variogram model fitted for the random component of the time series of hourly means; in other words, there is spatial homogeneity among hobo sites in that all the random components are essentially autocorrelated temporally. On a side note, it means that correlations could have been analyzed directly from the estimated random components, without passing by structural correlations estimated from sills as in the CRAD method.

The similarity or dissimilarity between hobo sites in the temporal fluctuations of temperature at small and larger scales generated hypotheses about the environmental factors on which they could be based. The resulting hypotheses for air vs. soil temperature differed to some degree.

8.3 State-space models

In well-defined conditions, the multivariate discrete-time stochastic process underlying observed multiple time series can be viewed as a dynamic system regenerating itself, following a model similar to a vector AR(1) and called "Markovian representation" (Akaike, 1974b, 1975) or "state-space representation" (Priestley, 1981, p. 797). In fact, any multivariate linear process is defined by a system of stochastic equations that is linked to the auto- and cross-correlations of and between the univariate components or processes; the normality distribution is usually assumed, so that the probability properties are derived from the second-order ones. In addition [see the example of the vector AR(2) process below], any vector discrete-time ARMA(p, q) model can be rewritten as a vector discrete-time AR(1) after a "state space" has been appropriately defined (Aoki, 1990, pp. 21–38). In this section, we will see how state-space models can be used for spatio-temporal heterogeneity analysis; the general methodology is first introduced in the following two paragraphs.

Assume that the p-variate discrete-time stochastic process $\{\boldsymbol{u}(t)|t \in \mathbb{Z}\}$ with $\boldsymbol{u}(t) \in \mathbb{R}^p$ is stationary up to order 2 – that is, the population mean and variance of each univariate process $\{U_j(t)|t \in \mathbb{Z}\}$ ($j = 1, \ldots, p$) are constant over time, their population autocorrelations are a function of the time lag $|t' - t|$ instead of the two times t and t' (Section 2.3), and so are their population cross-correlations (Section 2.5). If $\boldsymbol{v}(t)$ and $\boldsymbol{e}(t)$ with $t \in \mathbb{Z}$ denote, respectively, the "state vector" and the "innovation vector," the state-space model for the process $\{\boldsymbol{u}(t)|t \in \mathbb{Z}\}$ can then be written as a system of two multivariate stochastic equations:

$$\boldsymbol{v}(t) = \mathbf{F}\boldsymbol{v}(t-1) + \mathbf{G}\boldsymbol{e}(t) \tag{8.1}$$

and

$$\boldsymbol{u}(t) = \mathbf{H}\boldsymbol{v}(t). \tag{8.2}$$

Equation (8.1) resembles the model of a univariate discrete-time AR(1) process with zero population mean, $U(t) = \varepsilon(t) = \phi\varepsilon(t-1) + \eta(t)$ [equation (6.10)], except that each random term [i.e., $\boldsymbol{v}(t)$, $\boldsymbol{v}(t-1)$, $\boldsymbol{e}(t)$] is a vector and accordingly the autoregressive coefficient $\mathbf{F}$ is a matrix instead of a scalar. The innovation vector $\boldsymbol{e}(t)$ in equation (8.1) can be seen as a multivariate white noise or an error term that is not autocorrelated, $\mathrm{Corr}[\boldsymbol{e}(t), \boldsymbol{e}(t')] = \mathbf{0}$ if $t \neq t'$, but can have non-zero simultaneous cross-correlations; $\mathrm{Corr}[\boldsymbol{e}(t), \boldsymbol{e}(t)]$ is different from the identity matrix when this is the case. Equation (8.2) expresses how the vector of observations $\boldsymbol{u}(t)$ is related to the state vector $\boldsymbol{v}(t)$, which is of dimension greater than p in general. The state vector can be defined in various ways, using different criteria (Aoki, 1990, p. 51 and p. 99), but in all cases, it is aimed at containing the necessary common factors between the past and the future, with the present at the interface, to predict the future fluctuations of the multiple time series with some error but adequately over a short time span. A classical way to define the state vector is through canonical correlations (Akaike, 1974b, 1975). There exist iterative and non-iterative algorithms to estimate the system matrices $\mathbf{F}$ and $\mathbf{G}$ (Aoki, 1990, pp. 78–79 and Chapter 9).

Models for vector ARMA(p, q) processes can be rewritten in state-space representation form. For example, consider the vector AR(2) process with zero population mean, which is usually defined by the model $\boldsymbol{u}(t) = \mathbf{A}_1\boldsymbol{u}(t-1) + \mathbf{A}_2\boldsymbol{u}(t-2) + \boldsymbol{n}(t)$ [i.e., the p-variate extension of equation (6.11)]. By defining a state vector in two parts, $\boldsymbol{v}^{(1)}(t) = \boldsymbol{u}(t)$

and $\boldsymbol{v}^{(2)}(t) = \mathbf{A}_2\boldsymbol{u}(t-1)$, one obtains a state-space model [equations (8.1) and (8.2)], where

$$\mathbf{F} = \begin{pmatrix} \mathbf{A}_1 & \mathbf{I}_p \\ \mathbf{A}_2 & \mathbf{0} \end{pmatrix}, \quad \mathbf{G} = \begin{pmatrix} \mathbf{I}_p \\ \mathbf{0} \end{pmatrix} \text{ and } \mathbf{H} = \begin{pmatrix} \mathbf{I}_p & \mathbf{0} \end{pmatrix}.$$

It must be noted that (i) in the absence of a moving-average component in vector AR(p) processes, there is no parameter to be estimated in **G**; (ii) the dimension of the state vector is directly proportional to the order of the autoregressive component of the process, and is minimum for the vector AR(1) process because it naturally satisfies the Markovian representation [equation (8.1)] – in this case, $\boldsymbol{v}(t) = \boldsymbol{v}^{(1)}(t) = \boldsymbol{u}(t)$.

The application of state-space models to analyze the heterogeneity of spatio-temporal surface patterns is relatively simple: at a given time $t = 1, \ldots, n$, the vector of p observations $\boldsymbol{u}(t)$ is provided by the observation of one variable at p sampling locations or of 2, 3, etc. variables at $\frac{p}{2}$, $\frac{p}{3}$, etc. sampling locations. The use of these models in this context provides unique information about the dynamics of fluctuations of the random variable(s) of interest in space and time: the entries of matrices **F** and **G**, whether they correspond to the same sampling location, time, or variable, or not, reflect a certain relationship in space, time, space-time, or between variables. Unsurprisingly, state-space models have limitations: the spatio-temporal heterogeneity studied is related to autocorrelation and cross-correlations, the former being temporal while the latter are spatial and spatio-temporal; the application of equation (8.1) assumes stationarity up to order 2, so the EGLS residuals from CRAD Phase I will be used in the examples below; space, which is continuous in general, is here "discretized" because the number p of sampling locations is small relative to the number n of times, which prevents probabilistic modeling in space.

Prior to discussing the results of the example below, it is important to make the following note. While equations (8.1) and (8.2) facilitated the passage from some of the models of discrete-time stochastic processes covered in Chapter 6 to state-space models (see also Priestley, 1981, p. 797), the first of the two equations defining a state-space model is usually written as $\boldsymbol{v}(t+1) = \mathbf{F}\boldsymbol{v}(t) + \mathbf{G}\boldsymbol{e}(t+1)$, with **F** the transition matrix and **G** the input matrix for innovations, to indicate that the model can be used for prediction purposes. This is the writing privileged in Aoki (1990) as well as in SAS PROC STATESPACE, which was used for data analysis hereafter. Similarly, the state vector is then obtained by

Table 8.4. *State-space model analysis of air and soil temperature hourly means at Gault Nature Reserve in Mont-Saint-Hilaire (Québec, Canada) over the period May 1–15, 2004; three time series of temperature data of each type (A = air, S = soil), collected by pairs at three hobo sites (i.e., Hobo46, Hobo63, and Hobo64; see Table 8.1), were analyzed; (a) the estimated (transition) matrix* **F** *and (b) the estimated (input for innovations) matrix* **G**

(a)	Hobo46A(*t*;*t*)	Hobo63A(*t*;*t*)	Hobo64A(*t*;*t*)	Hobo46S(*t*;*t*)	Hobo63S(*t*;*t*)	Hobo64S(*t*;*t*)	Hobo46A(*t*+1;*t*)	Hobo46S(*t*+1;*t*)	Hobo64S(*t*+1;*t*)
Hobo46A(*t*+1;*t*)	0	0	0	0	0	0	1	0	0
Hobo63A(*t*+1;*t*)	**−0.656**	**0.691**	**−0.118**	0.003	−0.029	0.036	**1.072**	0	0
Hobo64A(*t*+1;*t*)	**−1.049**	0.016	**0.590**	0.038	**−0.089**	0.092	**1.403**	0	0
Hobo46S(*t*+1;*t*)	0	0	0	0	0	0	0	1	0
Hobo63S(*t*+1;*t*)	0.023	−0.033	**0.100**	0.234	**0.719**	**0.262**	−0.037	−0.248	0
Hobo64S(*t*+1;*t*)	0	0	0	0	0	0	0	0	1
Hobo46A(*t*+2;*t*)	−0.276	−0.008	**0.227**	0.202	0.142	0.347	**1.133**	−0.230	−0.563
Hobo46S(*t*+2;*t*)	**−0.324**	−0.010	−0.104	−0.143	**−0.121**	−0.271	**0.458**	**1.068**	**0.458**
Hobo64S(*t*+2;*t*)	**−0.431**	0.041	0.027	−0.236	−0.048	−0.218	**0.393**	0.276	**1.227**

(b)	Hobo46A(*t*+1;*t*)	Hobo63A(*t*+1;*t*)	Hobo64A(*t*+1;*t*)	Hobo46S(*t*+1;*t*)	Hobo63S(*t*+1;*t*)	Hobo64S(*t*+1;*t*)
Hobo46A(*t*+1;*t*)	1	0	0	0	0	0
Hobo63A(*t*+1;*t*)	0	1	0	0	0	0
Hobo64A(*t*+1;*t*)	0	0	1	0	0	0
Hobo46S(*t*+1;*t*)	0	0	0	1	0	0
Hobo63S(*t*+1;*t*)	0	0	0	0	1	0
Hobo64S(*t*+1;*t*)	0	0	0	0	0	1
Hobo46A(*t*+2;*t*)	**1.048**	−0.119	**0.378**	−0.048	0.053	0.017
Hobo46S(*t*+2;*t*)	0.138	−0.032	−0.021	**0.939**	0.030	**0.223**
Hobo64S(*t*+2;*t*)	0.034	0.042	0.074	**0.106**	**0.143**	**0.595**

The state vector is composed of the six original variables, denoted Hobo46A(*t*;*t*), Hobo63A(*t*;*t*), Hobo64A(*t*;*t*), Hobo46S(*t*;*t*), Hobo63S(*t*;*t*), and Hobo64S(*t*;*t*), plus three conditional predictions of future observations, Hobo46A(*t*+1;*t*), Hobo46S(*t*+1;*t*), and Hobo64S(*t*+1;*t*). Coefficients in bold are significantly ($P < 0.05$) different from zero.

adding (conditional predictions of) future observations instead of past observations to the current vector of observations $\boldsymbol{u}(t)$.

8.3.1 MSH air–soil temperature example (state-space model analysis)

This MSH example is a continuation of the coregionalization analyses with a drift performed on hourly mean temperature data in Section 8.2. The objectives here are to understand and explain the small-scale temporal fluctuations of air and soil temperatures at the reserve in the first half of May 2004 (i.e., $n = 360$) and to correlate them spatially using a subset of sampling locations, namely the three hobo sites common to Fig. 8.1(a) and (b) (i.e., Hobo46, Hobo63, and Hobo64; $p = 6$). To achieve these objectives and satisfy the stationarity assumption, a state-space model was built on the EGLS residuals of CRAD Phase I. Results are expected to provide insight into the dynamics of air and soil temperatures through their intra- and inter-relationships spatially and temporally. Special attention therefore will be paid to the composition of the state vector and the estimated coefficients of system matrices and their statistical significance.

Different measures of autocorrelation and cross-correlation are reported in Table 8.4. (Note that the statistical significance of estimated coefficients is dependent on their standard error, so two coefficients may have the same value but different standard errors, and thus have different probabilities of significance in the end.) First of all, the state vector is composed of nine components – that is, three more than the initial six of the vector of observations, and two of those three are for soil temperature against one for air temperature, which suggests that temporal autocorrelation (i.e., for the same hobo site at different times) is stronger in the former and that spatial autocorrelation (i.e., for different hobo sites at the same time) is very strong in the latter. Secondly, the statistically significant, positive diagonal entries of the estimated **F** matrix [Table 8.4(a)] indicate positive temporal autocorrelation at lag 1 (between times t and $t + 1$ and times $t + 1$ and $t + 2$) for the two temperature variables, with stronger coefficients for soil temperature overall. For air temperature, positive spatial autocorrelation at a given time can be seen in the statistically significant coefficients 1.072, between Hobo46A(t+1;t) and Hobo63A(t+1;t), and 1.403, between Hobo46A(t+1;t) and Hobo64A(t+1;t); there is no such positive spatial autocorrelation for soil temperature.

Key note: The results above are in accordance with those obtained with CRAD Phase II and presented in Section 8.2. The two sets of results are not directly comparable, though, since the correlations here are lag-based whereas those of Section 8.2 are structure-based, or structural, and are associated with a spherical variogram model that includes all the lags, leaving no room for asymmetrical cross-correlations.

Still in Table 8.4(a), the temporal cross-correlations are spatio-temporal autocorrelations (i.e., at different hobo sites and times for the same temperature variable). At the time lag of 1 h, they are negative for air temperature (see the off-diagonal entries of the top left 3×3 sub-matrix in the estimated **F** matrix), and positive for soil temperature (see the statistically significant coefficients 0.262 and 0.458); the sign may be related to the nature (gaseous vs. solid) of the medium. For their part, spatio-temporal cross-correlations between air and soil temperatures (i.e., for different hobo sites, times and temperature variables) are positive; see the coefficients 0.458 and 0.393. From the estimated **G** matrix [Table 8.4(b)], the moving-average component of soil temperature processes appears to be stronger than that of air temperature processes, with five statistically significant innovation coefficients for soil against two for air; no air–soil innovation coefficient is significantly different from zero.

Key note: The different temporal autocorrelation structures for air and soil temperatures, suggested in Section 8.2 and confirmed above by the composition of the state vector, the diagonal entries of the estimated **F** *matrix and the moving average components through the estimated* **G** *matrix, preclude the use of CRAD for a spatio-temporal joint analysis of both temperature processes.*

In the state-space model analyzed here, coefficients at the time lag of 2 h in the estimated **F** matrix, whether they involve the two temperature variables or not, are more difficult to interpret, and require a "bigger picture" to be discussed. This is the aim of the cross-correlogram analysis presented below.

8.3.2 MSH air–soil temperature example (cross-correlogram analysis)

While numerical information on autocorrelation and cross-correlation is condensed at a few time lags in the system matrices of a state-space model, the temporal cross-correlograms introduced in Subsection 6.6.1 [equations (6.33) and (6.34)] allow an analysis at a larger number of lags, and can be discussed and interpreted as spatio-temporal auto- and cross-correlograms. In fact, temporal cross-correlograms can be computed at positive and negative lags, and can be asymmetrical [Fig. 6.12(d)]. They also have the advantage of graphical representation, and can thus open our eyes to temporal and spatial relationships within and between variables. (Sample cross-correlation coefficients at time lags 2 and more are "true" simple correlation coefficients, unlike coefficients at time lag 2 in the estimated **F** matrix of Table 8.4, which are rather partial correlation coefficients; cf. the autoregressive coefficient ϕ_2 in equation (6.11) for the discrete-time AR(2) process.) Prior to computing temporal cross-correlograms here, prewhitening was performed with an appropriate ARMA(p, q) model, based on the same criteria as in Subsection 6.6.1 (i.e., information criteria, temporal autocorrelation of residuals, statistical significance of estimated coefficients). SAS PROC ARIMA was used for the data analysis.

Clearly, from the temporal cross-correlograms displayed in Fig. 8.4(a1)–(a3), there is a very strong and positive, simultaneous (i.e., at time lag 0) correlation in air temperature between hobo sites, and correlation is practically absent at other time lags including ± 1 h. In other words, spatio-temporal autocorrelation in air temperature is essentially spatial after purely temporal autocorrelation is removed, or there is no space-time interaction in the heterogeneity due to autocorrelation for air temperature. By contrast, temporal cross-correlations for soil temperature [Fig. 8.4(b1)–(b3)] are much weaker, down to 0.2–0.4 at most (from 0.8 for air temperature), but statistically significant sample cross-correlation coefficients (SCCCs) are not mainly limited to time lag 0. In Fig. 8.4(b1), SCCCs are positive and significantly different from zero at 0, +2, +3, and +6 h, indicating that a change in soil temperature at Hobo46 can be related to a similar change in soil temperature at Hobo63 up to 6 h later. For Hobo46 and Hobo64, changes in soil temperature appear to be almost synchronized [Fig. 8.4(b2)]. Accordingly, significant positive SCCCs are observed at negative time lags, from 0 to –4 h, for Hobo63 and Hobo64 [Fig. 8.4(b3)]. Such shifts or synchronization in soil temperature changes between hobo sites may be related to soil properties and local environmental factors such as orientation on the hill and tree canopy openness, the altitude at the three hobo sites being similar. Two main comments can be made on Fig. 8.4(c1)–(c9): (i) spatio-temporal cross-correlations between air and soil temperatures are essentially the same, whether soil temperature is cross-correlated with air temperature from the same or a different hobo site – this is likely a consequence of the strong and positive correlations in air temperature between hobo sites (see above); (ii) the highest SCCCs are observed at positive time lags (i.e., +2 or +6 h, depending on the pair of hobo sites) – this indicates a delay in soil temperature changes relative to air temperature changes over time.

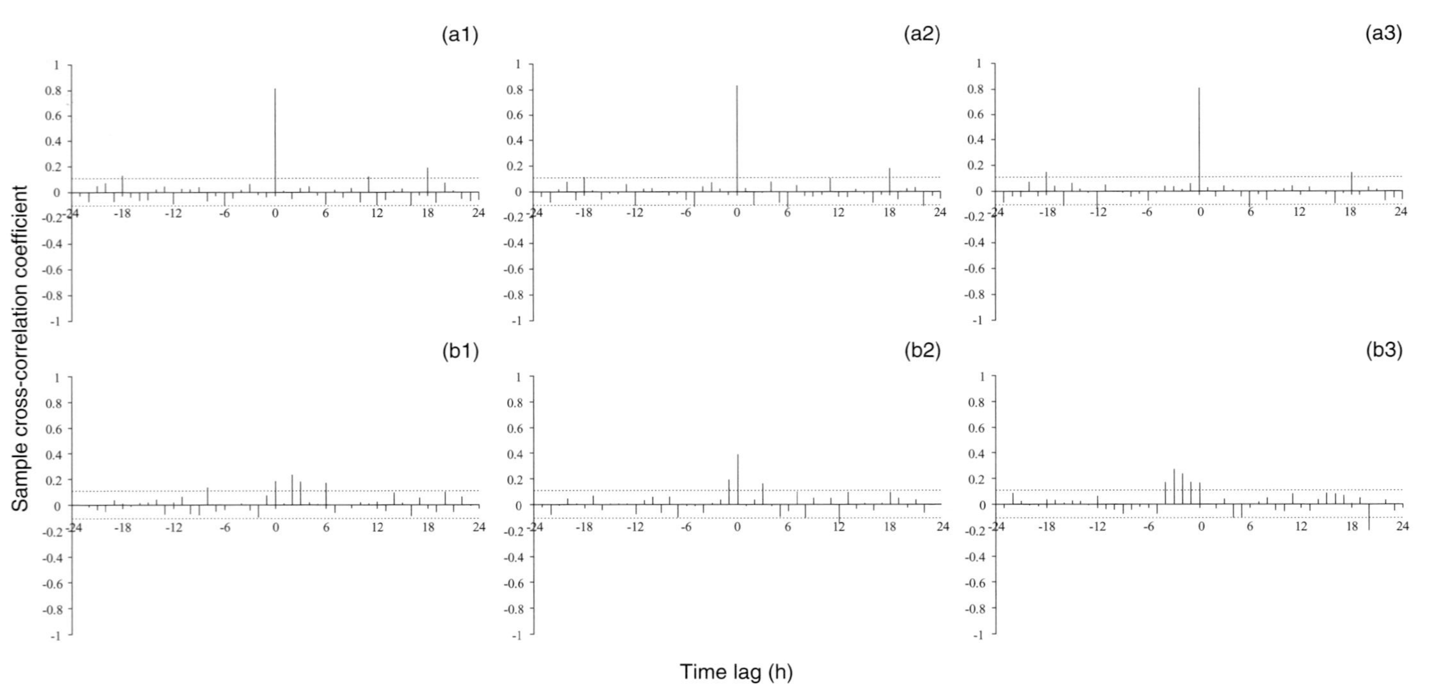

Fig. 8.4. Sample cross-correlograms for hourly mean air and soil temperatures over the period May 1–15, 2004 ($n = 360$), at the three sampling locations common to Fig. 8.1(a) and (b); see Hobo46, Hobo63, and Hobo64 in Table 8.1. Cross-correlations estimated from lag −24 to 24, by increments of 1 h (a1)–(a3) between air temperatures at different sampling locations; (b1)–(b3) between soil temperatures at different sampling locations; and (c1)–(c9) between air and soil temperatures at the same sampling location [diagonal panels (c1), (c5), and (c9)] and at different sampling locations (off-diagonal panels). The EGLS residuals of Fig. 8.2(a2)–(g2) were used for hourly mean air temperature because an orthogonal polynomial of degree 3 and four cosine and sine waves with periods of 24, 12, 8, and 6 h were sufficient to model and remove temporal heterogeneity of the mean in this case; there remained some trace of temporal heterogeneity of the mean in some of the residual series of Fig. 8.3(a2)–(g2) for hourly mean soil temperature, and the CRAD Phase I EGLS residuals were finally used in that case.

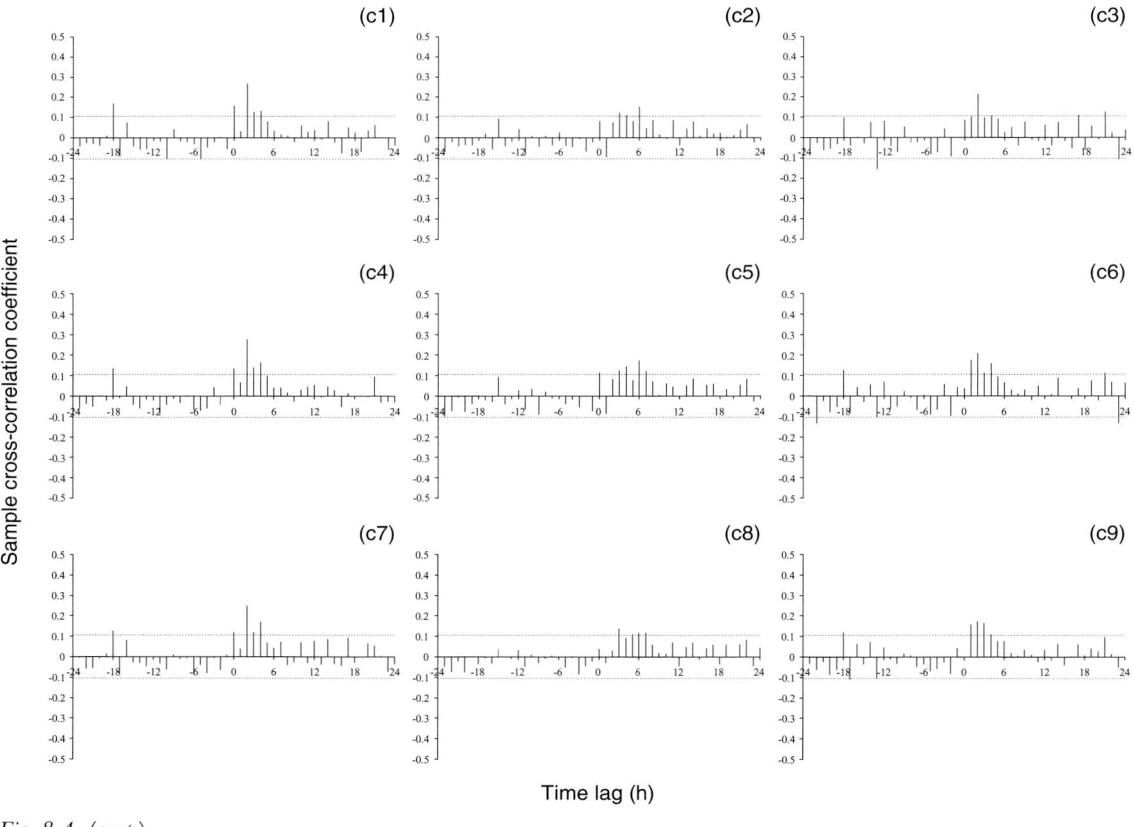
(c1)
(c2)
(c3)
(c4)
(c5)
(c6)
(c7)
(c8)
(c9)
0.5
0.4
0.3
0.2
0.1
0
-0.1
-0.2
-0.3
-0.4
-0.5
-24
-18
-12
-6
0
6
12
18
24
Sample cross-correlation coefficient
Time lag (h)

Fig. 8.4. (cont.)

Summary: When spatio-temporal surface patterns are observed in the form of multiple time series collected at a small number of sampling locations, whether for one or several random variables, state-space models offer the opportunity of analyzing spatio-temporal heterogeneity due to autocorrelation as well as relationships in space and between variables when there are several. In a preliminary step, spatio-temporal heterogeneity of the mean must be appropriately captured and removed. In the continued MSH example, the sample cross-correlograms computed between hobo sites and temperature variables at hourly lags -24 to $+24$ provided an important, easy-to-interpret complement to the system matrices of the state-space model, which are not a function of time per se although they involve time-lagged variables via the state vector.

8.4 Mantel analysis with spatial and temporal distances

In this section, heterogeneity is investigated in spatio-temporal surface patterns via distances. As it was the case for purely spatial surface patterns in Section 7.4, the Mantel test will be used with raw data vs. estimated drifts and the corresponding residuals – estimated drifts will represent the heterogeneity of the mean, and residuals will be susceptible to be autocorrelated. As mentioned in Section 7.4, distances, or dissimilarities, are measures of dispersion, as are variances. Thus, all three types of heterogeneity will be involved in the proposed Mantel analysis. By comparison with the spatio-temporal version of CRAD in Section 8.2 and the state-space models of Section 8.3, the number of sampling locations here is potentially larger, so the spatial data available for analysis are likely to form wider or denser surfaces. After a brief literature review, the specific objectives of this section are stated below.

Historically, the generalized (though simple linear) regression between two distance matrices, later called the "Mantel test," was introduced by Mantel (1967) in the context of spatio-temporal *point* pattern analysis. In his original *Cancer Research* article, Nathan Mantel proposed a statistical method to address the question: Is closeness in space related to closeness in time for people suffering from cancer? In terms of distances, this question can be re-expressed as: Does a relationship exist between (i) the geographical distances between cities, where the people with cancer live (spatial distances), and (ii) the lengths of time intervals between dates when cancer was diagnosed in them (temporal distances)? The Mantel test was then introduced in ecology and applied to distances that were not necessarily spatial and temporal and were evaluated from quantitative instead of point pattern data (see Legendre and Legendre, 1998,

Sections 10.5 and 13.6 for a review of ecological applications). Special attention must be paid when interpreting the outcome of a Mantel test: the tested relationship is about distances, and applying the outcome to the data from which the distances were derived is not without risk (Dutilleul *et al.*, 2000b; see Section 7.4 here). To underline the fact that spatial and temporal distances will be analyzed below, let us rewrite statistic (7.19) as

$$\text{Normalized Mantel statistic} = \frac{\sum\sum (d_{space,ii'} - \bar{d}_{space})(d_{time,ii'} - \bar{d}_{time})}{\sqrt{\sum\sum (d_{space,ii'} - \bar{d}_{space})^2 \sum\sum (d_{time,ii'} - \bar{d}_{time})^2}}. \quad (8.3)$$

In the following examples, the spatial distances $d_{space,ii'}$ $(i \neq i' = 1, \ldots, n)$ are either geographical distances calculated from MTM coordinates or absolute values of differences in altitude between sampling locations (i.e., hobo sites). By working with two different types of spatial distance (i.e., evaluated horizontally in 2-D vs. vertically in 1-D), we will explore the possibility that relationships with temporal distances change with the type of spatial distance. Below, the temporal distances $d_{time,ii'}$ $(i \neq i' = 1, \ldots, n)$, too, are of two types. They consist either in differences in absolute value between maxima (or minima) of the response over a given period at two sampling locations, or in amounts of time separating the occurrence of maxima (or minima) of the response at two sampling locations. The main objective is to test whether spatial distances (horizontal or vertical) are related to temporal distances associated with temperature extrema (maxima or minima).

8.4.1 MSH air–soil temperature example (first Mantel analysis)

In this example, (i) the number of sampling locations, n, is 65 for air temperature [Fig. 8.1(c)] and 16 for soil temperature [Fig. 8.1(b)] – the 16 hobo sites for soil temperature are the same as in Section 8.2, while the 65 hobo sites for air temperature cover most of the reserve; (ii) temperature data are hourly means over the period May 1–15, 2004; (iii) temporal distances are calculated by taking the absolute value of differences between temperature maxima (or minima) at two hobo sites – these distances are "temporal" because they are calculated from temporal extrema in air or soil temperature, but are expressed in Celsius degrees for the same reason; and (iv) Mantel tests are performed on raw data and on estimated drifts and the corresponding EGLS residuals obtained with a joint polynomial–trigonometric model (cf. Section 8.2). Thus, the total number of Mantel tests is 24, with 2 temperature variables (air, soil) × 2 temporal extrema (maximum, minimum) × 3 types of data (raw, drifts, residuals) × 2 types of spatial distance (2-D horizontal, 1-D vertical).

Results are presented in the form of scatter plots in Fig. 8.5(a1)–(d6). For air temperature, it appears that the Mantel statistic changes of sign whether it is evaluated with temporal maxima [panels (a1)–(a6)] or minima [panels (b1)–(b6)] and whether 2-D horizontal [panels (a1)–(a3) and (b1)–(b3)] or 1-D vertical [panels (a4)–(a6) and (b4)–(b6)] spatial distances are used. In (a1)–(a3), the positive values of the statistic (all three significant) indicate that dispersion in the temporal maximum of air temperature increases with increasing horizontal spatial distance – hobo sites that are more distant horizontally show greater differences in temporal maximum of air temperature. In (a4) and (a5), the two significant negative values of the statistic show that hobo sites located at about the same altitude had different temporal maxima in air temperature, whereas hobo sites could have similar temporal maxima in air temperature although they were at low and high altitudes – this might be related to the presence of three hills, the circulation of air between them, and the direction of dominant winds. The sign of relationships is reversed for temporal minima [panels (b1)–(b6)], with only one significant value of the Mantel statistic for 2-D horizontal spatial distances. Differences in results between raw data, estimated drifts (i.e., 24-h periodic components plus large-scale trends), and residuals (i.e., temporally autocorrelated, small-scale random components) are moderate for air temperature, except for temporal minima and 2-D spatial horizontal distances. In other words, there is little evidence for scale-dependent relationships. Note that relationships with temporal distances are the same for 2-D horizontal and 3-D spatial distances because the extent of the sampling domain is much wider in the horizontal plane than in altitude.

For soil temperature, results are different [Fig. 8.5(c1)–(d6)], possibly because of the smaller value of n (i.e., 16 vs. 65 for air temperature), which resulted in the analysis of 120 spatial and temporal distances instead of 2080, or simply because relationships in space and time are different for soil temperature, as suggested by results in previous sections or chapters. Here, significant results for soil temperature are obtained with temporal distances between maxima and 1-D vertical spatial distances. While three times negative for air temperature, the association between these temporal and spatial distances is negative on residuals, but positive on raw data

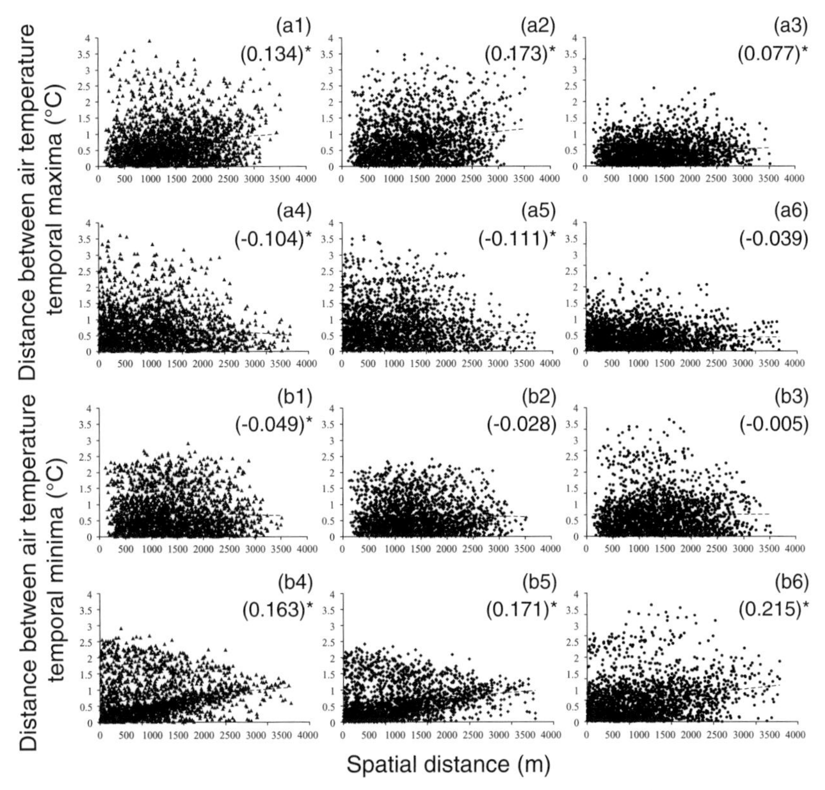

Fig. 8.5. Mantel analyses with spatial distances between sampling locations and distances between temporal extrema (i.e., maxima and minima) in hourly mean air and soil temperatures over the period May 1–15, 2004, at the 65 sampling locations of Fig. 8.1(c) for air temperature and at the 16 sampling locations of Fig. 8.1(b) for soil temperature. Results for air temperature here are presented in the (a) and (b) panels, and those for soil temperature in the (c) and (d) panels. Spatial distances between sampling locations are Euclidean distances of two types: horizontal, calculated from the MTM easting and northing coordinates [panels (a1)–(a3), (b1)–(b3), (c1)–(c3), and (d1)–(d3)], and vertical, calculated from the altitude values (other panels). The calculation of distances between temporal extrema was based on the observed time series (panels numbered 1), the predicted time series resulting from the joint EGLS fitting of an orthogonal polynomial of degree 3 and four cosine and sine waves with periods of 24, 12, 8, and 6 h (panels numbered 2), and the corresponding EGLS residuals (panels numbered 3). Observed values of the normalized Mantel statistics are reported in parentheses, and followed by an asterisk (∗) when significantly ($P < 0.05$) different from zero.

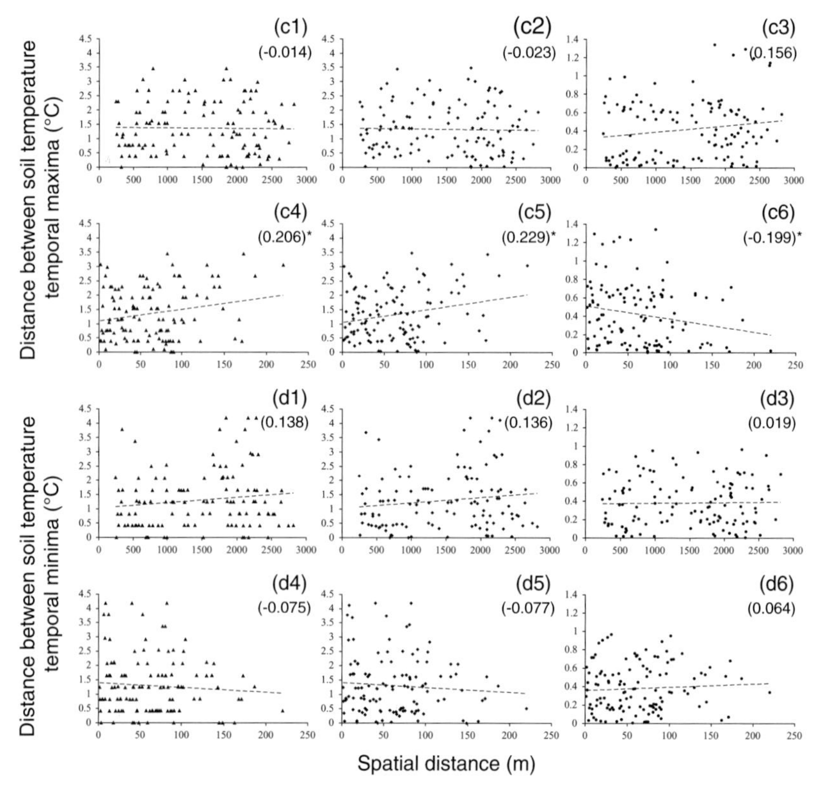

Fig. 8.5. (cont.)

and estimated drifts for soil temperature [Fig. 8.5(c4)–(c6)]. The three relationships are clear from graphs and may be interpreted as follows: the two positive ones indicate that the larger the difference in altitude, the wider the dispersion in temporal maximum with raw data and the mean signal, which includes the 24-h periodicity and large-scale trend – such results are related to the distinct nature of the soil medium compared to air; the negative relationship shows that the dispersion in temporal maximum of the autocorrelated small-scale random component is lower for hobo sites located at opposite altitudes, and vice versa – this may reflect an influence of air in soil temperature changes at small scale; see the three negative relationships found for air temperature [Fig. 8.5(a4)–(a6)]. Several non-significant values of the Mantel statistic (e.g., 0.156, 0.138, and 0.136, with 2-D horizontal spatial distances) would have been declared significant if they had been observed for air temperature (i.e., $n = 65$).

8.4.2 MSH air temperature example (second Mantel analysis)

This Mantel analysis was performed on raw data of air temperature for two days of May 2004 treated separately, using mean temperatures over 15-min intervals in order to refine the location of temporal extrema. Data were collected at the 65 hobo sites of Fig. 8.1(c). A temporal distance here is defined as the amount of time separating the temporal maxima (or minima) within a day at two hobo sites, and is thus expressed in multiples of 15 min. This type of temporal distance is closer to that analyzed in the original article of Mantel (1967), because it is related to times of occurrence (of temperature extrema); these times, however, are derived from surface patterns instead of point patterns. Since nights are generally colder than days, the period of time was centered, which ensured continuity in the data analyzed; a day was a 24-h period from midnight to midnight for temporal maxima and from noon to noon for temporal minima. Eventually, the number of Mantel tests is 8, with 2 days × 2 temporal extrema (maximum, minimum) × 2 types of spatial distance (2-D horizontal, 1-D vertical).

Two specific hypotheses were tested about the relationships between spatial and temporal distances: (I) they are consistent whether temporal distances are calculated from the temperature extrema themselves or from their times of occurrence; and (II) those relationships are stable from day to day. For the main part, Hypothesis (I) is accepted, since Fig. 8.6(a1)–(d2) repeats the relationships seen in Fig. 8.5(a1), (a4), (b1), and (b4), with the exceptions of the lack of statistical significance in Fig. 8.6(b1) and the positive sign in Fig. 8.6(c1). Concerning Hypothesis (II), the change of sign of the relationship in Fig. 8.6(c1) and (c2) provides a counter-example to the use of days as replicates in this type of analysis.

Summary: The Mantel analysis with spatial and temporal distances offers the possibility of studying the three types of heterogeneity in spatio-temporal surface patterns, by calculating distances (i.e., measures of dispersion) from estimated small-scale and large-scale components of surface patterns (i.e., the former component is likely to be autocorrelated, and the latter, to show heterogeneity of the mean; see first example) and by investigating relationships between spatial and temporal distances empirically. When the spatial index is 3-D, this investigation may take place in 1-D and 2-D spaces and show associations of different signs whether vertical or horizontal spatial distances are analyzed, as in the two MSH examples. The statistical procedure is particularly interesting for the analysis of

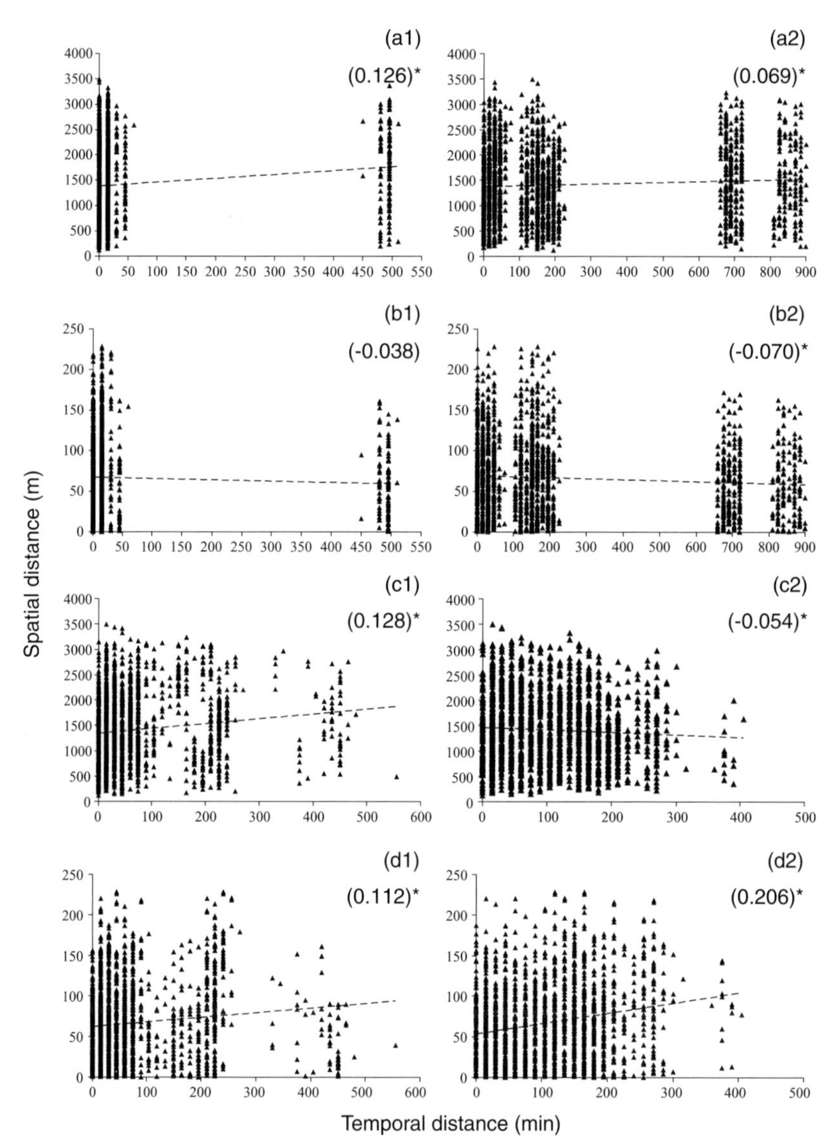

Fig. 8.6. Mantel analyses with spatial distances between sampling locations and temporal distances between 15-min intervals in which an extremum (i.e., maximum or minimum) in air temperature was observed on May 4–5, 2004 [panels numbered 1], and May 9–10, 2004 (panels numbered 2), at 61 of the 65 sampling locations of Fig. 8.1(c). Spatial distances between sampling locations are Euclidean distances calculated horizontally from the MTM easting and northing coordinates [panels (a1), (a2), (c1), and (c2)] and vertically from the altitude values (other panels). Using 15-min intervals, air temperature maxima were determined from midnight to midnight on May 5 and May 10, 2004, while air temperature minima were determined from noon to noon on May 4–5 and May 9–10, 2004. Observed values of the normalized Mantel statistics are reported in parentheses, and followed by an asterisk (*) when significantly ($P < 0.05$) different from zero.

extrema (maxima vs. minima), including their times of occurrence within different observational periods (see second example).

8.5 Recommended readings

On a fundamental basis, Chapter 2 ("Spatiotemporal Geometry") in the book of Christakos (2000) provides the reader with a thoughtful discussion on spatio-temporal metrics and distances and the spatio-temporal framework in general, its physical and mathematical bases and what they may mean for the natural sciences. On a more philosophical note, with a minimum number of mathematical expressions, the book of Sklar (1984), especially the second chapter in it, treats similar subjects and the Minkowski space, or space-time, in particular. Even if the framework is purely spatial instead of spatio-temporal, the articles by Pelletier *et al.* (2009a, 2009b) are the reference for the method of coregionalization analysis with a drift (CRAD), of which a spatio-temporal version was presented here in Section 8.2. The articles by Akaike (1974b, 1975) and the book of Aoki (1990), together with Priestley (1981, Sections 10.4 and 10.5), are key references for state-space model analysis; simply note that the writing style in them is more technical than in Section 8.3. When he proposed his statistical method to cancer researchers, Mantel (1967) did not know that it would become so popular outside of his field of investigation; the reading of the original publication is recommended, as is that of Dutilleul *et al.* (2000b) for guidelines in applications. Volumes on statistical methods for spatio-temporal data and their applications are not numerous, and the volume edited by Finkenstädt *et al.* (2007) is one in which the chapters by Jensen *et al.* "Spatio-temporal modelling – with a view to biological growth" and by Chandler *et al.* "Space-time modelling of rainfall for continuous simulation" should trigger the interest of biologists and environmentalists.

9 · *Sampling and study design aspects in heterogeneity analysis of surface patterns*

This chapter focuses on surface patterns. It has a threefold objective: (i) to discuss sampling principles in relation to the analysis of the three types of heterogeneity studied in this book, particularly in Chapters 6 to 8; (ii) to present theoretical measures of heterogeneity of the variance and heterogeneity due to autocorrelation in experimental designs; and (iii) to follow an applied approach and provide examples with results of real data analyses. Using four sampling grids representing main types of sampling design in 2-D space, sampling aspects are discussed in Section 9.1 in the context of the analysis of heterogeneity of the mean and trend surface analysis and in that of the semivariance and autocorrelation analyses. After a reminder of the ANOVA method and models for classical experimental designs and an introduction to repeated measures designs with one and two repeated measures factors, two theoretical measures of heteroscedasticity and autocorrelation are defined and evaluated in Section 9.2 in a total of six models for various variance–covariance structures. The measures are the effective numbers of degrees of freedom associated with expected mean squares in the ANOVA method and Box's "epsilon," due to Box (1954a, 1954b). Such an approach is particularly interesting, since it replaces millions of data simulations. The perspective is definitely more applied in Section 9.3, where the nature, fixed or random, of terms in the ANOVA model is discussed in relation to the heterogeneity of the mean on one hand and heteroscedasticity and autocorrelation on the other. This section is completed with three examples using real data, two in wood science (in time and space-time) and one in the environmental sciences (in space-time). As usual, recommended readings are given in the last section of the chapter (Section 9.4). Mathematical and statistical details required to understand the foundations of theoretical results in Section 9.2 are grouped in Appendix A9. The seven SAS codes used to calculate effective numbers of degrees of freedom and theoretical values of Box's "epsilon" are on the CD-ROM in a folder called "Expectation," because they are all related in some way to the expected mean squares which are at the basis of the ANOVA method; SAS codes used for the examples with real data are also available on the CD-ROM.

9.1 Sampling aspects

The statistical literature includes many books entirely devoted to sampling methods. My objective below is to introduce readers to a number of sampling aspects and discuss these aspects in relation to heterogeneity analysis of surface patterns. The interested reader will find information regarding other sampling aspects in the recommended readings of Section 9.4.

9.1.1 Basic sampling designs

The four sampling grids ($n = 100$) of Fig. 9.1 represent as many main types of sampling design in 2-D space:

- simple systematic [panel (a)] – the sampling grid is square (10×10) here, but it could be rectangular, circular, or elliptic, depending on the shape of the sampling domain (i.e., a square of side length 10 in Fig. 9.1);
- multiple systematic, with one grid that covers the sampling domain and on which a number of smaller grids are superimposed, each of the grids corresponding to a systematic sampling in the entire sampling domain or a portion of it [panel (b)] – the former grid is 6×6 and the latter are four smaller 4×4 grids here, but one 8×8 grid and four smaller 3×3 grids would have been possible too ($n = 100$);
- simple random [panel (c)], for which the generator of uniform pseudo-random numbers in a computer program (e.g., SAS, Matlab) can be used to define randomly the spatial coordinates of sampling locations, without any constraint within the sampling domain – the expressions "completely random" and "completely randomized" are reserved for a given type of point pattern (Chapters 3 to 5) and of experimental design (Section 9.2), respectively;
- stratified random [panel (d)], in which the sampling domain is divided into a number of plots of same shape and size, or strata, and a given number of sampling locations are randomly defined in each stratum – 100 squares of side length 1, with one sampling location in each of them, were used as strata here, but 25 squares of side length 2, with four sampling locations randomly defined inside each square, could have been used instead ($n = 100$).

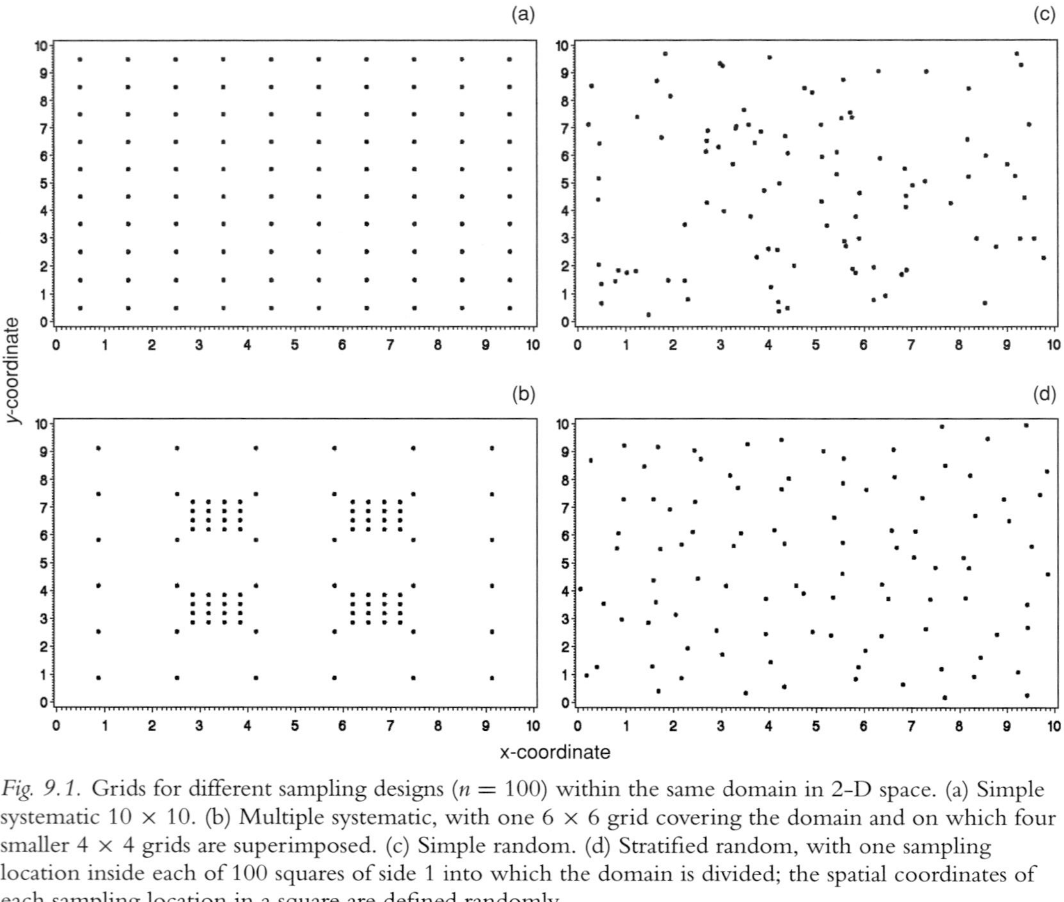

Fig. 9.1. Grids for different sampling designs ($n = 100$) within the same domain in 2-D space. (a) Simple systematic 10×10. (b) Multiple systematic, with one 6×6 grid covering the domain and on which four smaller 4×4 grids are superimposed. (c) Simple random. (d) Stratified random, with one sampling location inside each of 100 squares of side 1 into which the domain is divided; the spatial coordinates of each sampling location in a square are defined randomly.

If one considers that the sampling grid of Fig. 9.1(b) is a "hybrid" because it is the result of the combination of two sorts of simple regular sampling grid, other "hybrid" sampling grids may be considered, for example, by combining a simple regular grid with a simple random one, in order to benefit, if possible, from the advantages of the two types of sampling grid (see below).

Key note: A simple random sampling grid can be generated in the same way a completely random point pattern in 2-D space is simulated (e.g., by using the SAS function RANUNI; see Section A3.2). There are several fundamental differences between the two contexts, though. Unlike a completely random point pattern, no stochastic process is directly behind a simple random sampling grid. A node in a simple random sampling grid represents a location at which the continuous quantitative random variable of interest has been or will be observed and all the observations thus made provide a surface pattern or partial realization of a stochastic process, whereas a point in a completely random point pattern is an observation by itself, and composes with other points a partial realization of the underlying point process.

The spatial distribution of sampling locations in the four grids of Fig. 9.1 can be described as follows. In the case of the simple systematic 10×10 grid [Fig. 9.1(a)], it is characterized by one sampling location per area unit and a 1-unit distance between neighboring nodes. In this case, each observation in the surface pattern "represents" a distinct square of side length 1, and the sampling scale is readily defined and "exact." Distances between any two sampling locations are easy to calculate, from $1, \sqrt{2}, 2, \sqrt{5}, \sqrt{8}, 3, \ldots$ to $\sqrt{162}$, and can be used as "true" distances, instead of through distance classes, in the semivariance and autocorrelation analyses. In Fig. 9.1(b)–(d), there is one sampling location per area unit *on average*. In Fig. 9.1(b), there are two "exact" sampling scales. The 6×6 grid defines one, based on the distance of $\frac{10}{6}$ units between its neighboring nodes, and each of the four smaller 4×4 grids defines another (smaller), with the distance of $\frac{1}{3}$ unit. Such a combination of systematic sampling grids is the beginning of a hierarchical or multi-scale sampling grid. The sampling scale is approximate in Fig. 9.1(c) and (d), where the distance between neighboring nodes of the grid varies from much smaller or smaller than 1, to larger or much larger than 1. More extreme distance values (close to 0 and of several units) are provided by the simple random sampling grid, compared to the random stratified one. In theory, the average distance between neighboring nodes is 1 for the stratified random sampling grid, because each sampling location in a square of side length 1 is *expected* to be at the centre by randomization. (The expected value of the uniform distribution over the [0, 1] interval is 0.5.) Empirically though, each sampling location in Fig. 9.1(d) has eight potential nearest neighbors (in the eight 1×1 squares touching the 1×1 square in which the sampling location in question is), so the mean distance between nearest neighbors happens to be smaller than 1 for the stratified random sampling grid, compared to the simple systematic one. This *might* be an advantage for small-scale autocorrelation analysis (see the discussion in Subsection 9.1.3). Note that (i) a stratified random sampling grid with 25 square strata of side length 2 and four sampling locations per stratum would cover the domain less well, with the possibility of providing more uneven distance classes than the grid of Fig. 9.1(d); providing more (ii) if one continues to increase the size of strata further, a stratified random sampling with only one stratum is nothing but a simple random sampling.

The basic differences among the four sampling grids and designs have a number of implications for 2-D spatial heterogeneity analysis, as we will see below. In the next two subsections, the ability to capture heterogeneity of different types in 2-D space will be discussed in relation to the type of sampling grid or design. In doing this, the sample size, or grid size, will be kept fixed (i.e., $n = 100$). Where appropriate, references from the literature will be used to support statements about effects related to the number of sampling locations.

9.1.2 Comparison of sampling designs when heterogeneity of the mean is of interest

The objective of the following discussion is to compare sampling designs in their ability to give access to heterogeneity of the mean before any data analysis. It is even assumed that the mean function $\mu(x, y)$ is directly observable without error. Thus, the contour maps in Fig. 9.2 were produced with PROC GPLOT of SAS/GRAPH (SAS Institute Inc., 2009), using the values of (polynomial, negative quadratic exponential, cosine) functions at the nodes of the sampling grids of Fig. 9.1. To achieve a resolution of 0.01×0.01, linear interpolation was applied to triplets of neighboring values with PROC G3GRID of SAS/GRAPH (SAS Institute Inc., 2009), but no statistical method of data analysis was really performed.

In Fig. 9.2, heterogeneity of the mean is at relatively large scale, since the spatial structure is not repeated or repeated only one time over the sampling domain [cubic trend, panels (a); one central patch, panels (b)], or at most four times [four patches, each occupying one quarter of the domain, panels (c)]. The "true," or exact, surface patterns (of pure heterogeneity of the mean) are mapped in Fig. 9.2(a1), (b1), and (c1). Since no statistical method of data analysis was performed, the other 12 surface patterns are not said to be estimated or predicted, but are called "approximations."

Two common points arise from the comparison of the three "true" surface patterns and their approximations, obtained with the four sampling grids of Fig. 9.1: (i) differences between sampling designs are not large, but (ii) systematic designs provide the basis for better approximations than random samplings, when heterogeneity of the mean at intermediate-to-large scales is of interest. More specifically, the contours drawn in Fig. 9.2(a2)–(a3), (b2)–(b3), and (c2)–(c3) are closer to the "true" ones in (a1), (b1), and (c1), than those in (a4)–(a5), (b4)–(b5), and (c4)–(c5), as some discrepancies (i.e., altered rates of change, deformed patches) are noticeable. These discrepancies are the result of gaps of varying size in the sampling grid, even with stratification (Subsection 9.1.1). With a planar trend surface (results not shown), there is practically no difference between the four sampling designs.

Key note: Random sampling is not necessarily the best in spatial statistics, at the least not when spatial heterogeneity of the mean at intermediate-to-large scales is of interest. In particular, if a simple random sampling design does not provide a better reconstruction of the true mean function when accessible without error, a fortiori when sample data are submitted to statistical analysis results are likely to be not as good as with a simple systematic sampling design, among others.

9.1.3 Comparison of designs when heterogeneity due to autocorrelation is of interest

At the basis of the following comparison between sampling designs is a characteristic (i.e., autocorrelation) that concerns the random component of a stochastic process, represented by $\varepsilon(s)$ in fundamental equation (2.1), by opposition to the deterministic component $\mu(s)$ (i.e., the mean function) that was the focus in the previous subsection. Accordingly, in 2-D space no map of data can represent an autocorrelation structure uniquely. Instead, the autocorrelation functions used here to compare designs are represented via the theoretical semivariance functions, for a given range value in a stationary and isotropic variogram model [Fig. 9.3(a1) and (b1); cf. Fig. 7.1]. For each of the four sampling grids

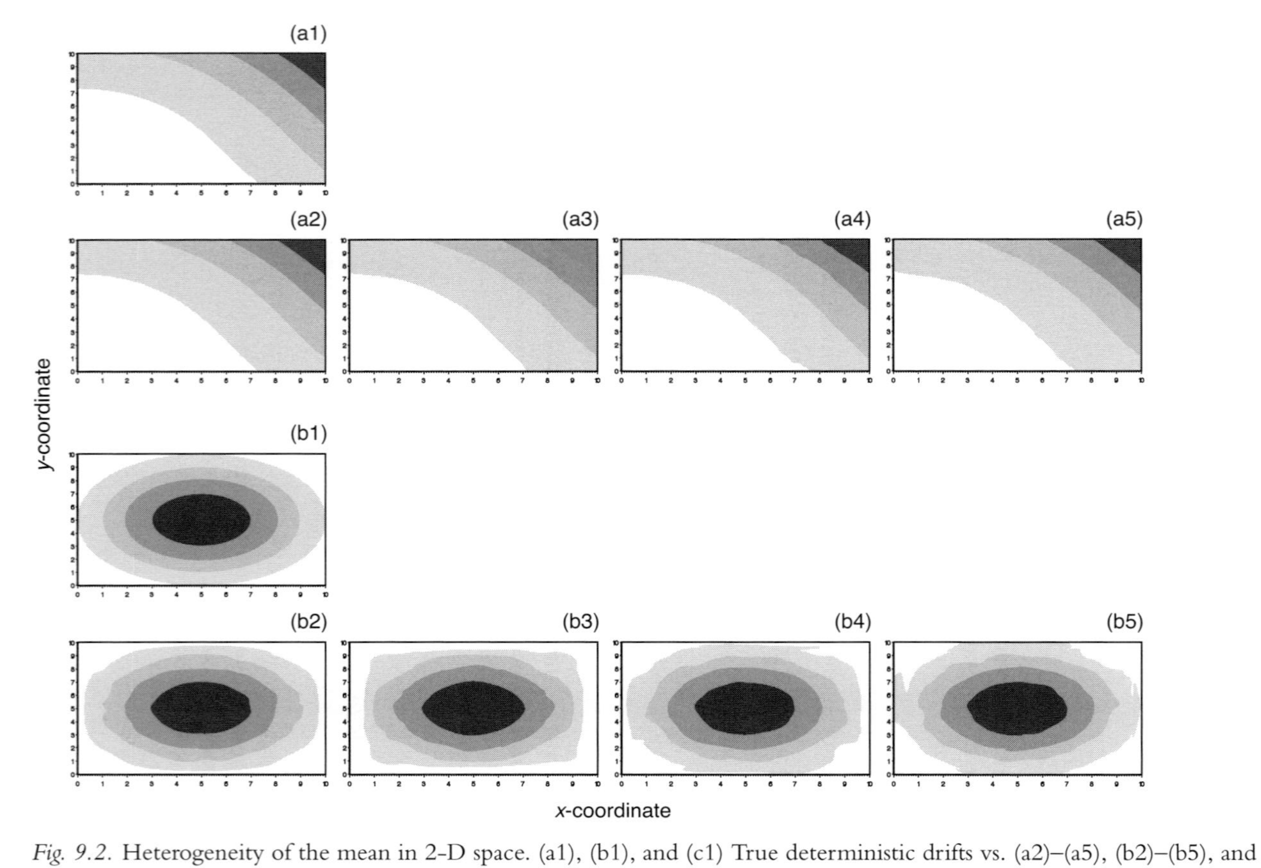

Fig. 9.2. Heterogeneity of the mean in 2-D space. (a1), (b1), and (c1) True deterministic drifts vs. (a2)–(a5), (b2)–(b5), and (c2)–(c5) approximations obtained by interpolation using the four sampling grids of Fig. 9.1. (a1) Purely cubic polynomial trend, $\mu(x, y) = x^3 + y^3 + x^2y + xy^2$. (b1) Bell-shaped trend surface, or single patch, $\mu(x, y) = \exp(-d(x, y)^2)$, with $d(x, y)^2 = \frac{(x-5)^2+(y-5)^2}{6.75}$. (c1) Cosine waves with a "period" of 5 in both directions, or multiple patches, $\mu(x, y) = \cos(\frac{2\pi x}{5}) + \cos(\frac{2\pi y}{5})$. Gray tones go from light (lower values) to dark (higher values).

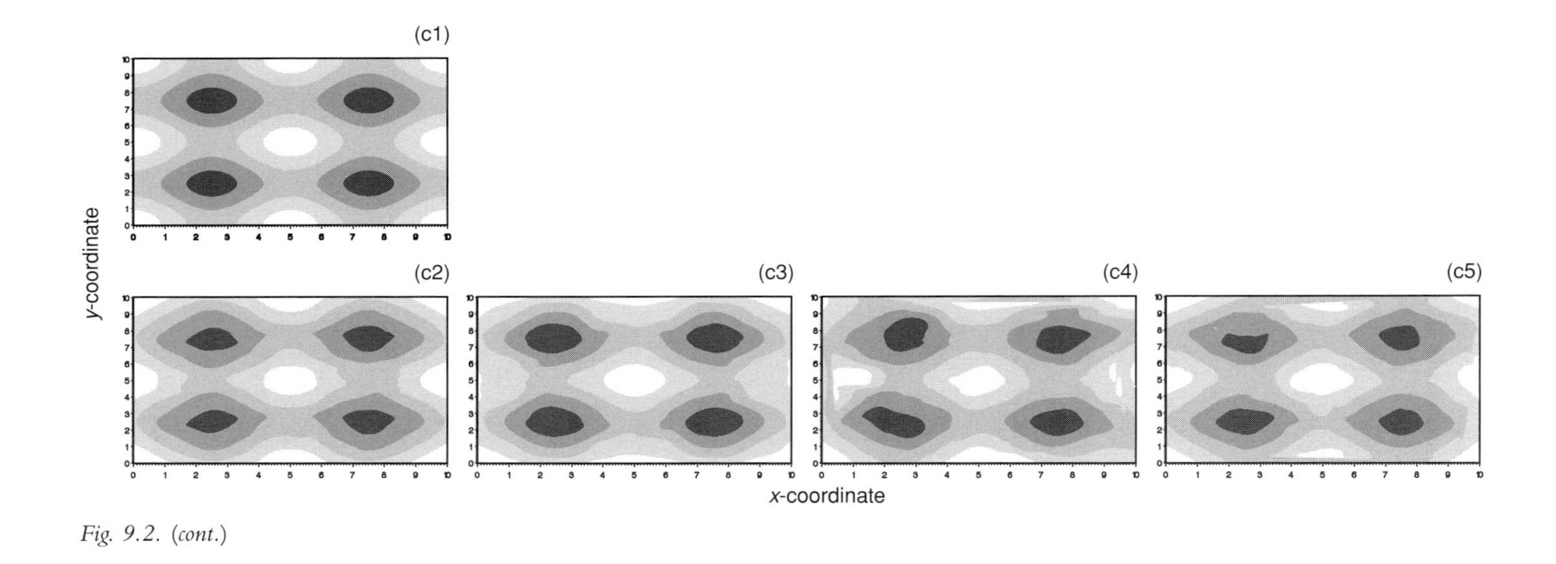

Fig. 9.2. (cont.)

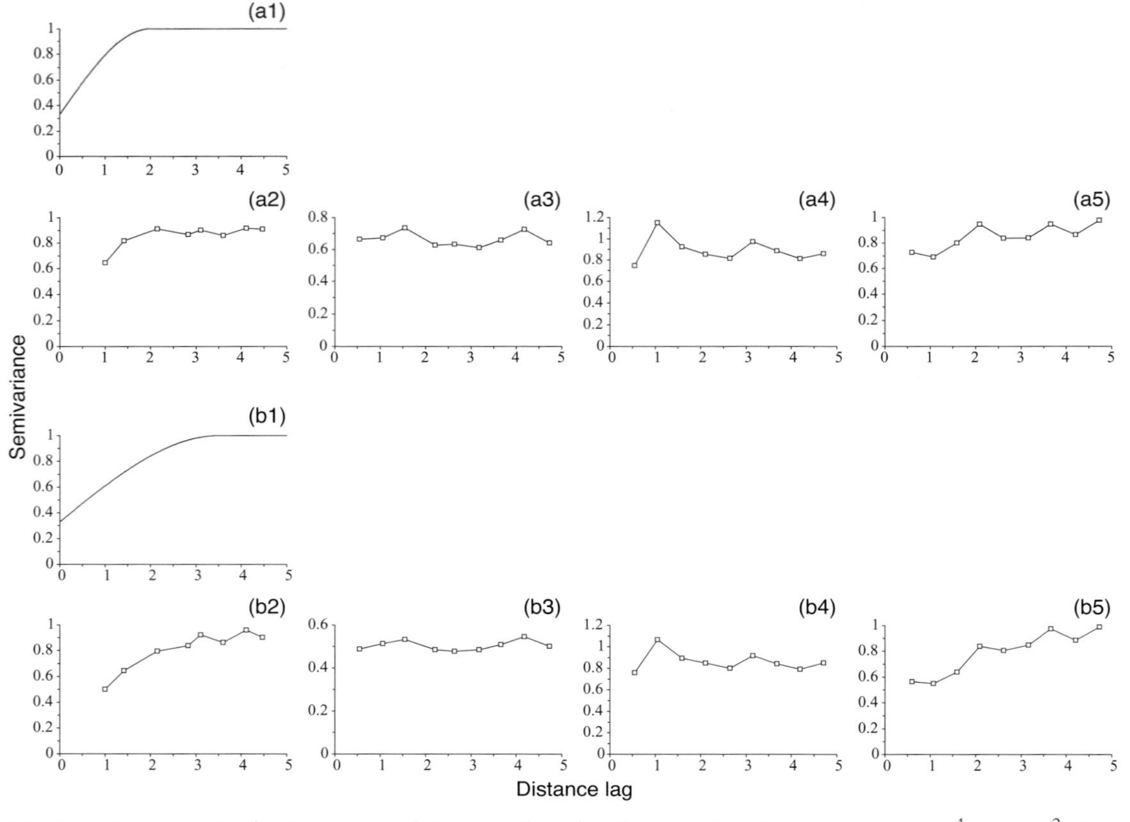

Fig. 9.3. Heterogeneity due to autocorrelation, combined with unequal variance components ($\frac{1}{3}$ against $\frac{2}{3}$) for the nugget effect (non-spatial) and the spherical basic function (small-scale, spatially autocorrelated) in the variogram model. (a1) Theoretical variogram model with a range of 2 for spatial autocorrelation and (a2)–(a5) experimental variograms computed from partial realizations obtained using the four sampling grids of Fig. 9.1 and the same "seed" in simulations. Similarly, the theoretical variogram model with range 3.5 and the four experimental variograms are plotted in (b1) and (b2)–(b5), respectively.

of Fig. 9.1, one partial realization of the 2-D spatial stochastic process, or surface pattern, was simulated, and each simulated surface pattern was used to compute one experimental variogram [Fig. 9.3(a2)–(a5) and (b2)–(b5)]. Such an approach is intended to be empirical and simple, leaving more complicated theoretical aspects for the next section. Despite the inherent variability between partial realizations, this will be enough to point out a number of differences among the four designs defined by the sampling grids of Fig. 9.1, when heterogeneity due to autocorrelation is of interest.

The computation of experimental variograms is the only statistical analysis of data performed, Matheron's semivariance estimator [equation (7.10)] being evaluated at distances or distance classes that can be compared as much as possible between sampling designs. Following up on the discussion about the spatial distribution of sampling locations in Subsection 9.1.1, there is no distance between sampling locations smaller than 1 in the simple systematic sampling grid [Fig. 9.1(a)], whereas the number of pairs of sampling locations in the first distance class (requested average distance: 0.525) varies a lot between the three other sampling grids, from 374 to 87 and 52 for the multiple systematic, simple random and stratified random sampling grids [Fig. 9.1(b)–(d)], respectively. The first two distances in experimental variograms computed from data simulated on the simple systematic sampling grid are not classes, as they correspond to nearest neighbors and diagonal neighboring nodes of the grid and are exactly equal to 1 and $\sqrt{2}$. No model is fitted to any experimental variogram because the objective is to discuss the quality of experimental variograms provided by different sampling designs as a basis for model fitting, instead of the quality of the model fitting itself. Note that, in terms of "minimum sample size," Webster and Oliver (1992) recommend 100 or more sampling locations for variogram analysis, but other factors must be taken into account, such as the ratio between the range of spatial autocorrelation and the extent of the sampling domain, and even greater numbers of sampling locations are required when parameter estimates obtained by variogram model fitting are used for multi-scale correlation analysis (Larocque *et al.*, 2007).

Relative to the extent of the 10×10 sampling domain in Fig. 9.1, the two values considered for the range of spatial autocorrelation (i.e., 2 and 3.5) in Fig. 9.3 are small and small-to-intermediate. In view of the theoretical semivariance functions and experimental variograms in Fig. 9.3(a1)–(a5) and (b1)–(b5), the following observations can be made. For both range values, and despite the fact that their evaluation started at

1 distance unit (see above), the experimental variograms for the simple systematic sampling design [Fig. 9.3(a2) and (b2)] represent the theoretical variance–covariance structure of the underlying 2-D spatial stochastic process very well, since one can easily imagine the theoretical curve across the semivariance estimates, including a nugget effect with a variance component smaller than that of the spatially autocorrelated structure (spherical here) and a longer range of autocorrelation in (b1)–(b2) than in (a1)–(a2). For both range values, the multiple systematic sampling design provides an experimental variogram that does not represent the theoretical semivariance function well [Fig. 9.3(a3) and (b3)], with a nugget effect that would be dominant over the spatially autocorrelated structure, no real evidence for a different range of autocorrelation from (a3) to (b3), and a plateau (supposedly reached at a height equal to the value of the sample variance) that is much too low, as a result of the strong positive autocorrelation at small distances for many of the simulated data. The two random sampling designs provide experimental variograms of intermediate quality, with a plateau that would be reached at good height and a nugget effect that would be exaggerated to moderate degrees, but with evidence for shorter ranges of autocorrelation than expected (simple random sampling) and longer ranges of autocorrelation than expected (stratified random sampling). The problems noticed for the two random sampling designs are related to the numbers of pairs of observations used in the evaluation of the experimental variogram at very small and small distance classes (i.e., those that matter for capturing the nugget effect and estimating the range of autocorrelation in model fitting). As mentioned above, those numbers (especially at first distance class) are small here, and Cressie (1985) showed that the variance of semivariance estimators is inversely proportional to the number of pairs of observations involved in their evaluation. Accordingly, the variance of experimental variograms at first distance class is large in Fig. 9.3(a4)–(a5) and (b4)–(b5), and the observed value of the semivariance statistic could have been much higher as well as much lower for another partial realization of the 2-D spatial stochastic process.

Key note: As in the analysis of spatial heterogeneity of the mean at intermediate-to-large scales, simple systematic sampling was found to be better than the three other designs when autocorrelation at small-to-intermediate scales is of interest, using tools (experimental variograms) that are in accordance with the nature of the characteristic studied. Again, simple random sampling was not found to be the best, although it may help capture the nugget effect and estimate the range of spatial autocorrelation when fitting a model, provided the number of sampling locations is sufficiently large and the range is short, thanks to a first class of distances that are smaller than in simple systematic sampling; the stratified random sampling design offers similar possibilities.

9.2 Study design: fundamental theoretical aspects

In this section, we will explore an avenue based on sound experimental design combined with the appropriate use of model equations and minimum mathematical development, to determine the importance of heterogeneity in spatial and temporal stochastic processes. Although at least one term is reserved for heterogeneity of the mean in most of the models, more emphasis will be put on the effects of heterogeneity of the variance and that due to autocorrelation on the results of statistical analysis.

To do this, we will proceed in a number of steps. In Subsection 9.2.1, six experimental designs will be presented and discussed, together with their respective ANOVA models (ANOVA: analysis of variance). The "expected mean squares," which are at the basis of the definition of ANOVA F-tests, will be defined in Subsection 9.2.2. Finally, two types of theoretical measures of heterogeneity that can be readily evaluated from a set of variance–covariance parameter values (i.e., without data simulation) will be established in Subsection 9.2.3. Important differences exist between the ANOVA-based approach to be followed here and the approach followed when using SAS PROC MIXED in Chapters 6 and 7; these differences are addressed and discussed mid-way of Subsection 9.2.2.

9.2.1 Experimental designs and analysis of variance: an overview

Among the six experimental designs studied here, four are in 2-D space and their field lay-outs are illustrated in Fig. 9.4(a)–(d), the other two being with simply and doubly repeated measures; temporal repeated measures (Chapter 6) provide an example of simply repeated measures. A general guideline in defining an experimental design is to take into account as many factors as possible in order to minimize the experimental error. As we will see below, the number of factors aimed at capturing spatial heterogeneity of the mean ranges from zero (i.e., when absent) to two (i.e., when present in perpendicular directions).

To discuss the field lay-outs of Fig. 9.4, imagine that four tree plantations were established with monoculture and mixed-species plots at

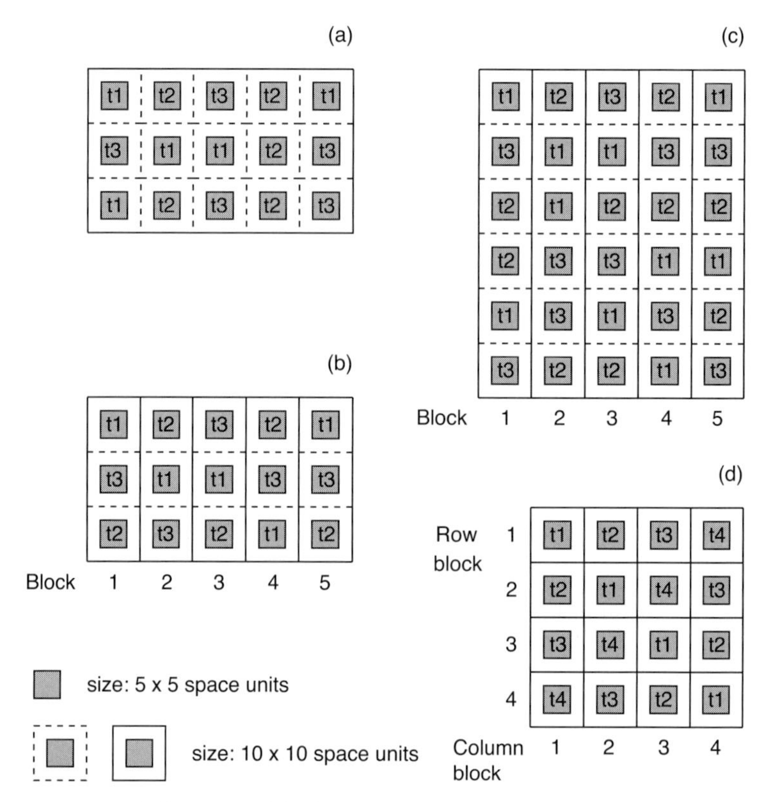

Fig. 9.4. Field lay-outs of the four experimental designs for which numerical results are presented in Fig. 9.5. (a) Completely randomized design. (b) Randomized complete block design with one replicate per treatment per block. (c) Randomized complete block design with two replicates per treatment per block. (d) 4 × 4 Latin square design. Buffer zones are in white; plots for which spatial data are actually collected are in gray; the number of treatment levels is three in (a)–(c) and four in (d).

sites characterized by various environmental conditions, to study the relationship between biodiversity and ecosystem functioning (cf. Potvin and Dutilleul, 2009). Height, basal diameter, and relative growth rate in diameter at breast height are examples of random variables that can be observed at the level of individual trees and averaged at the level of a plot. To avoid interference from trees of different plots, buffer zones are used between plots (cf. Fig. 9.4), so that only data from trees that are well inside the plot are used in the calculation of the average value.

In (a), the field, which can be divided into $3 \times 5 = 15$ plots, is spatially homogeneous, in that nothing appears to affect systematically the observed tree growth, except possibly the number and identity of species in a plot. Accordingly, a completely randomized design (CRD) is appropriate, and five replicates of each of three treatments (e.g., t1: monoculture; t2: three-species mixture; t3: six-species mixture) can be randomly assigned to the 15 plots. In (b), the field is of same shape and size as in (a), but there is a gradient in altitude (i.e., a hill) from left to right, which creates a gradient in soil moisture that could affect tree growth. Therefore, blocking the assignment of treatments is required, and a randomized complete block design with one replicate per treatment per block (RCBD 1 rep), with five blocks perpendicular to the gradient and 3 plots per block, is appropriate. Compared to the CRD, the randomization of treatments among plots must be repeated from block to block. In (c), the field is of double size than in (a) and (b), and there is a left-to-right gradient in altitude as in (b). The researcher is concerned by the fact that differences between biodiversity levels may change, in sign or in size, with altitude. The larger size of blocks, with six instead of three plots each, allows her to test for such an interaction between the treatment and block factors, by using a randomized complete block design with two replicates per treatment per block (RCBD >1 rep). In (d), a hill occupies 25% of the territory in the center of the field, so that two gradients in altitude perpendicular to each other can be defined, from left to right and from front to back. These two gradients require the use of column and row blocks. The square shape and slightly larger size of the field compared to (a) and (b) allow the application of a 4×4 Latin square design ("LSD"), with four column blocks, four row blocks, and four treatments (e.g., t4: 12-species mixture); randomization is even more constrained than in the RCBDs, with each treatment appearing once and only once in each row and each column; quotes are used in the abbreviation to distinguish it from that generally used for the Least Significant Difference in statistics. The four experimental designs above (i.e., CRD, RCBD 1 rep, RCBD >1 rep, "LSD") are important in the context of spatial heterogeneity analysis in the presence of treatment application (Dutilleul, 1993a).

At the basis of each version of the ANOVA method for a given experimental design, there is a linear model, or ANOVA model. The ANOVA models for the four experimental designs discussed so far in 2-D space are defined by equations (9.1)–(9.4) in Table 9.1. They are extensions of fundamental equation (2.1): $U(s) = \mu(s) + \varepsilon(s)$ and particular cases of the generalized form: $U_{f_1 \dots f_F r_1 \dots r_R k}(s) = \mu_{f_1 \dots f_F r_1 \dots r_R k}(s) + \varepsilon_{f_1 \dots f_F r_1 \dots r_R k}(s)$ [equation (2.14)]. With $\mu_{ij}(x, y) = \mu + a_i$, equation (9.1) reflects the spatial homogeneity of the mean in the CRD, the treatment main effect

Table 9.1. *ANOVA models for six experimental designs in which the three types of heterogeneity (i.e., mean, variance, autocorrelation) are taken into account in different degrees*

Experimental design	Framework	ANOVA model equation	
CRD	2-D space	$U_{ij}(x, y) = \mu + a_i + \varepsilon_{ij}(x, y) \quad (i = 1, \ldots, p; j = 1, \ldots, n)$	(9.1)
RCBD 1 rep	2-D space	$U_{ij}(x, y) = \mu + a_i + b_j + \varepsilon_{ij}(x, y) \quad (i = 1, \ldots, p; j = 1, \ldots, n)$	(9.2)
RCBD >1 rep	2-D space	$U_{ijk}(x, y) = \mu + a_i + b_j + (ab)_{ij} + \varepsilon_{ijk}(x, y) \quad (i = 1, \ldots, p; j = 1, \ldots, n; k = 1, \ldots, r)$	(9.3)
"LSD"	2-D space	$U_{ij_1j_2}(x, y) = \mu + a_i + b_{j_1}^{(col)} + b_{j_2}^{(row)} + \varepsilon_{ij_1j_2}(x, y) \quad (i, j_1, j_2 = 1, \ldots, p)$	(9.4)
CRD with simple RM	Time or space	$U_{ij}(s) = \mu + a_i + \varepsilon_{ij}^{(1)} + \Delta\mu(s) + \Delta a_i(s) + \varepsilon_{ij}^{(2)}(s)$ $(i = 1, \ldots, p; j = 1, \ldots, n; s = 1, \ldots, S^*)$	(9.5)
CRD with double RM	Time-time, space-space, or space-time	$U_{ij}(s_1, s_2) = \mu + a_i + \varepsilon_{ij}^{(1)} + \Delta\mu(s_1) + \Delta a_i(s_1) + \varepsilon_{ij}^{(2)}(s_1) + \Delta\mu(s_2) + \Delta a_i(s_2) + \varepsilon_{ij}^{(3)}(s_2)$ $+ \Delta\mu(s_1, s_2) + \Delta a_i(s_1, s_2) + \varepsilon_{ij}^{(4)}(s_1, s_2)$ $(i = 1, \ldots, p; j = 1, \ldots, n; s_1 = 1, \ldots, S_1^*; s_2 = 1, \ldots, S_2^*)$	(9.6)

Notations:

μ, overall population mean;
i, level of the treatment factor; a_i, main effect of treatment i; p, number of treatments; $i = 1, \ldots, p$; $\sum_{i=1}^{p} a_i = 0$;
j, level of the block factor; b_j, main effect of block j; n, number of blocks; $j = 1, \ldots, n$; $\sum_{j=1}^{n} b_j = 0$;

(x, y), 2-D spatial coordinates of sampling units (plot centers);
$(ab)_{ij}$, interaction effect between treatment i and block j ($i = 1, \ldots, p$; $j = 1, \ldots, n$); $\sum_{i=1}^{p} \sum_{j=1}^{p} (ab)_{ij} = 0$;
k, subscript of replicates; r, (even) number of replicates per treatment per block; $k = 1, \ldots, r$;
j_1, level of the column block factor, and j_2, level of the row block factor; $j_1, j_2 = 1, \ldots, p = n$;
s, level for either time or space, considered as a factor, in the analysis of simple repeated measures; $s = 1, \ldots, S^*$;
[*Note:* An asterisk is used to distinguish the number of repeated measures per individual here and the notation of the index space S used elsewhere in the book, especially in fundamental equation (2.1).]
s_1 and s_2, levels for time and time (different scales), space and space (different scales), or space and time, considered as factors, in the analysis of double repeated measures; $s_1 = 1, \ldots, S_1^*$ and $s_2 = 1, \ldots, S_2^*$;
$\Delta\mu(s)$, $\Delta\mu(s_1)$, and $\Delta\mu(s_2)$, main effects of the repeated measures factor;
$\Delta a_i(s)$, $\Delta a_i(s_1)$, and $\Delta a_i(s_2)$, interaction effects between the treatment factor and the repeated measures factor(s).

Abbreviations:
CRD = Completely Randomized Design;
RCBD = Randomized Complete Block Design;
1 rep = one replicate per treatment per block;
>1 rep = more than one replicate per treatment per block;
"LSD" = Latin Square Design;
(*Note:* Quotes are used not to confound the abbreviation here with the procedure for multiple comparisons of means, based on the Least Significant Difference.)
RM = repeated measures.

a_i representing the only possible (non-spatial) discrepancy from the overall population mean μ. Considering blocks as levels of a fixed factor, the ANOVA model for the RCBD 1 rep allows the expected value of the random variable to change from block to block, through the term b_j ($j = 1, \ldots, n$) in equation (9.2). In other words, $U_{ij}(x, y)$ for a given $i = 1, \ldots, p$ may have n different population means, one for each of the n blocks. With treatment-by-block interaction effects and column and row block main effects, the ANOVA models for the RCBD >1 rep [equation (9.3)] and the "LSD" [equation (9.4)] allow a greater diversification in the spatial heterogeneity of the mean in accordance with environmental conditions in the field. In particular, each of the p^2 cells of the square in Fig. 9.4(d) may have a different population mean prior to treatment application, through the expression $\mu + b_{j_1}^{(col)} + b_{j_2}^{(row)}$ ($j_1, j_2 = 1, \ldots, n = p$) in equation (9.4).

Key note: Considering blocks as levels of a random factor (normally distributed with a zero population mean and a variance component associated with it) is possible, but in this case the corresponding spatial heterogeneity is of random nature, instead of deterministic, which has several implications. One: Spatial heterogeneity is entirely related to the random term $\varepsilon(s)$ *in fundamental equation (2.1). Two: Spatial homogeneity of the mean must be assumed in the ANOVA model. Three: Under this assumption, the two remaining types of spatial heterogeneity are quantified via variance and covariance components (Dutilleul, 1993a). Considering blocks as levels of a fixed factor is consistent with the approach followed in Chapters 7 and 8 and the deterministic modeling of large-scale variability in the CRAD method (Pelletier* et al., *2009a, 2009b).*

The fifth and sixth experimental designs to be discussed and illustrated with datasets in this chapter are the CRDs with simple and double repeated measures. We have already worked with simple repeated measures in Chapter 6, in the frame of three forestry examples where ring width, fiber length and microfibril angle had been measured repeatedly in Norway spruces; each tree was the support for repeated measurements; a number of trees were fast-grown, and the others slow-grown, so growth category was the "treatment factor." Repeated measures were then simple and essentially temporal, since either the annual ring or the sampling site within a ring was the repeated measures factor, even if some spatial referencing might be associated with the latter. Purely spatial, simple repeated measures are collected, for example, when some mineral content (e.g., N, P, K) is measured at different depths in soil cores, each core providing the support for repeated measurements as an extension of paired observations. Double repeated measures data can be doubly temporal, doubly spatial, or spatio-temporal. Two spatio-temporal examples are presented in Subsection 9.3.2, with microfibril angle being measured in trees in a number of annual rings several times within a ring in the former and with turbidity and total suspended solids in water being monitored at given distances from the shore on different dates in the latter. Repeated measures data can be collected in the frame of other basic experimental designs than the CRD. For example, it suffices to imagine that growth (e.g., height, basal diameter) is measured more than once on trees in the plantation experiments discussed in view of Fig. 9.4(b)–(d), to obtain a RCBD 1 rep, a RCBD >1 rep and an "LSD" with temporal repeated measures (see also Dutilleul, 1998b).

To write the ANOVA model for any experimental design with simple repeated measures, one starts by writing the ANOVA model for the experimental design without repeated measures (including an error term without index), and one continues with a copy and paste of the first set of terms to which one adds a delta Δ and an index s for the level of the repeated measures factor (plus a second error term with index); the two sets of terms are called "between-subject effects" and "within-subject effects," respectively; see equation (9.5) in Table 9.1 for the ANOVA model for the CRD with simple repeated measures. The ANOVA model for an experimental design with double repeated measures is much longer to write, as it has four error terms and four sets of terms: one for the between-subject effects and three for three types of within-subject effects, depending on the repeated measures factor(s) involved; see equation (9.6) for the ANOVA model for the CRD with double repeated measures.

9.2.2 The key: the expected mean squares

The questions of field lay-out and modeling being covered, fundamental quantitative aspects can be addressed. Traditionally, special quantities called "expected mean squares" play a key role in the ANOVA method. We will not escape tradition in this subsection, but we will go beyond it by "preparing the terrain" for the two measures of heteroscedasticity and autocorrelation that will be defined in the next subsection, the perspective being that of the experimenter who wonders before starting: How many degrees of freedom can I expect to lose in my assessment of treatment effects, for example, because of heterogeneity of the variance and heterogeneity due to autocorrelation of this type and that strength? A

Table 9.2. *ANOVA decomposition for the first four experimental designs in Table 9.1*

Experimental design	Decomposition in terms of sums of squares (above) and with formulas (below)
CRD	Total SS = Treatment SS + Error SS
	$\sum_{i=1}^{p}\sum_{j=1}^{n}(U_{ij}-\bar{U})^2 = n\sum_{i=1}^{p}(\bar{U}_i-\bar{U})^2 + \sum_{i=1}^{p}\sum_{j=1}^{n}(U_{ij}-\bar{U}_i)^2$
RCBD 1 rep	Total SS = Treatment SS + Block SS + Error SS
	$\sum_{i=1}^{p}\sum_{j=1}^{n}(U_{ij}-\bar{U})^2 = n\sum_{i=1}^{p}(\bar{U}_i-\bar{U})^2 + p\sum_{j=1}^{n}(\bar{U}_j-\bar{U})^2 + \sum_{i=1}^{p}\sum_{j=1}^{n}(U_{ij}-\bar{U}_i-\bar{U}_j+\bar{U})^2$
RCBD >1 rep	Total SS = Treatment SS + Block SS + Trt × Block Interaction SS + Error SS
	$\sum_{i=1}^{p}\sum_{j=1}^{n}\sum_{k=1}^{r}(U_{ijk}-\bar{U})^2 = nr\sum_{i=1}^{p}(\bar{U}_i-\bar{U})^2 + pr\sum_{j=1}^{n}(\bar{U}_j-\bar{U})^2 + r\sum_{i=1}^{p}\sum_{j=1}^{n}(\bar{U}_{ij}-\bar{U}_i-\bar{U}_j+\bar{U})^2 + \sum_{i=1}^{p}\sum_{j=1}^{n}\sum_{k=1}^{r}(U_{ijk}-\bar{U}_{ij})^2$
"LSD"	Total SS = Treatment SS + Column Block SS + Row Block SS + Error SS
	$\sum_{j_1=1}^{p}\sum_{j_2=1}^{p}(U_{ij_1j_2}-\bar{U})^2 = p\sum_{i=1}^{p}(\bar{U}_i-\bar{U})^2 + p\sum_{j_1=1}^{p}(\bar{U}_{j_1}-\bar{U})^2 + p\sum_{j_2=1}^{p}(\bar{U}_{j_2}-\bar{U})^2 + \sum_{j_1=1}^{p}\sum_{j_2=1}^{p}(U_{ij_1j_2}-\bar{U}_i-\bar{U}_{j_1}-\bar{U}_{j_2}+2\bar{U})^2$

Notations: $\bar{U}$, the overall sample mean; $\bar{U}_i (i = 1, \ldots, p)$, Treatment sample means; $\bar{U}_j (j = 1, \ldots, n)$, Block sample means. Indices (x, y) are removed from the observations to simplify notations. In "LSD," a third summation over $i = 1, \ldots, p$ is not necessary, since once any two of the indices are fixed the third one is known and the total number of observations is p^2 instead of p^3; it follows that in the formula of the Error SS, the value of i is provided by the combination of j_1 and j_2 considered in the double summation.

part of the development is technical, so in order to follow the mainstream I will make an appropriate use of appendices.

First of all, the total variation in the sample data collected in the study is decomposed into the sources of variation defined in the ANOVA model. This decomposition is based on "sums of squares" (SS), which are functions of the observations and arithmetic means (e.g., the overall sample mean: $\bar{U}$; Treatment sample means: $\bar{U}_i$; and Block sample means: $\bar{U}_j$). The ANOVA decompositions for the CRD, RCBD 1 rep, RCBD >1 rep, and "LSD" are reported in Table 9.2, both in terms of SS and with formulas.

It must be noted that the arithmetic means in the ANOVA SS are the ordinary least-squares estimators of the corresponding population means (e.g., the overall population mean: μ; Treatment population means: μ_i; and Block population means: μ_j). In simple terms, homogeneity of the variance and the absence of autocorrelation are the classical conditions of application of this estimation method (McElroy, 1967; see OLS in Chapter 6). Thus, fixed effects are estimated in very different ways in the ANOVA method and in SAS PROC MIXED that was used in several examples in Chapters 6 and 7. In the latter, restricted maximum likelihood (REML; Patterson and Thompson, 1971) requires an estimated variance–covariance matrix to estimate fixed effects and thereby allows the presence of heteroscedasticity and autocorrelation. The equations of the two estimation procedures are available in Diggle *et al.* (1996, pp. 64–68). The advantage of REML of modifying the estimation of fixed effects by incorporating heteroscedasticity and autocorrelation has two counterparts here. One: The use of an estimated variance–covariance matrix precludes the application of properties of quadratic forms in normal random vectors, which is possible with the ANOVA sums of squares (see Section A9.2). Two: Instead of using the best estimator of fixed effects possible, the objective is to use an estimator that is best in the absence of the two types of heterogeneity, so that discrepancies due to their presence can be measured (see Subsection 9.2.3). The ANOVA-based approach was retained for these two reasons, in addition to the fact that the ANOVA remains a statistical method that cannot be circumvented in the analysis of experimental data.

What is then an expected mean square, and what are expected mean squares used for in the ANOVA? The two parts of this question are addressed separately below.

First, "expected mean square" is a shortening for the expected value of a mean square (MS), which results from applying the expectation operator E[·] (Section A2.1), to the random quantity (i.e., the MS) obtained by dividing an ANOVA SS by the corresponding number of degrees of freedom (df). In the classical conditions of application of the ANOVA (i.e., homoscedasticity and absence of autocorrelation), and in more general terms, when the circularity condition is satisfied (Huynh and Feldt, 1970; Rouanet and Lépine, 1970), the numbers of df associated with the Treatment SS and the Error SS in the CRD ANOVA are $p - 1$ and $p(n - 1)$; those associated with the Treatment SS, the Block SS and the Error SS in the RCBD 1 rep ANOVA are $p - 1$, $n - 1$ and $(p - 1)(n - 1)$; the number of df associated with the Treatment-by-Block Interaction SS in the RCBD >1 rep ANOVA is $(p - 1)(n - 1)$, leaving $pn(r - 1)$ df for the Error SS; all sums of squares have $p - 1$ df associated with them in the "LSD" ANOVA, except the Error SS which has $(p - 1)(p - 2)$ df.

(Circularity means homoscedasticity and absence of autocorrelation for orthonormal contrasts. It is the most general necessary and sufficient condition for valid unmodified ANOVA *F*-tests. More specifically, if $\Sigma = (\sigma_{ss'})$ denotes a variance–covariance matrix designed to model heterogeneity of the variance and heterogeneity due to autocorrelation in space or time, then Σ satisfies the circularity condition if $\sigma_{ss} + \sigma_{s's'} - 2\sigma_{ss'} = 2\lambda$ for any (s, s'), with $\lambda > 0$.)

Second, expected mean squares for the different sources of variation in the ANOVA models for the CRD, RCBD 1 rep, RCBD >1 rep, and "LSD" are reported in Table 9.3(a), whether the circularity condition is satisfied or not. Materials from Appendices A9.1 and A9.2 were used in the calculations, where each MS was given by a SS divided by the classical number of df; the expressions reported are not restricted to 2-D space. Differences between the situations Circularity Yes vs. No in Table 9.3(a) are fundamental. In the CRD ANOVA for example, E[Treatment MS] = E[Error MS] under the hypothesis of no treatment main effects (i.e., $\sum_{i=1}^{p} a_i^2 = 0$) when the circularity condition is satisfied. When the condition is not satisfied, the equality between expected mean squares no longer holds, which should affect the ANOVA *F*-test that is built on the ratio of the two mean squares. In Subsection 9.2.3, this kind of discrepancy is quantified through effective numbers of df in the CRD, RCBD 1 rep, RCBD >1 rep, and "LSD" ANOVAs and through a multiplicative factor to be applied to the numbers of df for within-subject effects in the repeated measures ANOVA.

(The nature of blocks, fixed or random, has an influence on the expression of expected mean squares in the RCBD 1 rep, RCBD >1 rep,

Table 9.3. *(a) Expected mean squares for fixed effects and the error term of ANOVA models (9.1)–(9.4) in Table 9.1, depending on whether the condition of circularity (i.e., restricted form of heterogeneity of variance and autocorrelation) is satisfied or not, and (b) theoretical expression of Box's "epsilon" (i.e., multiplicative factor for numbers of degrees of freedom) in modified* F*-tests for effects related to the repeated measures factor(s) in ANOVA models (9.5) and (9.6)*

(a) Experimental design	Fixed effect or error	Circularity (Yes/No)	Expected mean square
CRD	Treatment main effects	Yes	$\text{E[Treatment MS]} = \frac{n}{p-1}\sum_{i=1}^{p} a_i^2 + \sigma_\varepsilon^2$
	Error	Yes	$\text{E[Error MS]} = \sigma_\varepsilon^2$
CRD	Treatment main effects	No	$\text{E[Treatment MS]} = \frac{n}{p-1}\left\{\sum_{i=1}^{p} a_i^2 + \text{trace}(\mathbf{A}_{\text{Trt}}\boldsymbol{\Sigma}_{\text{Trt}})\right\}$
	Error	No	$\text{E[Error MS]} = \text{trace}(\mathbf{A}_{\text{Error}}\boldsymbol{\Sigma}_{\text{Error}})$
RCBD 1 rep	Treatment main effects	Yes	$\text{E[Treatment MS]} = \frac{n}{p-1}\sum_{i=1}^{p} a_i^2 + \sigma_\varepsilon^2$
	Block main effects	Yes	$\text{E[Block MS]} = \frac{p}{n-1}\sum_{j=1}^{n} b_j^2 + \sigma_\varepsilon^2$
	Error	Yes	$\text{E[Error MS]} = \sigma_\varepsilon^2$
RCBD 1 rep	Treatment main effects	No	$\text{E[Treatment MS]} = \frac{n}{p-1}\left\{\sum_{i=1}^{p} a_i^2 + \text{trace}(\mathbf{A}_{\text{Trt}}\boldsymbol{\Sigma}_{\text{Trt}})\right\}$

(cont.)

Table 9.3 *(cont.)*

(a) Experimental design	Fixed effect or error	Circularity (Yes/No)	Expected mean square
	Block main effects	No	$\text{E[Block MS]} = \frac{p}{n-1}\left\{\sum_{j=1}^{n} b_j^2 + \text{trace}(\mathbf{A}_{\text{Blk}}\boldsymbol{\Sigma}_{\text{Blk}})\right\}$
	Error	No	$\text{E[Error MS]} = \text{trace}(\mathbf{A}_{\text{Error}}\boldsymbol{\Sigma}_{\text{Error}})$
RCBD >1 rep	Treatment main effects	Yes	$\text{E[Treatment MS]} = \frac{nr}{p-1}\sum_{i=1}^{p} a_i^2 + \sigma_\varepsilon^2$
	Block main effects	Yes	$\text{E[Block MS]} = \frac{pr}{n-1}\sum_{j=1}^{n} b_j^2 + \sigma_\varepsilon^2$
	Treatment × Block interaction	Yes	$\text{E[Treatment} \times \text{Block MS]} = \frac{r}{(p-1)(n-1)}\sum_{i=1}^{p}\sum_{j=1}^{n}(ab)_{ij}^2 + \sigma_\varepsilon^2$
	Error	Yes	$\text{E[Error MS]} = \sigma_\varepsilon^2$
RCBD >1 rep	Treatment main effects	No	$\text{E[Treatment MS]} = \frac{nr}{p-1}\left\{\sum_{i=1}^{p} a_i^2 + \text{trace}(\mathbf{A}_{\text{Trt}}\boldsymbol{\Sigma}_{\text{Trt}})\right\}$
	Block main effects	No	$\text{E[Block MS]} = \frac{pr}{n-1}\left\{\sum_{j=1}^{n} b_j^2 + \text{trace}(\mathbf{A}_{\text{Blk}}\boldsymbol{\Sigma}_{\text{Blk}})\right\}$
	Treatment × Block interaction	No	$\text{E[Treatment} \times \text{Block MS]} = \frac{r}{(p-1)(n-1)}\left\{\sum_{i=1}^{p}\sum_{j=1}^{n}(ab)_{ij}^2 + \text{trace}(\mathbf{A}_{\text{Trt}\times\text{Blk}}\boldsymbol{\Sigma}_{\text{Trt}\times\text{Blk}})\right\}$

	Error	No	$E[\text{Error MS}] = \text{trace}(\mathbf{A}_{\text{Error}}\boldsymbol{\Sigma}_{\text{Error}})$
"LSD"	Treatment main effects	Yes	$E[\text{Treatment MS}] = \frac{p}{p-1}\sum_{i=1}^{p} a_i^2 + \sigma_\varepsilon^2$
	Column block main effects	Yes	$E[\text{Column block MS}] = \frac{p}{p-1}\sum_{j_1=1}^{p} (b_{j_1}^{(col)})^2 + \sigma_\varepsilon^2$
	Row block main effects	Yes	$E[\text{Row block MS}] = \frac{p}{p-1}\sum_{j_2=1}^{p} (b_{j_2}^{(row)})^2 + \sigma_\varepsilon^2$
	Error	Yes	$E[\text{Error MS}] = \sigma_\varepsilon^2$
"LSD"	Treatment main effects	No	$E[\text{Treatment MS}] = \frac{p}{p-1}\left\{\sum_{i=1}^{p} a_i^2 + \text{trace}(\mathbf{A}_{\text{Trt}}\boldsymbol{\Sigma}_{\text{Trt}})\right\}$
	Column block main effects	No	$E[\text{Column block MS}] = \frac{p}{p-1}\left\{\sum_{j_1=1}^{p} (b_{j_1}^{(col)})^2 + \text{trace}(\mathbf{A}_{\text{Col Blk}}\boldsymbol{\Sigma}_{\text{Col Blk}})\right\}$
	Row block main effects	No	$E[\text{Row block MS}] = \frac{p}{p-1}\left\{\sum_{j_2=1}^{p} (b_{j_2}^{(row)})^2 + \text{trace}(\mathbf{A}_{\text{Row Blk}}\boldsymbol{\Sigma}_{\text{Row Blk}})\right\}$
	Error	No	$E[\text{Error MS}] = \text{trace}(\mathbf{A}_{\text{Error}}\boldsymbol{\Sigma}_{\text{Error}})$

(cont.)

Table 9.3 *(cont.)*

(b) Experimental design	Theoretical expression of Box's "epsilon"
CRD with simple RM	$\varepsilon_{\mathrm{RM}} = \dfrac{\{\mathrm{trace}(\mathbf{C}'\boldsymbol{\Sigma}_{\mathrm{RM}}\mathbf{C})\}^2}{(S^* - 1)\mathrm{trace}\{(\mathbf{C}'\boldsymbol{\Sigma}_{\mathrm{RM}}\mathbf{C})^2\}}$
CRD with double RM	$\varepsilon_{\mathrm{RM}_j} = \dfrac{\{\mathrm{trace}(\mathbf{C}_j'\boldsymbol{\Sigma}_{\mathrm{RM}_j}\mathbf{C}_j)\}^2}{(S_j^* - 1)\mathrm{trace}\{(\mathbf{C}_j'\boldsymbol{\Sigma}_{\mathrm{RM}_j}\mathbf{C}_j)^2\}}$ if only one repeated measures factor is involved ($j = 1, 2$); $\varepsilon_{\mathrm{RM}_{12}} = \dfrac{\{\mathrm{trace}(\mathbf{C}_{12}'\boldsymbol{\Sigma}_{\mathrm{RM}_{12}}\mathbf{C}_{12})\}^2}{(S_1^* - 1)(S_2^* - 1)\mathrm{trace}\{(\mathbf{C}_{12}'\boldsymbol{\Sigma}_{\mathrm{RM}_{12}}\mathbf{C}_{12})^2\}}$ if both repeated measures factors are involved; where $\mathbf{C}$ and $\mathbf{C}_j$ ($j = 1, 2$) are matrices of orthonormal contrasts, that is, $\mathbf{C}'\mathbf{C} = \mathbf{I}_{S^*-1}$ and $\mathbf{C}\mathbf{C}' = \mathbf{I}_{S^*} - \frac{1}{S^*}\mathbf{J}_{S^*}$; $\mathbf{C}_j'\mathbf{C}_j = \mathbf{I}_{S_j^*-1}$ and $\mathbf{C}_j\mathbf{C}_j' = \mathbf{I}_{S_j^*} - \frac{1}{S_j^*}\mathbf{J}_{S_j^*}$; and $\mathbf{C}_{12} = \mathbf{C}_2 \otimes \mathbf{C}_1$, with $\otimes$, the Kronecker product (Graybill, 1983).

Notations: $\boldsymbol{\Sigma}$, variance–covariance matrix with dimensions $np \times np$ (CRD, RCBD 1 rep, "LSD" with $n = p$) or $npr \times npr$ (RCBD >1 rep);
$\boldsymbol{\Sigma}_{\mathrm{Trt}}$, $p \times p$ variance–covariance matrix of Treatment sample means (CRD, RCBD 1 rep and >1 rep, "LSD");
$\boldsymbol{\Sigma}_{\mathrm{Blk}}$, $n \times n$ variance–covariance matrix of Block sample means (RCBD 1 rep and >1 rep);

$\Sigma_{\text{Trt}\times\text{Blk}}$, $np \times np$ variance–covariance matrix of Treatment-by-Block sample means (RCBD >1 rep);
$\Sigma_{\text{Col blk}}$, $p \times p$ variance–covariance matrix of Column block sample means ("LSD");
$\Sigma_{\text{Row blk}}$, $p \times p$ variance–covariance matrix of Row block sample means ("LSD");
$\mathbf{A} = \mathbf{I}_{np} - \frac{1}{np}\mathbf{J}_{np}$ (CRD, RCBD 1 rep, "LSD" with $n = p$) or $\mathbf{I}_{npr} - \frac{1}{npr}\mathbf{J}_{npr}$ (RCBD >1 rep) (**I**: identity matrix, **J**: matrix of ones), matrix of centering of individual observations with respect to the overall sample mean;
$\mathbf{A}_{\text{Trt}} = \mathbf{I}_p - \frac{1}{p}\mathbf{J}_p$ (CRD, RCBD 1 rep and >1 rep, "LSD"), matrix of centering of Treatment sample means with respect to the overall sample mean;
$\mathbf{A}_{\text{Blk}} = \mathbf{I}_n - \frac{1}{n}\mathbf{J}_n$ (RCBD 1 rep and >1 rep), matrix of centering of Block sample means with respect to the overall sample mean;
$\mathbf{A}_{\text{Trt-Blk}} = \mathbf{I}_{np} - \mathbf{A}^*_{\text{Trt}} - \mathbf{A}^*_{\text{Blk}} + \frac{1}{np}\mathbf{J}_{np}$ (RCBD 1 rep and >1 rep), matrix of double centering with respect to Treatment and Block sample means;
(*Notes:* In the case of the RCBD >1 rep, $\mathbf{A}_{\text{Trt-Blk}} = \mathbf{A}_{\text{Trt}\times\text{Blk}}$. Stars are used to indicate that matrices $\mathbf{A}^*_{\text{Trt}}$ and $\mathbf{A}^*_{\text{Blk}}$ are different from $\mathbf{A}_{\text{Trt}}$ and $\mathbf{A}_{\text{Blk}}$ above; see Appendix A9 for details.)
$\mathbf{A}_{\text{Col blk}} = \mathbf{I}_p - \frac{1}{p}\mathbf{J}_p$ ("LSD"), matrix of centering of Column block sample means with respect to the overall sample mean;
$\mathbf{A}_{\text{Row blk}} = \mathbf{I}_p - \frac{1}{p}\mathbf{J}_p$ ("LSD"), matrix of centering of Row block sample means with respect to the overall sample mean;
(*Note:* $\mathbf{A}_{\text{Trt}} = \mathbf{A}_{\text{Col blk}} = \mathbf{A}_{\text{Row blk}}$ for "LSD.")
Different matrices $\mathbf{A}_{\text{Error}}$ are used depending on the experimental design; see Appendix A9 for details.
(*Note:* In the case of the RCBD 1 rep, $\mathbf{A}_{\text{Error}} = \mathbf{A}_{\text{Trt-Blk}}$.)
σ^2_ε, variance component of the error term if i.i.d. (i.e., particular case of circularity) or of orthonormal contrasts if circular but not i.i.d. (see text);
Σ_{RM}, $S^* \times S^*$ variance–covariance matrix of the random vector of repeated measures (CRD with simple RM);
Σ_{RM_j}, $S^*_j \times S^*_j$ variance–covariance matrix of one type of random vector of repeated measures, $j = 1, 2$ (CRD with double RM);
$\Sigma_{\text{RM}_{12}}$, $S^*_1 S^*_2 \times S^*_1 S^*_2$ variance–covariance matrix of the complete random vector of repeated measures (CRD with double RM).

and "LSD" ANOVAs. Expected mean squares were calculated with fixed blocks here. It is possible to calculate them with random blocks, using similar procedures. However, in addition to the question of modeling discussed above, it would mean more technicalities, without changing much to the take-home message.)

9.2.3 Effective df and Box's "epsilon": measures of spatial and temporal heterogeneity

In the previous subsection, we have learned that heteroscedasticity and autocorrelation may alter the number of df associated with an ANOVA SS. Since the number of df appears in the definition of the MS itself (i.e., $\mathrm{MS} = \frac{\mathrm{SS}}{\mathrm{df}}$), we will work with the SS instead of the MS below. The two measures of heterogeneity that will be presented are aimed at quantifying the same thing: the discrepancy from the classical number of df due to heteroscedasticity and autocorrelation. One measure is absolute and called the "effective number of df," and the other is relative and called "epsilon," due to Box (1954a, 1954b).

After a given ANOVA SS has been expressed as a quadratic form $\boldsymbol{u}'\mathbf{A}\boldsymbol{u}$ in the normal random vector $\boldsymbol{u}$ (Section A9.1), it is possible to calculate its expected value and population variance using matrix algebra tools (see Section A9.2). If the circularity condition is satisfied (e.g., $\mathrm{Var}[\boldsymbol{u}] = \sigma^2\mathbf{I}$), then $\frac{\{\mathrm{E}[\boldsymbol{u}'\mathbf{A}\boldsymbol{u}]\}^2}{0.5\mathrm{Var}[\boldsymbol{u}'\mathbf{A}\boldsymbol{u}]}$ is equal to the classical number of df. Otherwise,

$$\text{effective number of df} = \frac{\{\mathrm{E}[\boldsymbol{u}'\mathbf{A}\boldsymbol{u}]\}^2}{0.5\mathrm{Var}[\boldsymbol{u}'\mathbf{A}\boldsymbol{u}]} \tag{9.7}$$

and

$$\text{Box's "epsilon"} = \frac{\{\mathrm{E}[\boldsymbol{u}'\mathbf{A}\boldsymbol{u}]\}^2}{(q-1)0.5\mathrm{Var}[\boldsymbol{u}'\mathbf{A}\boldsymbol{u}]}, \tag{9.8}$$

which can be rewritten using the trace operator [Table 9.3(b); see also Section A9.2]; by definition, Box's "epsilon" cannot be greater than 1 and is constrained to be greater than or equal to $\frac{1}{q-1}$; in the case of one repeated measures factor, $q = S^*$, and when there are two repeated measures factors and both are involved in the interaction effect of interest, $q-1$ is replaced by $(S_1^* - 1)(S_2^* - 1)$.

The results presented in Figs. 9.5 and 9.6 are not the fruit of intensive simulations. Instead, they are the results of theoretical calculations made with equation (9.7) for Fig. 9.5 and equation (9.8) for Fig. 9.6, or with equivalent expressions written in terms of traces and variance–covariance matrices designed to model autocorrelation or heteroscedasticity, or both [Table 9.3(b)]. The standard error bars in Fig. 9.5(a)–(c) are the results of 1000 randomized assignments of treatments to plots in the cases of the CRD, the RCBD 1 rep and the RCBD >1 rep; the field lay-out of Fig. 9.4(d) was used for the "LSD," because of the restrictions in randomization in this case. The interested reader may use the Expectation software on the CD-ROM to study other scenarios than those considered here. Results presented in Figs. 9.5 and 9.6 can be discussed as follows.

Effective df

In Fig. 9.5, spatial autocorrelation decreases with increasing distance between observations (supposed to be made at the level of plots of size 5×5), following a spherical variogram model (Fig. 7.1) with a range varying from 5 to 50 space units by increments of 5. There is no heteroscedasticity, and because of buffer zones, the distance between plot centers is 10 space units (Fig. 9.4).

Whatever the experimental design considered, a decrease in the effective number of df with increasing value of the range of spatial autocorrelation is observed for all ANOVA SS. With no real surprise, the decrease is the strongest for the Error SS because spatial autocorrelation among observations is equivalent to spatial autocorrelation among errors in the ANOVA model, the error being the only random term on the right-hand side of equations (9.1)–(9.4) in Table 9.1. The Error df tends to decrease to half its classical value and even lower in the cases of the CRD and the RCBD >1 rep. Spatial autocorrelation also affects the Treatment df, since Treatment sample means are calculated from observations that are autocorrelated. Likely thanks to the randomization of treatment assignment, the loss in Treatment df is limited to about 12.5% when the range of spatial autocorrelation is the largest (50 space units), with the exception of the single "LSD" studied here. The Block SS, Treatment × Block Interaction SS, Column Block SS and Row Block SS experience intermediate losses in df because blocks are aligned by definition, so that sample means computed for neighboring blocks combine observations that are not very distant spatially.

The fact that spatial autocorrelation affects the df associated with an ANOVA SS even if this represents a fixed effect is a very important point. It makes much sense afterwards, but was not so evident beforehand.

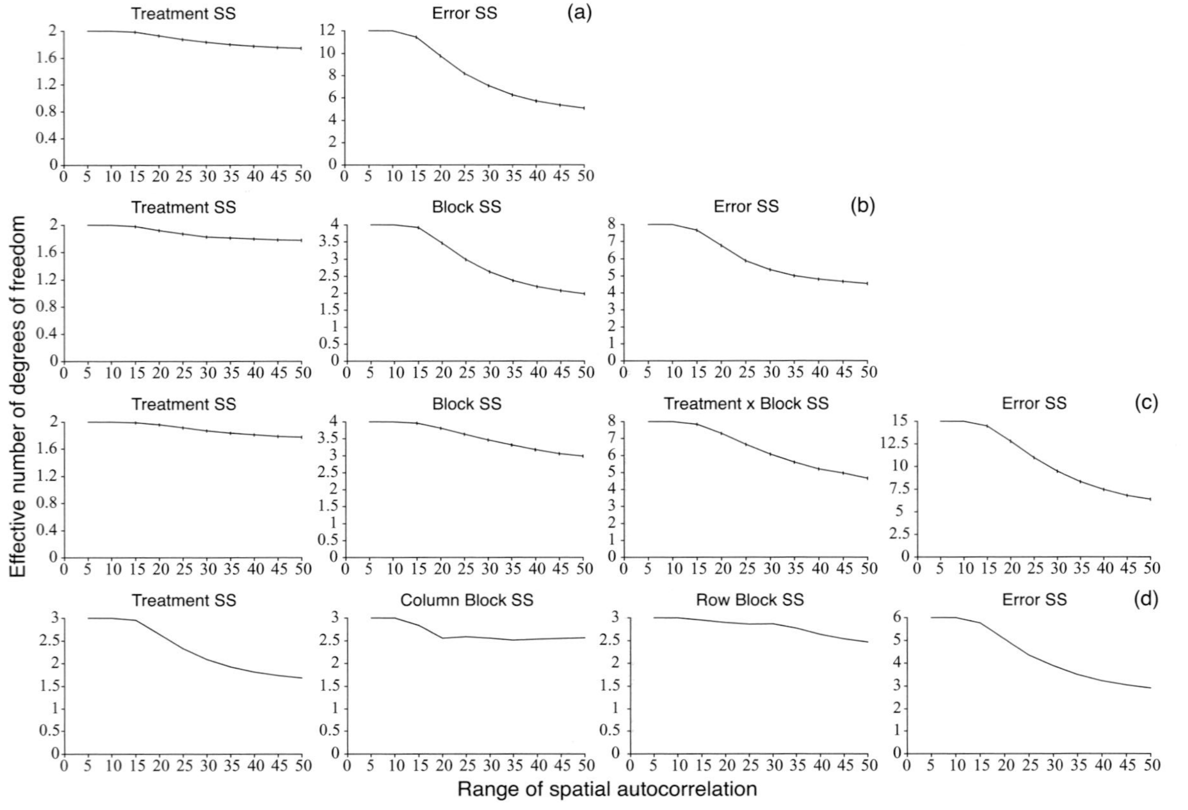

Fig. 9.5. Effective numbers of degrees of freedom associated with the expected mean squares for fixed effects and the error term in the ANOVA models for the four experimental designs outlined in Fig. 9.4, in relation to the range of spatial autocorrelation in a spherical variogram model for the error term. See also Tables 9.1 and 9.2.

In practical terms, the critical value ($\alpha = 0.05$) of the ANOVA F-test for Treatment main effects in the CRD is 3.89 when full df (2 and 12) are used; it is up to 5.86 when the range of spatial autocorrelation is 50, which represents an important loss of statistical power. The ANOVA F-test for Block main effects (i.e., spatial heterogeneity of the mean at large scale) would experience a stronger loss of statistical power in the RCBD 1 rep. To anticipate such losses, the experimenter could increase the number of replicates in the CRD while keeping the plot size the same in a larger field still homogeneous spatially, or increase the plot and field sizes in the RCBD 1 rep in order to make the range of spatial autocorrelation relatively smaller than the distance between plot centers while keeping blocks homogeneous.

Box's "epsilon"

All Box's "epsilon" values reported here were calculated directly from theoretical variance–covariance matrices; no simulation or randomization in treatment assignment was involved. These calculations were motivated by: the assessment of the respective importance of temporal autocorrelation (in sign and strength) and temporal heteroscedasticity on the reduction in df associated with SS for within-subject effects in the repeated measures ANOVA (objective 1); a similar evaluation of the respective importance of the (practical) range and model of autocorrelation (exponential, spherical, Gaussian; Fig. 7.1) in the case of spatial repeated measures (objective 2); and the assessment of the reduction in df for within-subject effects in the framework of spatio-temporal repeated measures ANOVA, by combining various temporal and spatial variance–covariance structures (objective 3); space-time separability of the variance–covariance structure is assumed [equation (2.9)]. *Note:* Any 2×2 variance–covariance matrix is circular because $\sigma_{11} + \sigma_{22} - 2\sigma_{12} = 2\lambda > 0$ is always true. It follows that Box's "epsilon" is equal to 1 when the number of repeated measures is two and there is one repeated measures factor ($S^* = 2$) and when there are two repeated measures factors and the number of repeated measures is 2×2 ($S_1^* = S_2^* = 2$).

Overall, the larger the number of repeated measures, the lower the Box's "epsilon" values in all the scenarios considered here (Fig. 9.6), because a larger number of repeated measures resulted in larger differences between the lowest and highest variances or between the lowest and highest autocorrelations.

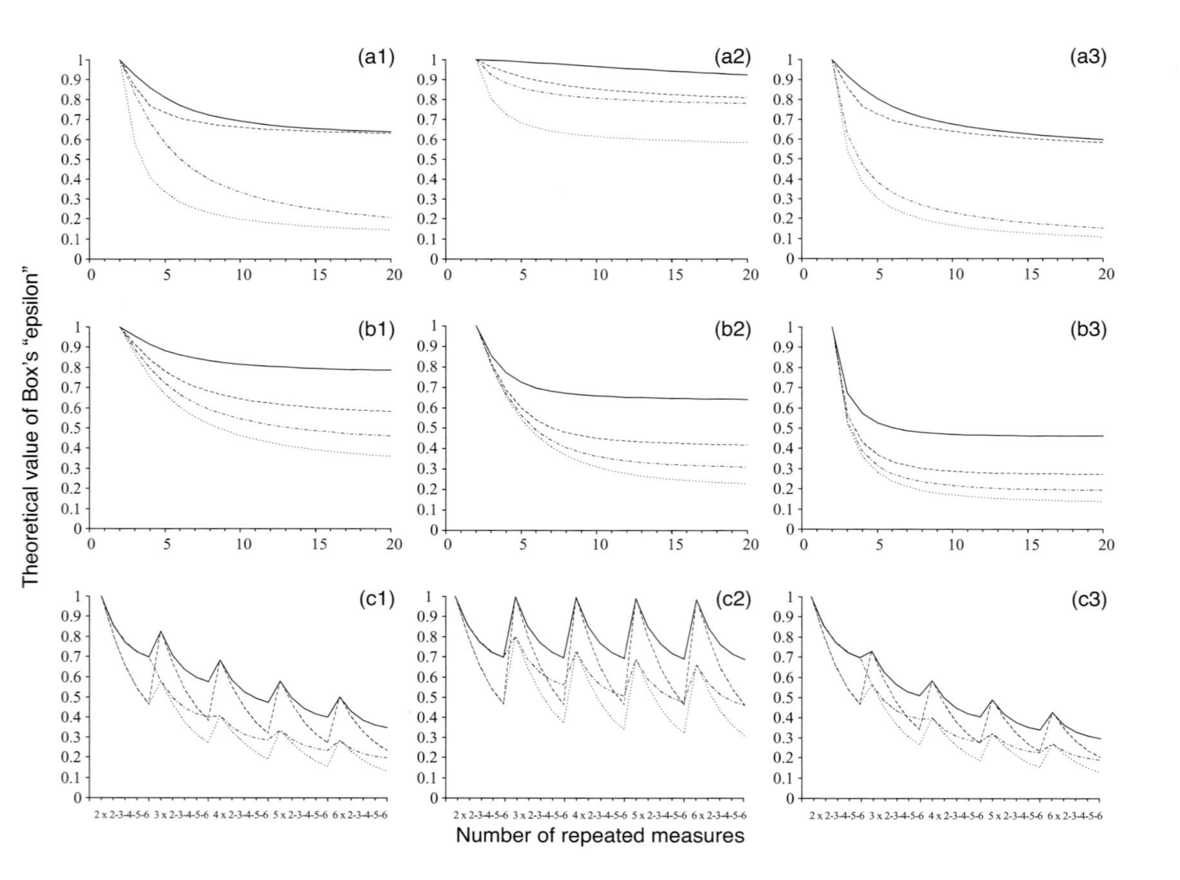
(a1)
(a2)
(a3)
(b1)
(b2)
(b3)
(c1)
(c2)
(c3)
Theoretical value of Box's "epsilon"
Number of repeated measures
2 x 2-3-4-5-6 3 x 2-3-4-5-6 4 x 2-3-4-5-6 5 x 2-3-4-5-6 6 x 2-3-4-5-6

Objective 1

At equal strength (i.e., same absolute value of the autoregressive coefficient ϕ), autocorrelation that alternates in sign has a stronger effect on the df [Fig. 9.6(a1)], because the alternation in sign (i.e., $\phi < 0$) creates a wider range of autocorrelation values. Temporal heteroscedasticity without temporal autocorrelation [Fig. 9.6(a2)] has a weaker effect on the df for within-subject effects than with temporal autocorrelation decreasing with increasing time lag [Fig. 9.6(a3)]. The stronger the temporal heteroscedasticity (i.e., the ratio between the lowest and highest variances is greater), the lower Box's "epsilon."

Objective 2

The longer the range of spatial autocorrelation, the lower Box's "epsilon," and the slower the decrease in spatial autocorrelation with increasing distance (i.e., slowest for the Gaussian model and fastest for the exponential, with the spherical in-between), the stronger the reduction in df [Fig. 9.6(b1)–(b3)].

←

Fig. 9.6. Theoretical value of Box's "epsilon" in relation to the number of repeated measures, for various autocorrelated or heteroscedastic variance–covariance structures in time (a1)–(a3), space (b1)–(b3), and space-time (c1)–(c3). (a1) Second-order stationary AR(1), with parameter ϕ value = –0.9 (dotted curve), –0.5 (dashed), 0.5 (continuous), and 0.9 (dashed-dotted). (a2) Heteroscedastic and independent: $\text{Var}[U(t)] = 1 + 0.1(t-1)$ (continuous curve), $\text{Var}[U(t)] = 1 + 0.5(t-1)$ (dashed), $\text{Var}[U(t)] = t$ (dashed-dotted), and $\text{Var}[U(t)] = t^2$ (dotted), $t = 1, \ldots, S^*$. (a3) Combinations of variance–covariance structures in (a1) and (a2): AR(1), $\phi = 0.5$ with $\text{Var}[U(t)] = 1 + 0.1(t-1)$ (continuous curve); AR(1), $\phi = -0.5$ with $\text{Var}[U(t)] = 1 + 0.1(t-1)$ (dashed); AR(1), $\phi = 0.9$ with $\text{Var}[U(t)] = t^2$ (dashed-dotted); and AR(1), $\phi = -0.9$ with $\text{Var}[U(t)] = t^2$ (dotted). (b1) Exponential, (b2) spherical, and (b3) Gaussian variogram models, with (practical) range of autocorrelation = 3 (continuous curve), 5 (dashed), 7 (dashed-dotted), and 10 (dotted). (c1)–(c3) Combinations of variance–covariance structures in (a1), (a2), and (b2). (c1) AR(1), $\phi = 0.9$ (in time; first number of repeated measures) with spherical variogram model, range 3 (in space; second number of repeated measures) (continuous curve); AR(1), $\phi = 0.9$ with spherical, range 10 (dashed); AR(1), $\phi = -0.9$ with spherical, range 3 (dashed-dotted); and AR(1), $\phi = -0.9$ with spherical, range 10 (dotted). (c2) $\text{Var}[U(t)] = 1 + 0.1(t-1)$, $t = 1, \ldots, S^*$ (in time) with spherical variogram model, range 3 (in space) (continuous curve); $\text{Var}[U(t)] = 1 + 0.1(t-1)$ with spherical, range 10 (dashed); $\text{Var}[U(t)] = t^2$ with spherical, range 3 (dashed-dotted); and $\text{Var}[U(t)] = t^2$ with spherical, range 10 (dotted). (c3) AR(1), $\phi = 0.9$ (in time) and spherical variogram model, range 3 (in space) (continuous curve); AR(1), $\phi = 0.9$ and spherical, range 10 (dashed); AR(1), $\phi = -0.9$ and spherical, range 3 (dashed-dotted); and AR(1), $\phi = -0.9$ and spherical, range 10 (dotted), all four with $\text{Var}[U(t)] = t$ ($t = 1, \ldots, S^*$).

Objective 3
Temporal autocorrelation and temporal heteroscedasticity combined with spatial autocorrelation [Fig. 9.6(c3)] tend to provide a stronger reduction in df than only temporal autocorrelation and spatial autocorrelation [Fig. 9.6(c1)] and temporal heteroscedasticity and spatial autocorrelation [Fig. 9.6(c2)] in this order. The number of repeated measures has a particularly important effect in the spatio-temporal framework. Note that, from the four peaks in Fig. 9.6(c2), only the first one is exactly at 1 ($S_1^* = S_2^* = 2$).

Summary: The expected value and population variance of ANOVA SS provide the basis for two very useful measures of spatial and temporal heterogeneity (variance and autocorrelation): the effective number of df and Box's "epsilon." Both quantities can be evaluated for different experimental designs, numbers of factors and levels per factor, and variance–covariance structures. Accordingly, it is possible for the experimenter to evaluate his/her situation before actually starting the study and possibly redirect initial choices that had been made about the number of replicates or blocks, the size of plots or blocks, or the number of repeated measures. With the effective number of df, we have learned in particular that spatial autocorrelation can affect the df associated with an ANOVA SS even if this represents a fixed effect. We have evaluated Box's "epsilon" in various scenarios (temporal, spatial, spatio-temporal), which provided us with insight into the "interaction" between heteroscedasticity and autocorrelation and the role of the model of spatial autocorrelation, among other things. We did not see it, but in very particular cases heteroscedasticity can decrease the discrepancy from circularity due to unequal autocorrelations at same lag in nonstationary models. In the next section, we will see how Box's "epsilon" can be estimated from the data in applications.

9.3 Study design: more applied aspects

After the foundations for the use of the ANOVA method in spatial and temporal heterogeneity analysis were established in the previous subsection, we can look at applications and illustrate the method with real datasets in this one. In doing this, we will focus on the version of the ANOVA for repeated measures (ANOVAR). More specifically, we will see how the between-subject and within-subject effects defined in equations (9.5) and (9.6) in Table 9.1 are assessed with *F*-tests of different types in Subsection 9.3.1. An introduction to the multivariate ANOVA

(MANOVA) will be given at the end of the subsection. Three examples with real datasets will be presented in Subsection 9.3.2. The first two are extensions or variants of forestry and wood science examples presented in Chapter 6, where SAS PROC MIXED was used for data analysis. The third example is environmental, without a treatment factor with several levels, and is inspired from the study of Tosic *et al.* (2009).

9.3.1 Assessment of between-subject and within-subject effects in the ANOVAR

As discussed in Subsection 9.2.1, the ANOVA model for the CRD with simple repeated measures, either temporal or spatial, is obtained by adding three terms to the ANOVA model for the CRD without repeated measures: $\varepsilon_{ij}^{(1)}$, a first error term that does not depend on the temporal or spatial index; $\Delta\mu(s)$, the main effects of the repeated measures factor; and $\Delta a_i(s)$, the interaction effects between the treatment factor and the repeated measures factor. While $\varepsilon_{ij}^{(1)}$ and $\varepsilon_{ij}^{(2)}(s)$ are random, the other four terms in equation (9.5) are fixed. It follows that

$$\mu(s) = E[\bar{U}(s)] = \mu + \Delta\mu(s),$$

where $\Delta\mu(s)$ represents any discrepancy from homogeneity of the mean (μ) that would exist on average over all treatment levels and replicates, and

$$\mu_i(s) = \mathrm{E}[\bar{U}_i(s)] = \mathrm{E}[U_{ij}(s)] = \mu + a_i + \Delta\mu(s) + \Delta a_i(s),$$

where $\Delta a_i(s)$ ($i = 1, \ldots, p$) indicate any discrepancies from homogeneity of the mean (represented here by $\mu + a_i$ for a given i) that would arise, in addition to $\Delta\mu(s)$, from different fluctuations in the expected value of the response in time or space depending on the level of the treatment factor. Clearly, the heterogeneity of the mean, if any, is in $\Delta\mu(s)$ and $\Delta a_i(s)$, because these two terms allow the expected value of $U_{ij}(s)$ to change with s, while the heterogeneity of the variance and the heterogeneity due to autocorrelation, if they exist, are in $\varepsilon_{ij}^{(2)}(s)$ ($s = 1, \ldots, S^*$) or the variance–covariance matrix of the random vector of repeated measures, Σ ($S^* \times S^*$).

Accordingly, the F-test for the treatment main effects a_i (i.e., the only between-subject effects in the case of the CRD), using $\varepsilon_{ij}^{(1)}$ as error term, requires no modification or adjustment of the number of df. *Note:* In the cases of the RCBD 1 rep, RCBD >1 rep, and "LSD," there are one, two, and two supplemental between-subject effects: RCBD 1 rep, b_j; RCBD >1 rep, b_j and $(ab)_{ij}$; and "LSD," $b_{j_1}^{(col)}$ and $b_{j_2}^{(row)}$. The statistical assessment of the within-subject effects, i.e., $\Delta\mu(s)$ and $\Delta a_i(s)$ for the CRD with simple repeated measures, is a different story. Modified *F*-tests with numbers of df adjusted with Box's "epsilon," that is, df_1 = Box's "epsilon" times the classical df_1 and df_2 = Box's "epsilon" times the classical df_2, are then likely to be required (see Subsection 9.2.3). The exception will be when the heterogeneity of the variance and the heterogeneity due to autocorrelation are of a type that satisfies the circularity condition, in which case Box's "epsilon" is one and the numbers of df of the *F*-tests for within-subject effects remain unchanged.

In applications, Box's "epsilon" is estimated from the data. First, Greenhouse and Geisser (1959) proposed an estimator in which the theoretical variance–covariance matrix Σ $(S^* \times S^*)$ [see equations (9.8), (A9.8) and (A9.9)] is replaced by the sample variance–covariance matrix computed from the vectors of repeated measures data. Then, Huynh and Feldt (1976) showed that the Greenhouse–Geisser estimator of Box's "epsilon" decreases the numbers of df excessively, leading to a loss of statistical power of the *F*-test, when the departure from circularity is moderate and the number of vectors of repeated measures data is small. Therefore, they designed an alternative estimator that performs better in these situations. In all cases, for a given estimator the same estimated value of Box's "epsilon" is applied to both numbers of df of the Fisher–Snedecor distribution of the test statistic for within-subject effects considered, prior to calculating the probability of significance of its observed value. Box's "epsilon" is never really applied to the *F*-ratio statistic itself because, if it was, it would be applied to the numerator and the denominator and would thus simplify in the end.

With double repeated measures, the statistical assessment of between-subject effects is exactly the same as with simple repeated measures. On the side of within-subject effects, things get complicated because these are of three types; for example, purely temporal, purely spatial, and spatio-temporal if one repeated measures factor is temporal and the other is spatial. Accordingly, there are three error terms, indexed differently in equation (9.6): $\varepsilon_{ij}^{(2)}(s_1)$ (e.g., purely temporal), $\varepsilon_{ij}^{(3)}(s_2)$ (e.g., purely spatial), and $\varepsilon_{ij}^{(4)}(s_1, s_2)$ (e.g., spatio-temporal), and three different Box's "epsilons" are estimated and used in the relevant modified *F*-tests for each set of within-subject effects [i.e., $\Delta\mu(s_1)$ and $\Delta a_i(s_1)$; $\Delta\mu(s_2)$ and $\Delta a_i(s_2)$; $\Delta\mu(s_1, s_2)$ and $\Delta a_i(s_1, s_2)$]. Among other things,

$$\begin{aligned}\mu(s_1, s_2) &= E[\bar{U}(s_1, s_2)] \\ &= \mu + \Delta\mu(s_1) + \Delta\mu(s_2) + \Delta\mu(s_1, s_2),\end{aligned}$$

where $\Delta\mu(s_1)$, $\Delta\mu(s_2)$, and $\Delta\mu(s_1, s_2)$ represent discrepancies from homogeneity of the mean (μ), on average over all treatment levels and replicates, purely in time (if $s_1 = t$), purely in 1-D space (if $s_2 = x$), and in both (if $s_1 = t$ and $s_2 = x$). This situation is illustrated by the second example in Subsection 9.3.2.

Performing the ANOVAR with SAS PROC GLM (SAS Institute Inc., 2009) requires that one writes the ANOVA model in the multivariate (MANOVA) form. Here is how this can be done from equation (9.5) (CRD with simple repeated measures) and equation (9.6) (CRD with double repeated measures); a similar rewriting procedure can be followed for other experimental designs with repeated measures (Dutilleul, 1998b).

In equation (9.5), let us regroup the terms without subscript, with subscript i only, and with subscripts i and j together. This provides three pairs of terms (i.e., one term without the index s and the other with it):

$$U_{ij}(s) = \mu + \Delta\mu(s) + a_i + \Delta a_i(s) + \varepsilon_{ij}^{(1)} + \varepsilon_{ij}^{(2)}(s).$$

Consider the equality above for $s = 1, \ldots, S^*$ (i.e., for all levels of the repeated measures factor) and define four $S^* \times 1$ vectors: $\boldsymbol{u}_{ij} = (U_{ij}(1), \ldots, U_{ij}(S^*))'$; $\boldsymbol{m} = (\mu + \Delta\mu(1), \ldots, \mu + \Delta\mu(S^*))'$; $\boldsymbol{a}_i = (a_i + \Delta a_i(1), \ldots, a_i + \Delta a_i(S^*))'$; and $\boldsymbol{e}_{ij} = (\varepsilon_{ij}^{(1)} + \varepsilon_{ij}^{(2)}(1), \ldots, \varepsilon_{ij}^{(1)} + \varepsilon_{ij}^{(2)}(S^*))'$. Then, the MANOVA model for the CRD with simple repeated measures can be written as

$$\boldsymbol{u}_{ij} = \boldsymbol{m} + \boldsymbol{a}_i + \boldsymbol{e}_{ij}.$$

A similar grouping of terms in equation (9.6) leads to

$$\begin{aligned}U_{ij}(s_1, s_2) &= \mu + \Delta\mu(s_1) + \Delta\mu(s_2) + \Delta\mu(s_1, s_2) + a_i + \Delta a_i(s_1) \\ &+ \Delta a_i(s_2) + \Delta a_i(s_1, s_2) + \varepsilon_{ij}^{(1)} + \varepsilon_{ij}^{(2)}(s_1) + \varepsilon_{ij}^{(3)}(s_2) + \varepsilon_{ij}^{(4)}(s_1, s_2),\end{aligned}$$

and finally, with $\boldsymbol{u}_{ij} = (U_{ij}(1, 1), \ldots, U_{ij}(1, S_2^*), U_{ij}(2, 1), \ldots, U_{ij}(2, S_2^*), \ldots, U_{ij}(S_1^*, 1), \ldots, U_{ij}(S_1^*, S_2^*))'$ ($S_1^* S_2^* \times 1$) and other similar notations,

$$\boldsymbol{u}_{ij} = \boldsymbol{m} + \boldsymbol{a}_i + \boldsymbol{e}_{ij}.$$

That is, a MANOVA model that looks the same as that for the CRD with simple repeated measures, but the order of the two repeated measures factors under the statement REPEATED (where one specifies how many repeated measures factors there are) is very important in SAS PROC GLM. It is dependent on how indices change inside $\boldsymbol{u}_{ij}$. Above, the factor with S_1^* repeated measures would be first, and the one with S_2^* repeated measures, second.

9.3.2 Examples in time, space and space-time

The three examples of ANOVAR presented below can be briefly described as follows. The first example (forestry) is related to the forestry examples of Chapter 6; the repeated measures data analyzed in it are restricted to time. The second example (wood science) is related to part (b) of the second forestry example in Section 6.4; the repeated measures data are spatio-temporal, with sampling site within a growth ring considered to be of spatial nature in 1-D. The third example (Barbados environmental) is new; the spatio-temporal repeated measures data in it are analyzed under space-time separability of the variance–covariance structure, due to low replication; space is 2-D; and no "treatment" is "applied." In these examples, time and space are considered as two fixed qualitative ANOVA factors with a finite number of levels. The main and interaction effects of these factors (i.e., the within-subject effects) correspond to temporal and spatial heterogeneity of the mean and can be assessed statistically, the two other types of heterogeneity (variance and autocorrelation) being quantified via Box's "epsilon."

Forestry example

In the study by Herman *et al.* (1998) (from which this example is directly inspired), the main objective was to test the hypothesis that there are differences in the mean values of fiber length, ring width and wood density, depending on the growth rate in circumference of Norway spruce (*Picea abies* (L.) Karst.). Therefore, 40 Norway spruces were first classified, independently of the three variables, as 20 "fast-grown" and 20 "slow-grown." For each tree, the three variables were measured on each identifiable growth ring. Here, we will concentrate on the first 12 years after thinning had started in the forest stand. With Growth category as the treatment factor and Year of growth ring as the temporal repeated measures factor, the experimental design followed is a CRD with simple repeated measures and the ANOVA model for it is given by equation

(9.5) in Table 9.1, so that the material of the previous subsection can be readily applied.

Figure 9.7(a1)–(a3) shows differences in the mean values of the three variables that could be related to Growth category or Year, or both, but are they statistically significant? *Note:* Error bars in Fig. 9.7 are based on one standard error only; approximate 95% confidence intervals would be obtained with two standard errors. The content of Table 9.4 provides an answer to the question above. For all three variables, Year main effects are highly significant ($P < 0.0001$), even with modified *F*-tests. *Note:* Unmodified *F*-tests (in which downward adjusted numbers of df are not used) tend to exaggerate the significance of within-subject effects when the observed value of the statistic is greater than 1, by producing probabilities of significance lower than they should be. For each variable, temporal heterogeneity of the mean appears to be similar for both growth categories, since the Growth category × Year interaction is not declared significant with the modified *F*-test. Growth category main effects are statistically significant for ring width and wood density but not for fiber length, although differences in mean between fast-grown and slow-grown trees keep the same sign and are almost constant in size over years for the three variables [Fig. 9.7(a1)–(a3)]. The non-significant result obtained for fiber length indicates that, across years, the difference between the two fiber length means is not large enough relative to a pooled estimate of the dispersion in the two growth categories (i.e., the basic principle for the comparison of two means in statistics).

The use of contrasts (i.e., functions of means that each takes 1 df from the number of df associated with the corresponding main effects) provides the following complements of information about temporal heterogeneity of the mean in the three variables. For fiber length [Table 9.4(a)], the POLYNOMIAL contrasts show that temporal heterogeneity of the mean is more in global fluctuations (see linear, quadratic and cubic contrasts 1–3) than in local ones (see contrasts 9–11). For ring width [Table 9.4(b)], the MEAN contrasts allow one to determine which years are closer to the average (essentially the very first year and the last two here). For wood density [Table 9.4(c)], the CONTRAST(1) contrasts compare the mean of the first year to the mean of each of the following years, and several years are found to be significantly different from the first one; the application of the Bonferroni correction (which consists in using $P < \frac{0.05}{11}$ as criterion of significance) does not change the conclusions drastically.

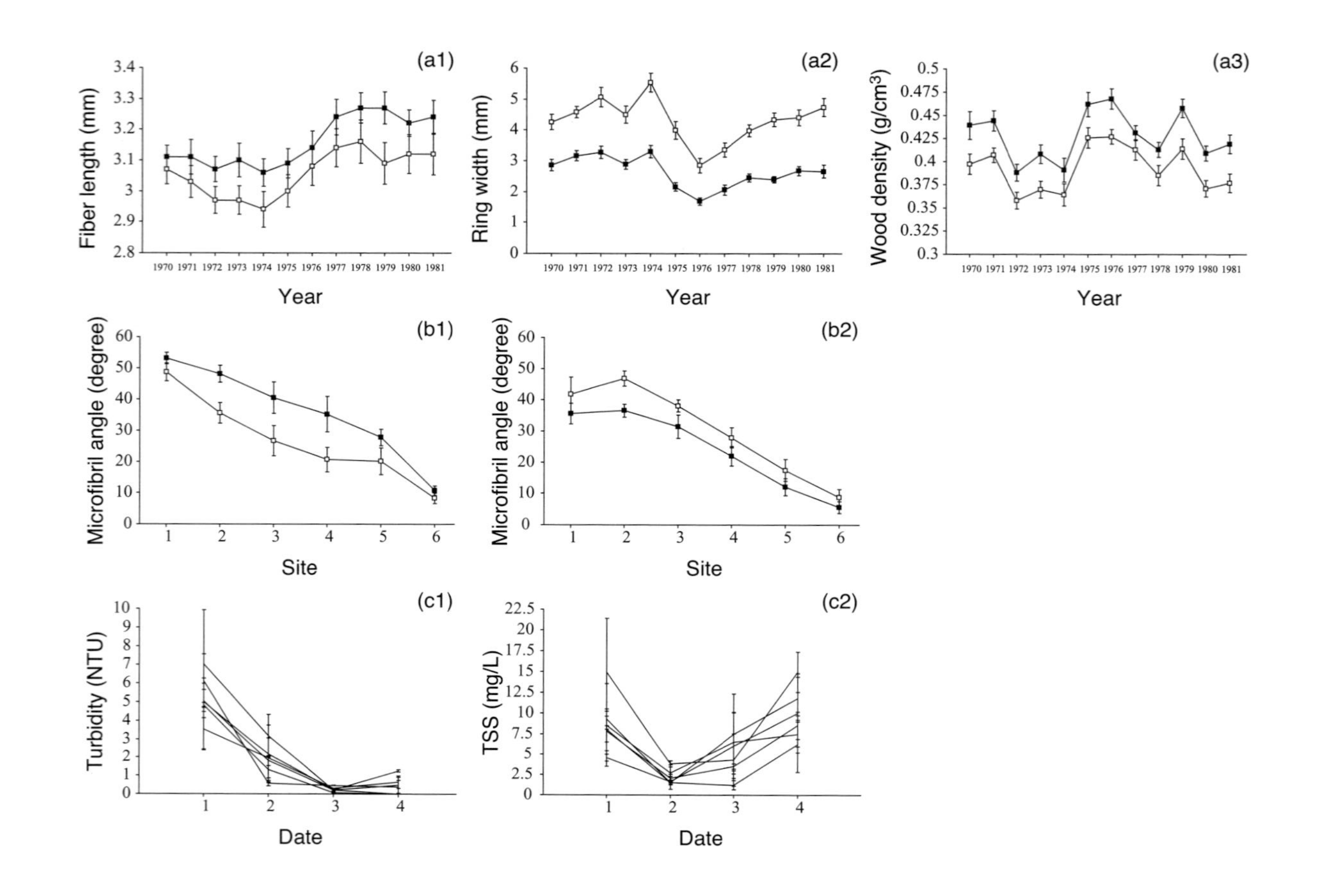

(a1)
Fiber length (mm)
Year
(a2)
Ring width (mm)
Year
(a3)
Wood density (g/cm^3)
Year
(b1)
Microfibril angle (degree)
Site
(b2)
Microfibril angle (degree)
Site
(c1)
Turbidity (NTU)
Date
(c2)
TSS (mg/L)
Date

The heterogeneity of the variance and the heterogeneity due to autocorrelation appear to be of same size for fiber length and wood density (Box's "epsilon" estimates: 0.5–0.6), and slightly stronger (i.e., the departure from circularity is slightly stronger) for ring width (Box's "epsilon" estimates: 0.4–0.5). With 12 repeated measures, the smallest value of Box's "epsilon" (which corresponds to the largest departure from circularity) is $\frac{1}{11}$ (0.0909) in theory, so the estimates obtained here are intermediate. Because of space constraints, it was not possible to include the 12 × 12 sample variance–covariance matrices and corresponding autocorrelations in Table 9.5, but they can be saved in the output of SAS PROC GLM (by using the option PRINTE under the statement MANOVA). While heteroscedasticity is moderate for each of the three variables (i.e., the ratio between the lowest and highest sample variances is about 2 to 3), autocorrelation is more constant and positive for fiber length and wood density (numerical results not reported), compared to ring width for which correlations between successive years can be around 0.8 while those between some of the first years and some of the last years can be negative. This difference is reflected by the estimates of Box's "epsilon."

Wood science example

In this example, the biological hypothesis is basically the same as in the previous example because it concerns growth category effects on the mean value of a tree variable, except that these effects are studied between and within growth rings and the variable is microfibril angle (Herman *et al.*, 1999). Two years of growth ring are considered, namely, the first (1959) and the last (1987) in Herman *et al.* (1998, 1999). Microfibril angle was measured on seven fast-grown and seven slow-grown Norway

←

Fig. 9.7. Mean values (± standard errors) of repeated measures in time, space, and space-time. (a1)–(a3) Forestry example, with three variables, two growth categories (fast: open symbols; slow: filled symbols) as levels of a treatment factor, and the year of growth ring as temporal repeated measures factor. (b1) and (b2) Wood science example, with one variable measured in two years, (b1) 1959 and (b2) 1987 (levels of a temporal repeated measures factor), at six sites within a growth ring (levels of a spatial repeated measures factor), for the same two growth categories as in (a1)–(a3). (c1) and (c2) Barbados environmental example, with two variables, and station (six levels) and date (four levels) as spatial and temporal repeated measures factors, respectively; there is no treatment factor and the two repeated measures factors are crossed; TSS = total suspended solids.

Table 9.4. *Results of ANOVAR (i.e., analysis of variance for repeated measures) for the forestry example in time, with two treatment levels (i.e., fast vs. slow growth), 20 replicates per treatment level, and 12 time levels (i.e., CRD with temporal repeated measures). Variables are: (a) fiber length; (b) ring width; and (c) wood density. The temporal repeated measures factor is the year of growth ring. Different types of time contrasts were used for the three variables: (a) polynomial; (b) comparison with the mean of all other years (MEAN in SAS PROC GLM); and (c) comparison with first year [CONTRAST(1) in GLM]. See text and Fig. 9.7(a) for details*

(a)

Unmodified F-tests for between-subject effects

Source	df	Type III SS	Mean square	Observed F	Pr > F
Growth category	1	1.2948	1.2948	2.67	0.1108
Error 1	38	18.4600	0.4858		

Unmodified and modified F-tests for within-subject effects

						Adjusted Pr > F	
Source	df	Type III SS	Mean square	Observed F	Pr > F	G – G†	H – F†
Year	11	2.4815	0.2256	11.29	< 0.0001	< 0.0001	< 0.0001
GC × Year	11	0.1247	0.0113	0.57	0.8554	0.7479	0.7827
Error 2	418	8.3496	0.0200				

† Adjustments made using Greenhouse–Geisser and Huynh–Feldt estimates of Box's "epsilon," which are respectively equal to 0.5196 and 0.6381 here; GC = Growth category

Unmodified two-tailed t-tests for polynomial contrasts

	Pr > | t |		Pr > | t |		Pr > | t |
Contrast 1	< 0.0001	Contrast 2	0.0901	Contrast 3	< 0.0001
	Pr > | t |		Pr > | t |		Pr > | t |
Contrast 4	0.7782	Contrast 5	0.0008	Contrast 6	0.2193

	$Pr > \|t\|$		$Pr > \|t\|$		$Pr > \|t\|$
Contrast 7	0.4059	Contrast 8	0.0417	Contrast 9	0.4131
	$Pr > \|t\|$		$Pr > \|t\|$		
Contrast 10	0.3173	Contrast 11	0.6246		

Contrast 1: linear; Contrast 2: quadratic, etc.

(b)

Unmodified F-tests for between-subject effects

Source	df	Type III SS	Mean square	Observed F	$Pr > F$
Growth category	1	334.3174	334.3174	80.31	< 0.0001
Error 1	38	158.1914	4.1629		

Unmodified and modified F-tests for within-subject effects

						Adjusted $Pr > F$	
Source	df	Type III SS	Mean square	Observed F	$Pr > F$	G – G†	H – F†
Year	11	156.3982	14.2180	22.29	< 0.0001	< 0.0001	< 0.0001
GC × Year	11	11.6947	1.0632	1.67	0.0784	0.1464	0.1325
Error 2	418	266.5942	0.6378				

† Greenhouse–Geisser and Huynh–Feldt estimates of Box's "epsilon": 0.4431 and 0.5296

Unmodified two-tailed t-tests for MEAN contrasts

	$Pr > \|t\|$		$Pr > \|t\|$		$Pr > \|t\|$
Contrast 1	0.4823	Contrast 2	0.0006	Contrast 3	< 0.0001
	$Pr > \|t\|$		$Pr > \|t\|$		$Pr > \|t\|$
Contrast 4	0.0640	Contrast 5	< 0.0001	Contrast 6	0.0040
	$Pr > \|t\|$		$Pr > \|t\|$		$Pr > \|t\|$
Contrast 7	< 0.0001	Contrast 8	< 0.0001	Contrast 9	0.0186
	$Pr > \|t\|$		$Pr > \|t\|$		
Contrast 10	0.3093	Contrast 11	0.4407		

(cont.)

Table 9.4. *(cont.)*

(c)

Unmodified F-tests for between-subject effects

Source	df	Type III SS	Mean square	Observed F	Pr > F
Growth category	1	0.1606	0.1606	10.13	0.0029
Error 1	38	0.6025	0.0159		

Unmodified and modified F-tests for within-subject effects

						Adjusted Pr > F	
Source	df	Type III SS	Mean square	Observed F	Pr > F	G – G†	H – F†
Year	11	0.2981	0.0271	28.53	< 0.0001	< 0.0001	< 0.0001
GC × Year	11	0.0069	0.0006	0.66	0.7747	0.6756	0.7077
Error 2	418	0.3971	0.0009				

† Greenhouse–Geisser and Huynh–Feldt estimates of Box's "epsilon": 0.5282 and 0.6506

Unmodified two-tailed t-tests for CONTRAST(1) contrasts

	Pr > | t |		Pr > | t |		Pr > | t |
Contrast 1	0.1536	Contrast 2	< 0.0001	Contrast 3	0.0006
	Pr > | t |		Pr > | t |		Pr > | t |
Contrast 4	< 0.0001	Contrast 5	0.0047	Contrast 6	0.0009
	Pr > | t |		Pr > | t |		Pr > | t |
Contrast 7	0.6013	Contrast 8	0.0146	Contrast 9	0.0294
	Pr > | t |		Pr > | t |		
Contrast 10	0.0018	Contrast 11	0.0159		

Table 9.5. *Results of ANOVAR for the wood science example in time, space and space-time, with 2 treatment levels (i.e., fast vs. slow tree growth), 7 replicates per treatment level, and 2 time levels × 6 space levels (i.e., CRD with spatio-temporal repeated measures). The variable is microfibril angle. The repeated measures factors are the year of growth ring and the sampling site within a growth ring. Polynomial contrasts were tested for the spatial repeated measures factor. See text and Fig. 9.7(b) for details*

Unmodified *F*-tests for between-subject effects

Source	df	Type III SS	Mean square	Observed *F*	Pr > *F*
Growth category	1	93.2293	93.2293	0.32	0.5849
Error 1	12	3550.4967	295.8747		

Unmodified† *F*-tests for purely temporal within-subject effects

Source	df	Type III SS	Mean square	Observed *F*	Pr > *F*
Year	1	774.6462	774.6462	5.89	0.0319
GC × Year	1	2520.6879	2520.6879	19.17	<0.0001
Error 2	12	1577.5386	1577.5386		

† There is no need to modify the ANOVA *F*-tests in this case because the circularity condition is always satisfied when there are only two repeated measures (i.e., paired observations); see text. GC = growth category

Unmodified and modified *F*-tests for purely spatial within-subject effects

						Adjusted Pr > *F*	
Source	df	Type III SS	Mean square	Observed *F*	Pr > *F*	G – G†	H – F†
Site	5	27019.3292	5403.8658	91.37	<0.0001	<0.0001	<0.0001
GC × Site	5	150.8070	30.1614	0.51	0.7676	0.6869	0.7595
Error 3	60	3548.6315	59.1439				

† Adjustments made using Greenhouse–Geisser and Huynh–Feldt estimates of Box's "epsilon," which are respectively equal to 0.6314 and 0.9543 here

(*cont.*)

Table 9.5. *(cont.)*

Unmodified two-tailed t-tests for purely spatial polynomial contrasts

	Pr > | t |		Pr > | t |		Pr > | t |
Contrast 1	< 0.0001	Contrast 2	0.0209	Contrast 3	0.7620
	Pr > | t |		Pr > | t |		
Contrast 4	0.2333	Contrast 5	0.9156		

Contrast 1: linear; Contrast 2: quadratic, etc.

Unmodified and modified F-tests for spatio-temporal within-subject effects

						Adjusted Pr > F	
Source	df	Type III SS	Mean square	Observed F	Pr > F	G – G†	H – F†
Year × Site	5	974.1176	194.8235	3.97	0.0035	0.0103	0.0035
GC × Year × Site	5	404.9467	80.9893	1.65	0.1603	0.1850	0.1603
Error 4	60	2942.0967	49.0349				

† Greenhouse–Geisser and Huynh–Feldt estimates of Box's "epsilon": 0.7059 and 1.1195

Unmodified two-tailed t-tests for spatio-temporal contrasts†

	Pr > | t |		Pr > | t |		Pr > | t |
Contrast 1	0.4216	Contrast 2	0.0733	Contrast 3	0.0015
	Pr > | t |		Pr > | t |		
Contrast 4	0.4199	Contrast 5	0.7479		

† Polynomial contrasts for the spatial component

spruces, so the number of sampling sites within a growth ring is limited to six (i.e., the original sites numbered 1, 3, 5, 7, 9, and 11), in order to have more trees (individuals) than repeated measures (i.e., $14 > 2 \times 6$). Since a sampling site within a growth ring in a given direction at a given height corresponds to a spatial location as well as a time, Site will be called a spatial repeated measures factor here, contrary to Section 6.4. This has no implication for statistical analysis. Just the interpretation of results will be simplified by having one temporal and one spatial repeated measures factor (instead of two temporal repeated measures factors). In all cases, the experimental design followed is a CRD with double repeated measures, for which the ANOVA model is given by equation (9.6) in Table 9.1, and the relevant material from Subsection 9.3.1 is readily applicable.

Across the two years and the six sampling sites within a growth ring, Growth category main effects (i.e., the between-subject effects) are not statistically significant (Table 9.5, first part). This is because one mean curve is systematically above the other in one year, and the reverse is observed the other year [Fig. 9.7(b1) and (b2)], so the difference between the two microfibril angle means is small overall. Concerning temporal heterogeneity of the mean (i.e., the purely temporal within-subject effects), the difference between yearly means is significant ($P < 0.05$), but the Growth category × Year interaction (of crossing type) is highly significant ($P < 0.0001$), because of the change in sign of the difference in mean between growth categories from one year to the other [Table 9.5, second part; see also Fig. 9.7(b1) and (b2)]. *Note:* Lower tensile and tear strengths of fibers as well as some modifications of the mechanical properties of solid wood and paper can be expected when microfibril angle is higher. As for spatial and spatio-temporal heterogeneities of the mean (i.e., the purely spatial and spatio-temporal within-subject effects), Site main effects and the Year × Site interaction are both significant (at different levels), but interactions with Growth category are not (Table 9.5, third and fourth parts), although the triple interaction shows a tendency to be significant with a probability approaching 0.15.

More specifically, polynomial contrasts show that site effects, when averaged over growth categories and years, are mostly linear and quadratic (Table 9.5, third part), while the same type of contrasts point out that the Year × Site interaction is significant because of differences in the quadratic and cubic curvatures between years (Table 9.5, fourth part); see also the little "peaks" present in (b2) but absent from (b1) in Fig. 9.7. Since Box's "epsilon" is one when there are only two repeated measures, calculations were made outside of SAS PROC GLM, and showed moderate heteroscedasticity and weak negative correlation between years. Concerning spatial and spatio-temporal heterogeneities of the variance and heterogeneities due to autocorrelation, estimates of Box's "epsilon" close to 1 (and even greater than 1 with the Huynh–Feldt estimator, in which case the estimate is rounded to 1) indicate almost no departure from circularity. *Reminder:* The Huynh–Feldt estimator of Box's "epsilon" is recommended when the departure from circularity appears moderate and the number of vectors of repeated measures data is small (Crowder and Hand, 1990, p. 55).

Barbados environmental example

In Tosic *et al.* (2009), the authors studied the contributions of runoff to seawater quality degradation off Holetown, Barbados. In particular, they analyzed seawater quality following flow events in the rainy season (June–December) of 2006. Samples (300 mL) were then taken from the runoff's surface water, and among the variables known as indicators of eutrophication and sedimentation, turbidity and total suspended solids (TSS) were measured. Periodic runoff events create plumes of nutrient-rich, sediment-laden freshwater which can extend over 1 km offshore to the island's bank reefs. The data submitted to ANOVAR in this example were collected at six sampling stations located in a reef area about 500 m from the watershed outlet. Temporally, four flow events were studied, on the dates of August 24, October 16 and 27, and November 14, 2006. Three water samples (replicates) were taken at each station on each date.

In the absence of a treatment factor with several levels (i.e., there was a flow event on each of the four dates), the experimental design is not a CRD with double repeated measures, even though there are spatio-temporal repeated measures. This experimental situation is similar to that in the study of Dutilleul and Pinel-Alloul (1996), where the water column was sampled at three thermal strata on five dates in a small number of lakes. Accordingly, equation (9.6) of the ANOVA model simplifies into

$$U_j(s_1, s_2) = \mu + \varepsilon_j^{(1)} + \Delta\mu(s_1) + \varepsilon_j^{(2)}(s_1) + \Delta\mu(s_2) + \varepsilon_j^{(3)}(s_2) + \Delta\mu(s_1, s_2) + \varepsilon_j^{(4)}(s_1, s_2),$$

with $j = 1, \ldots, n$; μ, $\Delta\mu(s_1)$, $\Delta\mu(s_2)$, and $\Delta\mu(s_1, s_2)$, the overall population mean, the Date main effects, the Station main effects, and the Date $\times$ Station interaction, respectively; and their respective error terms.

Due to low replication (i.e., $n = 3$), as in Dutilleul and Pinel-Alloul (1996), the variance–covariance structure here is analyzed under space-time separability (Section 2.3). In the frame of this example, it means that for a given j, $U_j(s_1, s_2)$ ($s_1 = 1, \ldots, S_1^*$; $s_2 = 1, \ldots, S_2^*$) defines a random matrix with S_1^* rows (time) and S_2^* columns (space); using the algorithm of Dutilleul (1999), a temporal variance–covariance matrix and a spatial variance–covariance matrix are first estimated from the n replicates; and the estimated spatio-temporal variance–covariance matrix is obtained by Kronecker (direct) product (Graybill, 1983).

Figure 9.7(c1) and (c2) suggests important fluctuations in time and space for turbidity and TSS, but some of the error bars associated with mean values calculated from replicates are huge. The ANOVAR (Table 9.6) indicates that heterogeneity of the mean is more temporal than spatial; only the Date main effects are declared significant or highly significant with the modified F-test. Thus, despite some variability, water quality in the reef area is not more affected at near-shore stations than at off-shore stations, or the entire reef is affected similarly; actually, the effect of pollution on the reef area is dependent on whether or not the plume's dispersion reaches the reef or not, and the likeliness of the plume reaching the reef is dependent on the prevailing winds at the time of an event (Tosic *et al.*, 2009). Moreover, Box's "epsilon" estimates and the estimated variance–covariance matrices show a much greater heterogeneity of the variance in time than in space, while autocorrelation estimates vary from strong negative to strong positive in both frameworks for both variables (Table 9.6). With 24 repeated measures, the smallest theoretical value of Box's "epsilon" is $\frac{1}{23}$ (0.0435). The value estimated for spatio-temporal within-subject effects in TSS is close to it, indicating a strong departure from circularity.

9.4 Recommended readings

When entering the world of sampling techniques, the monograph by Cochran (1963) is one that should be considered. Although the author does not focus on spatial heterogeneity, he shows how the expectation operator can be used to obtain general results on sampling statistics, beyond a single partial realization of the stochastic process. Haining (1990, Sections 5.2 and 5.3) presents results in 1-D space and references in 2-D space that confirm the results presented here in Section 9.1, in particular that simple random sampling is not the best sampling design in spatial statistics. Mead (1988) provides a very interesting, conceptual and

Table 9.6 *Results of ANOVAR for the Barbados environmental example in time, space and space-time, with 4 time levels × 6 space levels and 3 replicates per space-time level; there is no treatment factor with several levels. The variables are (a) turbidity and (b) total suspended solids. The temporal and spatial repeated measures factors are the date of a rainfall event and a station in the reef area. Results are of two types: unmodified and modified ANOVA* F-*tests, and Box's "epsilon" estimates and the corresponding variance–covariance matrices estimated under space-time separability*

(a)

Unmodified and modified F-tests for purely temporal, purely spatial and spatio-temporal effects†

Source	df	Type III SS	Mean square	Observed F	Pr > F	Adjusted Pr > F
Date	3	290.5971	96.8657	34.26	<0.0001	0.0002
Station	5	14.9309	2.9862	1.06	0.3963	0.3845
Station × Date	15	21.4543	1.4303	0.51	0.9253	0.6569
Error	48	135.7203	2.8275			

† Heterogeneity of the mean; Box's "epsilon" estimates change with effects (see below)

Box's "epsilon" estimates and estimated variance–covariance matrices†

$$\hat{\varepsilon}_{\mathrm{RM}_1} = 0.5594 \text{ (purely temporal) and } \hat{\Sigma}_{\mathrm{RM}_1} = \begin{pmatrix} 11.512 & -0.349 & -0.319 & 0.326 \\ -0.349 & 12.674 & 0.221 & 2.447 \\ -0.319 & 0.221 & 0.050 & 0.037 \\ 0.326 & 2.447 & 0.037 & 0.525 \end{pmatrix}$$

$$\hat{\varepsilon}_{\mathrm{RM}_2} = 0.3696 \text{ (purely spatial) and } \hat{\Sigma}_{\mathrm{RM}_2} = \begin{pmatrix} 0.244 & 0.542 & 0.027 & -0214 & 0.022 & -0.344 \\ 0.542 & 1.506 & -0.046 & -0.552 & 0.021 & -1.042 \\ 0.027 & -0.046 & 0.559 & -0.407 & 0.081 & -0.033 \\ -0.214 & -0.552 & -0.407 & 0.732 & -0.098 & 0.525 \\ 0.022 & 0.021 & 0.081 & -0.098 & 0.160 & -0.164 \\ -0.344 & -1.042 & -0.033 & 0.525 & -0.164 & 0.921 \end{pmatrix}$$

$\hat{\varepsilon}_{\mathrm{RM}_{12}} = 0.1670$ (spatio-temporal) and $\hat{\Sigma}_{\mathrm{RM}_{12}} = \hat{\Sigma}_{\mathrm{RM}_2} \otimes \hat{\Sigma}_{\mathrm{RM}_1}$

† The diagonal and off-diagonal entries of these matrices are indicative of the presence of heterogeneity of the variance and autocorrelation, respectively; ⊗ denotes the Kronecker product between two matrices (Graybill, 1983).

(b)

Unmodified and modified F-tests for purely temporal, purely spatial and spatio-temporal effects

Source	df	Type III SS	Mean square	Observed F	$\Pr > F$	Adjusted $\Pr > F$
Date	3	685.9911	228.6637	9.36	<0.0001	0.0405
Station	5	204.6628	40.9325	1.68	0.1584	0.2907
Station × Date	15	211.2772	14.0851	0.58	0.8781	0.5154
Error	48	1172.0800	24.4183			

Box's "epsilon" estimates and estimated variance–covariance matrices

$$\hat{\varepsilon}_{\mathrm{RM}_1} = 0.4037 \text{ (purely temporal) and } \hat{\Sigma}_{\mathrm{RM}_1} = \begin{pmatrix} 1.170 & -0.090 & 1.238 & -0.561 \\ -0.090 & 0.085 & -1.137 & 0.369 \\ 1.238 & -1.137 & 15.762 & -4.649 \\ -0.561 & 0.369 & -4.649 & 2.851 \end{pmatrix}$$

$$\hat{\varepsilon}_{\mathrm{RM}_2} = 0.2961 \text{ (purely spatial) and } \hat{\Sigma}_{\mathrm{RM}_2} = \begin{pmatrix} 59.172 & 43.624 & 55.676 & -28.982 & 19.736 & 80.076 \\ 43.624 & 34.539 & 40.098 & 17.065 & 12.946 & 61.632 \\ 55.676 & 40.098 & 60.521 & 39.733 & 22.447 & 83.632 \\ -28.982 & -17.065 & -39.733 & 39.727 & -12.016 & -43.181 \\ 19.736 & 12.946 & 22.447 & -12.016 & 21.079 & 30.421 \\ 80.076 & 61.632 & 83.632 & -43.181 & 30.421 & 140.600 \end{pmatrix}$$

$$\hat{\varepsilon}_{\mathrm{RM}_{12}} = 0.0765 \text{ (spatio-temporal) and } \hat{\Sigma}_{\mathrm{RM}_{12}} = \hat{\Sigma}_{\mathrm{RM}_2} \otimes \hat{\Sigma}_{\mathrm{RM}_1}$$

methodological introduction to experimental designs, with many exercises for the reader. By comparison, Milliken and Johnson (1992) treat more applied aspects of the analysis of experimental data in the ANOVA approach, while Winer *et al.* (1991) is a key reference for ANOVA expected mean squares, which are derived in a variety of experimental situations therein. Crowder and Hand (1990) is one of a few books dedicated to the analysis of repeated measures. More fundamental references are Scheffé's (1959) *Analysis of Variance* and Searle's (1971) *Linear Models*, Chapter 2 in the latter being particularly useful for properties of quadratic forms in normal random vectors. A very large number of applications of matrices in statistics, including in ANOVA and other linear models, can be found in Graybill (1976, 1983). More ecologically oriented references (on topics related to sampling and study design) are the following: Dutilleul (1993a, spatial heterogeneity and the design of experiments); Dutilleul (1998a, 1998b, study design and statistical methods in scale analysis); Dutilleul and Legendre (1993, heteroscedasticity vs. spatial heterogeneity); Filion *et al.* (2000, experimental design with FACE rings); Potvin *et al.* (1990) and Herman *et al.* (1998, 1999, analysis of repeated measures).

Appendix A9: Matrix algebra complements

A9.1 How to write ANOVA sums of squares as quadratic forms

In a preliminary step, consider the sample variance $S^2 = \frac{1}{n-1}\sum_{i=1}^{n}(U(s_i) - \bar{U})^2$ in the spatial or temporal context, where $U(s_i)$ $(i = 1, \ldots, n)$ are the sample data and $\bar{U}$ denotes the sample mean. The sum of squares $\sum_{i=1}^{n}(U(s_i) - \bar{U})^2$ represents the squared Euclidean distance (defined by the classical scalar product, without any estimated matrix) between the vector of observations and the sample mean in the n-dimensional space of the sample.

By definition, $\bar{U} = \frac{1}{n}\sum_{i=1}^{n} U(s_i)$, so in vector notation,

$$\bar{U} = \frac{1}{n}\mathbf{1}_n'\boldsymbol{u}, \qquad \text{(A9.1)}$$

where $\mathbf{1}_n$ denotes the $n \times 1$ column vector of ones (so its transpose, $\mathbf{1}_n'$, is the $1 \times n$ row vector of ones), and $\boldsymbol{u}$ is the $n \times 1$ random vector of observations, $(U(s_1), \ldots, U(s_n))'$.

It follows that, for a given i, $U(s_i) - \bar{U}$ can be written as

$$\boldsymbol{b}_i'\boldsymbol{u} - \frac{1}{n}\mathbf{1}_n'\boldsymbol{u}, \qquad \text{(A9.2)}$$

where all the components of $\boldsymbol{b}_i'(1 \times n)$ are zero, except the i-th component which is one. Doing this for $i = 1, \ldots, n$ leads to the following matrix expression:

$$\mathbf{I}_n \boldsymbol{u} - \frac{1}{n} \mathbf{J}_n \boldsymbol{u} = (\mathbf{I}_n - \frac{1}{n} \mathbf{J}_n)\boldsymbol{u}, \tag{A9.3}$$

where $\mathbf{I}_n$ is the $n \times n$ identity matrix and $\mathbf{J}_n$ is the $n \times n$ matrix of ones. In other words, equation (A9.3) represents the $n \times 1$ random vector whose i-th component is $U(s_i) - \bar{U}$, as the linear transformation of $\boldsymbol{u}$ by the matrix $\mathbf{I}_n - \frac{1}{n}\mathbf{J}_n$.

Thus, the sum of squares $\sum_{i=1}^{n} (U(s_i) - \bar{U})^2$ can be written as

$$\boldsymbol{u}'(\mathbf{I}_n - \frac{1}{n} \mathbf{J}_n)\boldsymbol{u}, \tag{A9.4}$$

because matrix $\mathbf{I}_n - \frac{1}{n}\mathbf{J}_n$ is idempotent, that is, $(\mathbf{I}_n - \frac{1}{n}\mathbf{J}_n)' = \mathbf{I}_n - \frac{1}{n}\mathbf{J}_n$ and $(\mathbf{I}_n - \frac{1}{n}\mathbf{J}_n)^2 = \mathbf{I}_n - \frac{1}{n}\mathbf{J}_n$. In conclusion, expression (A9.4) is a quadratic form $\boldsymbol{u}'\mathbf{A}\boldsymbol{u}$ in the random vector $\boldsymbol{u}$, defined by matrix $\mathbf{A} = \mathbf{I}_n - \frac{1}{n}\mathbf{J}_n$. To write the sample variance in matrix notation, it suffices to multiply $\mathbf{I}_n - \frac{1}{n}\mathbf{J}_n$ by $\frac{1}{n-1}$.

In a second step, let us see how Total SS, Treatment SS and Error SS can be written as quadratic forms in the spatial or temporal context in the case of the CRD. In this case,

- Total SS $= \sum_{i=1}^{p} \sum_{j=1}^{n} (U_{ij}(s) - \bar{U})^2$, with $\bar{U} = \frac{1}{N}\boldsymbol{1}'_N\, \boldsymbol{u}$ where $N = np$, so that

$$\text{Total SS } = \boldsymbol{u}'(\boldsymbol{I}_N - \frac{1}{N}\mathbf{J}_N)\boldsymbol{u}; \tag{A9.5}$$

- Treatment SS $= n\sum_{i=1}^{p} (\bar{U}_i - \bar{U})^2$, with $\bar{U}_i = \frac{1}{n}\boldsymbol{b}'_{(i)}\boldsymbol{u}$ where $\boldsymbol{b}'_{(i)}$ $(1 \times N)$ has all its components equal to zero, except components $(i-1)n + 1, \ldots, in$ which are equal to one, so that $\bar{U}_i - \bar{U} = \frac{1}{n}\boldsymbol{b}'_{(i)}\boldsymbol{u} - \frac{1}{N}\boldsymbol{1}'_N\boldsymbol{u}$ $= \left(\frac{1}{n}\boldsymbol{b}'_{(i)} - \frac{1}{N}\boldsymbol{1}'_N\boldsymbol{u}\right)$ and

$$\text{Treatment SS } = \boldsymbol{u}'(\mathbf{B} - \frac{1}{N}\mathbf{J}_N)\boldsymbol{u}, \tag{A9.6}$$

  where $\mathbf{B}$ $(N \times N)$ is obtained by copying and pasting each $\boldsymbol{b}'_{(i)} n$ times for $i = 1, \ldots, p$;
- finally, by subtraction,

$$\text{Error SS } = \boldsymbol{u}'(\mathbf{I}_N - \mathbf{B})\boldsymbol{u}. \tag{A9.7}$$

For a given experimental design, each and every ANOVA SS can be written as a quadratic form $\boldsymbol{u}'\mathbf{A}\boldsymbol{u}$ defined by a symmetrical $N \times N$ matrix $\mathbf{A}$, where N denotes the total number of observations (e.g., $N = npr$ for the RCBD >1 rep and p^2 for the "LSD"). Writing the Total SS and Treatment SS like that should not be more difficult for other experimental designs than for the CRD above. Some of the other ANOVA SS may require some subtleties, though. For example, the Error SS $= \sum_{i=1}^{p} \sum_{j=1}^{n} \sum_{k=1}^{r} (U_{ijk}(x, y) - \bar{U}_{ij})^2$ for the RCBD >1 rep and the Error SS $= \sum_{j_1=1}^{p} \sum_{j_2=1}^{p} (U_{ij_1j_2}(x, y) - \bar{U}_i - \bar{U}_{j_1} - \bar{U}_{j_2} + 2\bar{U})^2$ for the "LSD," both in 2-D space. It is then critical, but not straightforward, to use the correct sample means in relation to the indices (x, y) indicating the spatial location of plots. Such subtleties can be found in the relevant SAS codes of the Expectation software (on the CD-ROM).

A9.2 Properties of quadratic forms in normal random vectors

Quadratic (and bilinear) forms in random vectors have several useful properties to offer. Below, the focus is on first-order and second-order moment properties (i.e., the expected value and the population variance) and their application to ANOVA SS in direct relation to equations (9.7) and (9.8), but probability distributional properties are interesting, too. Salient material from Searle (1971, Chapter 2) is used. Note that the normality of random vectors (of observations) is not required for the first-order moment property.

If (i) the random vector $\boldsymbol{u}$ $(N \times 1)$ follows a probability distribution such that $\mathrm{E}[\boldsymbol{u}] = \boldsymbol{m}$ and $\mathrm{Var}[\boldsymbol{u}] = \Sigma$ and (ii) $\boldsymbol{u}'\mathbf{A}\boldsymbol{u}$ denotes a quadratic form in $\boldsymbol{u}$ (Section A9.1), then

$$\mathrm{E}[\boldsymbol{u}'\mathbf{A}\boldsymbol{u}] = \mathrm{tr}(\mathbf{A}\Sigma) + \boldsymbol{m}'\mathbf{A}\boldsymbol{m}, \tag{A9.8}$$

where $\mathrm{tr}(\cdot)$ denotes the trace operator, which applies to any square matrix and returns a value equal to the sum of diagonal entries of the matrix.

If $\boldsymbol{u}$ follows an N-variate normal distribution $N_N(\boldsymbol{m}, \Sigma)$, which implies that $\mathrm{E}[\boldsymbol{u}] = \boldsymbol{m}$ and $\mathrm{Var}[\boldsymbol{u}] = \Sigma$, then

$$\mathrm{Var}[\boldsymbol{u}'\mathbf{A}\boldsymbol{u}] = \mathrm{tr}\{(\boldsymbol{A}\Sigma)^2\} + \boldsymbol{m}'\mathbf{A}\Sigma\mathbf{A}\boldsymbol{m}, \tag{A9.9}$$

Equations (A9.8) and (A9.9) can be used in a variety of applications, including the following two as a follow-up to Section A9.1.

Example 1

If $\boldsymbol{m} = \mu \boldsymbol{1}_n$ (i.e., there is homogeneity of the mean) and $\Sigma = \sigma^2 \mathbf{I}_n$ (i.e., there is homogeneity of the variance and no heterogeneity due to autocorrelation), which means $\mathrm{E}[U(s_i)] = \mu$ for $i = 1, \ldots, n$, $\mathrm{Var}[U(s_i)] = \sigma^2$ for $i = 1, \ldots, n$ and $\mathrm{Corr}[U(s_i), U(s_{i'})] = 0$ for $i \neq i' = 1, \ldots, n$, then

$$\mathrm{E}[S^2] = \mathrm{tr}\left\{\frac{1}{n-1}\left(\mathbf{I}_n - \frac{1}{n}\mathbf{J}_n\right)\sigma^2\mathbf{I}_n\right\} + \frac{\mu^2}{n-1}\mathbf{1}_n'\left(\mathbf{I}_n - \frac{1}{n}\mathbf{J}_n\right)\mathbf{1}_n, \tag{A9.10}$$

which simplifies into σ^2 because $\mathrm{tr}\{\frac{1}{n-1}(\mathbf{I}_n - \frac{1}{n}\mathbf{J}_n)\,\sigma^2\mathbf{I}_n\} = \sigma^2$ and $(\mathbf{I}_n - \frac{1}{n}\mathbf{J}_n)\mathbf{1}_n = \mathbf{0}$. It follows that in these conditions, S^2 is an unbiased estimator of σ^2.

If $\boldsymbol{m} = \mu \boldsymbol{1}_n$ and $\mathrm{Var}[U(s_i)] = \sigma^2$ for $i = 1, \ldots, n$ but $\mathrm{Corr}[U(s_i), U(s_{i'})] \neq 0$ for $i \neq i' = 1, \ldots, n$ (i.e., there is heterogeneity due to autocorrelation) in a way that Σ is not circular, then

$$\mathrm{E}[S^2] = \mathrm{tr}\left\{\frac{1}{n-1}\left(\mathbf{I}_n - \frac{1}{n}\mathbf{J}_n\right)\Sigma\right\} + 0 = \frac{n}{n-1}\sigma^2 - \frac{1}{n(n-1)}\mathrm{tr}(\mathbf{J}_n\Sigma), \tag{A9.11}$$

which can be used to measure the bias due to autocorrelation (Legendre and Dutilleul, 1991).

Example 2

Under H_0: $a_i = 0$ ($i = 1, \ldots, p$) in equation (9.1) in Table 9.1 for the CRD, and if $\boldsymbol{u}$ follows an N-variate normal distribution $N_N(\mu \boldsymbol{1}_N, \sigma^2 \boldsymbol{I}_N)$ with $N = np$, then

$$\mathrm{E}[\text{Error SS}] = \mathrm{tr}\{(\mathbf{I}_N - \mathbf{B})\,\sigma^2\mathbf{I}_N\} + \mu^2 \boldsymbol{1}_N'(\mathbf{I}_N - \mathbf{B})\boldsymbol{1}_N, \tag{A9.12}$$

which simplifies into $p(n-1)\,\sigma^2 + 0$, and

$$\mathrm{Var}[\text{Error SS}] = 2\mathrm{tr}[\{(\mathbf{I}_N - \mathbf{B})\,\sigma^2\mathbf{I}_N\}^2] + \mu^2 \boldsymbol{1}_N'(\mathbf{I}_N - \mathbf{B})\,\Sigma\,(\mathbf{I}_N - \mathbf{B})\boldsymbol{1}_N, \tag{A9.13}$$

which simplifies into $2\,p(n-1)\,\sigma^2 + 0$.

The results above are consistent with two main properties of the chi-square distribution $W \sim \chi^2(k)$: $\mathrm{E}[W] = k$ and $\mathrm{Var}[W] = 2k$. (*Note:* The ANOVA F-test is built on the ratio of independent chi-square distributions.) They also explain the definition of the effective number of df and Box's "epsilon" in equations (9.7) and (9.8): when the circularity condition is satisfied, $\mathrm{E}[\boldsymbol{u}'\mathbf{A}\boldsymbol{u}]$ is equal to some k and $\mathrm{Var}[\boldsymbol{u}'\mathbf{A}\boldsymbol{u}]$ is equal to $2k$, so that $\frac{\{\mathrm{E}[\boldsymbol{u}'\mathbf{A}\boldsymbol{u}]\}^2}{0.5\mathrm{Var}[\boldsymbol{u}'\mathbf{A}\boldsymbol{u}]} = k$ and Box's "epsilon" is one.

In closing, if $\boldsymbol{m} \neq \mu \boldsymbol{1}_N$ and Σ is not circular in the ANOVA for the CRD or another experimental design, then the expressions of expected mean squares (i.e., $\mathrm{MS} = \frac{\mathrm{SS}}{\mathrm{df}}$) become complicated [Table 9.3(a)], the effective numbers of df are different from the classical ones, and Box's "epsilon" is smaller than 1 (Figs. 9.5 and 9.6).

10 · *Conclusions*

This relatively short chapter is "closing the loop" on the thoughts made about the concept of spatio-temporal heterogeneity in previous chapters and on the application of the statistical approaches and methods available for its analysis in the biological and environmental sciences. First, a condensed summary will be presented in the form of a compilation of concluding remarks (Section 10.1). Then, this summary will be completed with final thoughts and a word aimed at being helpful to readers in their future work (Section 10.2). So doing, key points will be highlighted, and sometimes inter-related, with references to biological and environmental examples.

10.1 Concluding remarks

Spatio-temporal heterogeneity takes different forms depending on the family of patterns, point (Chapters 3–5) or surface (Chapters 6–8). For both families of patterns, however, it is possible to distinguish three types of heterogeneity based on the mean, the variance, and autocorrelation. This distinction is the key to sound spatio-temporal heterogeneity analysis (Chapter 2). For surface patterns, the population mean, variance and autocorrelations are readily defined for the underlying stochastic process, thanks to the expectation operator (Appendix A2). For point patterns, integrals of the first-order and second-order intensity functions must be solved to obtain the expected number of points in a given portion of space or time and the population variance of and population covariance between counts (Appendix A3). The latter calculations plead in favor of the approach based on counts in heterogeneity analysis of point patterns, instead of a pure distance-based approach using nearest-neighbor distances (Chapters 3 and 4). Nevertheless, the definition of quadrats of a given size corresponds to a discretization of continuous space, with potentially different results depending on quadrat size; something similar happens when intervals of a given length are defined in continuous time and events of interest are counted therein. In a sense, Ripley's K and L functions (Chapter 5) fall between the two approaches, as they are evaluated at a series of distances h and the K function is related to an expected number of points while $L(h) = \sqrt{\frac{K(h)}{\pi}}$, but one overall mean number of points per space or time unit is used in the definition and estimation of K, which implicitly assumes homogeneity of the mean.

For each family of patterns, the three key types of heterogeneity (of the mean, of the variance, and due to autocorrelation) are or should be inter-related, when they exist, in the statistical analysis. For point patterns, the sample variance of counts (i.e., the numerator in the variance-to-mean ratio, Chapter 3) is distributed differently depending on the sign of autocorrelation, the sample mean being the same. Still for point patterns, the EGLS fitting of a trend surface to counts and the EGLS estimation of a local mean of counts within a moving window are efficient because they incorporate autocorrelation (see the Arisaema 5 example in 2-D space, Chapter 5), whereas kernel-based estimators of the first-order intensity function are classically adjusted for edge effects but not for autocorrelation.

For surface patterns, any heteroscedasticity or autocorrelation should be taken into account whenever possible in the analysis of heterogeneity of the mean, and vice versa. Examples of this were given by (i) the fitting of polynomial and trigonometric regression models with autocorrelated errors to time series of atmospheric CO_2 concentration in Chapter 6; (ii) the EGLS fitting of the LMC to experimental variograms of EGLS residuals (i.e., after the removal of EGLS drift estimates) in the CRAD method (see, e.g., the plant diversity study at Molson Ecological Reserve in Chapter 7); (iii) the application of state-space models to multiple time series or spatio-temporal datasets after the removal of EGLS drift estimates (see the example of air and soil temperatures at Gault Nature Reserve in Chapter 8); and (iv) the assessment of within-subject effects, using modified F-tests adjusted for heteroscedasticity and autocorrelation in the repeated measures ANOVA (see the forestry and wood science examples in time, space, and space-time in Chapter 9).

Distances play a key role in the statistical analysis of spatial and temporal heterogeneity, but this role changes with the family of patterns. At the n points of an observed point pattern, be those points in space or time, the binary response is 1, and for a given point, the nearest-neighbor distance is the smallest distance, in space or time, between the point and the $n - 1$ other points. Frequency distributions, cumulative or not, of nearest-neighbor distances are used to characterize aggregated and regular point patterns vs. the completely random one (Chapter 3). An excess of small nearest-neighbor distances, relative to the completely random point pattern, is indicative of an aggregated point pattern (see, e.g., the California earthquakes in 2-D space, Chapter 3), whereas a lack is typical of a regular point pattern (see, e.g., the Arctic terrestrial and Martian polygonal terrain networks of connected trough-like features, Chapter 5). Ripley's K and L statistics are evaluated as functions of the distance h from a point of the observed pattern, but there is not necessarily a point at that distance from each point and the interest lies in the number of extra points to be expected within the distance (Chapter 5). In the absence of a "true" spatio-temporal distance, spatial (or temporal) distances are used in the analysis of spatio-temporal point patterns (Chapter 4).

In the heterogeneity analysis of surface patterns, the main use of distances is for defining and estimating auto- and cross-correlations under the appropriate assumptions (i.e., stationarity and isotropy). The response is not restricted to be binary (1/0) in surface patterns, so distances calculated from the data can be related to spatial or temporal distances. This is the case in variograms (including Geary's c statistic), since semivariance estimates are nothing but mean squared (Euclidean) distances calculated between data as a function of distance (Chapter 7). In correlograms (including sample auto- and cross-correlation coefficients and Moran's I statistic), the squared distance (sum of squares) is replaced by a scalar product (sum of cross-products), allowing correlation statistics to take positive or negative values (Chapters 6–8). Separate Mantel analyses for the deterministic mean component and the autocorrelated random component were presented in Chapter 7 (see the Lake Erie example), and spatial distances were related to temporal distances in the original version of the Mantel test in Chapter 8 (see the continued example of air and soil temperatures at Gault Nature Reserve).

Despite basic differences (i.e., presence/absence of an order, dimensionality), there exist many resemblances between space and time as frameworks for heterogeneity analysis. Among the few analytical differences, (i) the presence of an order and the uni-dimensionality of time imply that the isotropy condition is automatically met once the assumption of stationarity at order 2 is satisfied in time series analysis, but this is not the case in 2-D space, where isotropy and stationarity have distinct implications for autocorrelation analysis (Chapter 2); (ii) the autoregressive models (Chapter 6) are more popular in time than in 2-D space, and to a lesser degree than in 1-D space, because of the presence of an order in time and the equal spacing of observations in time series in general; and (iii) from the spatio-temporal applications of CRAD (Chapter 8), it seems that the nugget effect is not a common thing in the variogram modeling of time series data. The third difference must be taken with care because temperature data collected remotely instead of on-site are likely to present measurement errors, which should be reflected by a discontinuity at the origin in the variogram.

Scale aspects are very important for spatial and temporal heterogeneity analysis. In 2-D space, for example, the size of sampling plots and the ratio of the range of autocorrelation to the extent of the sampling domain are critical for the correct application of statistical methods such as CRAD (Chapter 7) and for questions regarding sample size (Chapter 9). Space and time may interact, and the heterogeneity arising from their interaction can be analyzed by treating spatio-temporal point patterns as bivariate spatial point patterns (Chapters 3 and 4) and by using CRAD, state-space models, and the Mantel test with spatio-temporal surface pattern data, depending on the numbers of sampling locations and times (Chapter 8). When both these numbers are small, the repeated measures ANOVA (Chapter 9) offers the possibility of testing for interaction and main effects of space and time factors, provided there is sufficient replication.

Key note: The exploration of the Space-Time Response Cube (Fig. 1.1) is thus complete.

10.2 Final thoughts and closing word

What are the implications of a third spatial dimension for spatio-temporal heterogeneity analysis? As far as the calculation of distances is concerned, working in 3-D space is not more difficult than in 2-D space, but analyzing heterogeneity in 3-D space goes beyond calculating distances. As seen with spatio-temporal surface patterns in Chapter 8 and as illustrated by the spatio-temporal point patterns in Fig. 10.1(a) and (b), the horizontal distances are generally larger than the vertical ones. In both families of patterns, this asymmetry in spatial distances follows from a combination of factors, including the shape, size and substance of our planet and the consequence (i.e., gravity) of its rotational movement around the Sun. (Exceptions, i.e., very large vertical distances, arise when data are collected by satellites or at the bottom of oceans.)

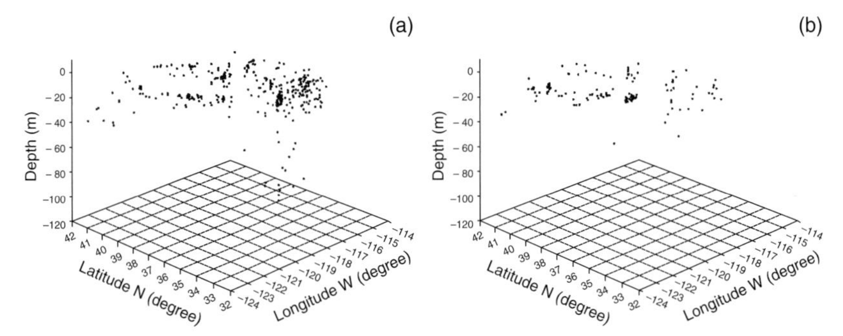

Fig. 10.1. Epicenters of California earthquakes of magnitude 4 or more, located in 3-D space (longitude, latitude, depth): (a) 1990–1999 and (b) 2000–2009.

The depth of epicenters in the example of California earthquakes provides a third spatial axis for heterogeneity analysis. In view of Fig. 10.1(a) and (b), the 3-D spatial distribution of earthquakes appears very different between the two periods of 10 years. While some deep earthquakes occurred in a limited part of California over the period 1990–1999, no such thing happened over 2000–2009. Given the different order of magnitude of distances (i.e., horizontal, km, vs. vertical, m), it is recommended to proceed separately (i.e., in 2-D horizontally and in 1-D vertically), instead of globally (i.e., in 3-D), both in the purely spatial heterogeneity analyses and in the spatio-temporal heterogeneity analysis (cf. the surface pattern analyses in Chapter 8). Otherwise, the existing patterns and differences are likely to come out improperly.

New technologies are giving researchers access to spatial and temporal datasets that were non-existent before. While the availability of these datasets opens up new avenues for discovery, the assessment of the heterogeneity contained in them offers new challenges to analysts, and may raise fundamental questions about the definitions of Space and Time. For example, consider computed tomography (CT) scanning, a technique originally designed for medical diagnostics and recently adapted for the study of plant structures (Dutilleul *et al.*, 2005). CT scanning a 1-cm long segment of tree branch, with CT images constructed every 0.1 mm, provides a set of more than 25 million observations that can be plotted in 3-D (Fig. 10.2); a CT image is a map of indirect measures of material density, which is produced from a 512 × 512 matrix of CT scan data called "CT numbers." (The segment of tree branch in Fig. 10.2 could be replaced by a segment of trunk of a diameter approaching that of a human thoracic cage.)

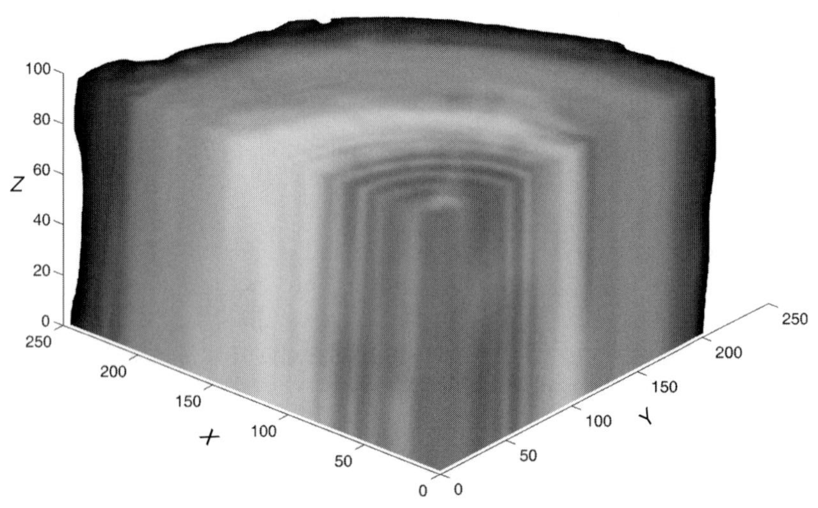

Fig. 10.2. Three-dimensional image of a section of tree branch segment, reconstructed from computed tomography (CT) scan data. The X and Y axes correspond to the plane perpendicular to the couch of the CT scanner, on which the tree branch was installed; the Z axis is parallel to the couch. A segment of 1 cm (along the Z axis) was CT scanned, and a quarter of the CT scan dataset was used to produce (on computer) the sectional image presented here. CT scan data (or CT numbers) are indirect measures of density, and the convention for image contrasts is: light gray = dense material; dark gray = light material.

Such a volumetric approach to the study of tree growth presents several advantages over the traditional tree-ring series analyses performed in 1-D in dendrochronology, but a number of analytical challenges must be met before it becomes real. First of all, the boundaries of annual growth increments must be drawn in 3-D space, horizontally and vertically, so that volumes can be evaluated. It is also possible to compute average annual ring widths, by averaging ring widths in all directions at a given height and by averaging again over heights; the associated variability can be quantified at an unprecedented level of precision, by taking the autocorrelation between observations into account. Furthermore, wood density can be estimated from CT scan data, at the level of the ring as well as within a ring, since CT numbers show fluctuations in density within growth rings (see the light gray vs. dark gray bands in Fig. 10.2). Wood density estimates can be located in 3-D space by using an index $s = (x, y, z)$, where x and y are the horizontal spatial coordinates, and z the vertical one. (By convention, the three axes are represented with capitals in the world of CT scanning.) For ring widths, a spatio-temporal index

$s = (z, \theta, t)$, where θ denotes the angle from a given half-line starting from the center of the wood sample at height z and t is the year of growth, can be used. In these cases, an index with four components, (x, y, z, t), is not necessary because the three spatial components determine the temporal one. All of this is consistent with the discussions and choices made in the tree, forestry, and wood science examples of Chapters 1, 2, 6, and 9.

In closing, I hope that this book provides readers with answers to a large number of questions regarding the analysis of heterogeneity in space and time. With minimum mathematical development, it gives insight into the various aspects of the concepts in the statistical perspective, while illustrating the application of appropriate analytical methods with relevant biological and environmental examples. Accordingly, it should help "open eyes" as well as "prevent mistakes." It suffices to think of the distinction between pattern and process – hypotheses are stated about population parameters and processes, instead of patterns which are the sample data from which statistics are computed – and of the potential effects of heterogeneity on the outcome of correlation analysis – different relationships can be found on raw data vs. drifts, autocorrelated errors and distances. Thereafter, I am confident that the parametric triumvirate mean–variance–autocorrelation, to be used in a model-based approach, will prove helpful in many spatio-temporal heterogeneity studies in the future, and the computer programs on the CD-ROM will definitively facilitate the applications.

References

Akaike, H. (1974a). A new look at the statistical model identification. *IEEE Transactions on Automatic Control*, **AC-19**, 716–722.

Akaike, H. (1974b). Markovian representation of stochastic processes and its application to the analysis of autoregressive moving average processes. *Annals of the Institute of Mathematical Statistics*, **26**, 363–387.

Akaike, H. (1975). Markovian representation of stochastic processes by canonical variables. *SIAM Journal on Control*, **13**, 162–173.

Allen, T.F.H. and Hoekstra, T.W. (1991). Role of heterogeneity in scaling of ecological systems under analysis. *Ecological Heterogeneity*, eds. Kolasa, J. and Pickett, S.T.A., pp. 47–68. New York: Springer-Verlag.

Alpargu, G. and Dutilleul, P. (2001). Efficiency analysis of ten estimation procedures for quantitative linear models with autocorrelated errors. *Journal of Statistical Computation and Simulation*, **69**, 257–275.

Alpargu, G. and Dutilleul, P. (2006). Stepwise regression in mixed quantitative linear models with autocorrelated errors. *Communications in Statistics – Simulation and Computation*, **35**, 79–104.

Anderson, T.W. (1971). *The Statistical Analysis of Time Series*. New York: Wiley.

Aoki, M. (1990). *State Space Modeling of Time Series,* Second, Revised and Enlarged Edition. Berlin: Springer-Verlag.

Bell, G. and Lechowicz, M.J. (1994). Spatial heterogeneity at small scales and how plants respond to it. *Exploitation of Environmental Heterogeneity by Plants: Ecophysiological Processes above and below Ground*, eds. Caldwell, M.M. and Pearcy, R.W., pp. 391–414. San Diego: Academic Press.

Box, G.E.P. (1954a). Some theorems on quadratic forms applied in the study of analysis of variance problems. I. Effect of inequality of variance in the one-way classification. *Annals of Mathematical Statistics*, **25**, 290–302.

Box, G.E.P. (1954b). Some theorems on quadratic forms applied in the study of analysis of variance problems. II. Effects of inequality of variance and of correlation between errors in the two-way classification. *Annals of Mathematical Statistics*, **25**, 484–498.

Box, G.E.P. and Cox, D.R. (1964). An analysis of transformations. *Journal of the Royal Statistical Society Series B*, **26**, 211–243.

Box, G.E.P., Jenkins, G.M., and Reinsel, G.C. (1994). *Time Series Analysis: Forecasting and Control*, Third Edition. Englewood Cliffs: Prentice Hall.

Brillinger, D.R. (1981). *Time Series: Data Analysis and Theory*, Expanded Edition. New York: McGraw-Hill.

Brillinger, D.R. (1994). Time series, point processes and hybrids. *Canadian Journal of Statistics*, **22**, 177–206.

Brownie, C., Hines, J.E., Nichols, J.D., Pollock, K.H., and Hestbeck, J.B. (1993). Capture–recapture studies for multiple strata including non-Markovian transitions. *Biometrics*, **49**, 1173–1187.

Burrough, P.A. (1981). Fractal dimensions of landscapes and other environmental data. *Nature*, **294**, 240–242.

Carrière, Y., Ellers-Kirk, C., Sisterson, M., *et al.* (2003). Long-term regional suppression of pink bollworm by *Bacillus thuringiensis* cotton. *Proceedings of the National Academy of Sciences, USA*, **100**, 1519–1523.

Carrière, Y., Dutilleul, P., Ellers-Kirk, C., *et al.* (2004). Sources, sinks, and the zone of influence of refuges for managing insect resistance to Bt crops. *Ecological Applications*, **14**, 1615–1623.

Chatfield, C. (2004). *The Analysis of Time Series: An Introduction*, Sixth Edition. Boca Raton: Chapman and Hall/CRC.

Chilès, J.P. and Delfiner, P. (1999). *Geostatistics: Modeling Spatial Uncertainty*. New York: Wiley.

Christakos, G. (2000). *Modern Spatiotemporal Geostatistics*. Oxford: Oxford University Press.

Clapham, A.R. (1936). Over-dispersion in grassland communities and the use of statistical methods in plant ecology. *Journal of Ecology*, **24**, 232–251.

Clark, P.J. and Evans, F.C. (1954). On some aspects of spatial pattern in biological populations. *Science*, **121**, 397–398.

Cliff, A.D. and Ord, J.K. (1981). *Spatial Processes: Models and Applications*. London: Pion.

Cochran, W.G. (1963). *Sampling Techniques*, Second Edition. New York: Wiley.

Cressie, N.A.C. (1985). Fitting variogram models by weighted least squares. *Mathematical Geology*, **17**, 563–586.

Cressie, N.A.C. (1993). *Statistics for Spatial Data*, Revised Edition. New York: Wiley.

Cressie, N.A.C. and Hawkins, D.M. (1980). Robust estimation of the variogram, I. *Mathematical Geology*, **12**, 115–125.

Crowder, M.J. and Hand, D.J. (1990). *Analysis of Repeated Measures*. London: Chapman and Hall.

Daley, D.J. and Vere-Jones, D. (1988). *Introduction to the Theory of Point Processes*. New York: Springer-Verlag.

Diggle, P.J. (1990). *Time Series: A Biostatistical Introduction*. Oxford: Clarendon Press.

Diggle, P.J. (2003). *Statistical Analysis of Spatial Point Patterns*, Second Edition. London: Arnold Publishers.

Diggle, P.J., Liang, K.-Y. and Zeger, S.L. (1996). *Analysis of Longitudinal Data*. Oxford: Clarendon Press.

Donnelly, K. (1978). Simulation to determine the variance and edge-effect of total nearest neighbour distance. *Simulation Methods in Archaeology*, ed. Hodder, I.R., pp. 91–95. London: Cambridge University Press.

Du Rietz, G.E. (1930). Classification and nomenclature of vegetation. *Svensk Botanisk Tidskrift*, **24**, 489–503.

Dutilleul, P. (1990). "Apport en analyse spectrale d'un périodogramme modifié et modélisation des séries chronologiques avec répétitions en vue de leur comparaison en fréquence". Doctoral dissertation (D.Sc. in Mathematics), Université catholique de Louvain, Belgium.

Dutilleul, P. (1993a). Spatial heterogeneity and the design of ecological field experiments. *Ecology*, **74**, 1646–1658.

Dutilleul, P. (1993b). Modifying the *t*-test for assessing the correlation between two spatial processes. *Biometrics*, **49**, 305–314.

Dutilleul, P. (1995). Rhythms and autocorrelation analysis. *Biological Rhythm Research*, **26**, 173–193.

Dutilleul, P. (1998a). Incorporating scale in ecological experiments: Study design. *Ecological Scale: Theory and Applications*, eds. Peterson, D.L. and Parker, V.T., pp. 369–386. New York: Columbia University Press.

Dutilleul, P. (1998b). Incorporating scale in ecological experiments: Data analysis. *Ecological Scale: Theory and Applications*, eds. Peterson, D.L. and Parker, V.T., pp. 387–425. New York: Columbia University Press.

Dutilleul, P. (1999). The MLE algorithm for the matrix normal distribution. *Journal of Statistical Computation and Simulation*, **64**, 105–123.

Dutilleul, P. (2001). Multi-frequential periodogram analysis and the detection of periodic components in time series. *Communications in Statistics – Theory and Methods*, **30**, 1063–1098.

Dutilleul, P. (2008). A note on sufficient conditions for valid unmodified *t* testing in correlation analysis with autocorrelated and heteroscedastic sample data. *Communications in Statistics – Theory and Methods*, **37**, 137–145.

Dutilleul, P. and Carrière, Y. (1998). Among-environment heteroscedasticity and the estimation and testing of genetic correlation. *Heredity*, **80**, 403–413.

Dutilleul, P. and Legendre, P. (1992). Lack of robustness in two tests of normality against autocorrelation in sample data. *Journal of Statistical Computation and Simulation*, **42**, 79–91.

Dutilleul, P. and Legendre, P. (1993). Spatial heterogeneity against heteroscedasticity: An ecological paradigm versus a statistical concept. *Oikos*, **66**, 152–171.

Dutilleul, P. and Pinel-Alloul, B. (1996). A doubly multivariate model for statistical analysis of spatio-temporal environmental data. *Environmetrics*, **7**, 551–566.

Dutilleul, P. and Potvin, C. (1995). Among-environment heteroscedasticity and genetic autocorrelation: implications for the study of phenotypic plasticity. *Genetics*, **139**, 1815–1829.

Dutilleul, P. and Till, C. (1992). Evidence of periodicities related to climate and planetary behaviors in ring-width chronologies of Atlas cedar (*Cedrus atlantica*) in Morocco. *Canadian Journal of Forest Research*, **22**, 1469–1482.

Dutilleul, P., Cumming, B., and Lontoc-Roy, M. (2011). Assessing shifts in drought conditions using autocorrelogram and periodogram analyses of paleolimnological temporal series from lakes in central and western North America. *Tracking Environmental Change Using Lake Sediments: Data Handling and Statistical Techniques*, eds. Birks, H.J.B., Juggins, S., Lotter, A. and Smol, J.P.. New York: Springer.

Dutilleul, P., Deswysen, A.G., Fischer, V., and Maene, D. (2000a). Time-dependent transition probabilities and the assessment of seasonal effects on within-day variations in chewing behaviour of housed sheep. *Applied Animal Behaviour Science*, **68**, 13–37.

Dutilleul, P., Haltigin, T.W., and Pollard, W.H. (2009). Analysis of polygonal terrain landforms on Earth and Mars through spatial point patterns. *Environmetrics*, **20**, 206–220.

Dutilleul, P., Herman, M., and Avella-Shaw, T. (1998). Growth rate effects on correlations among ring width, wood density and mean tracheid length in Norway spruce (*Picea abies* (L.) Karst). *Canadian Journal of Forest Research*, **28**, 56–68.

Dutilleul, P., Lontoc-Roy, M., and Prasher, S.O. (2005). Branching out with a CT scanner. *Trends in Plant Science*, **10**, 411–412.

Dutilleul, P. and Pelletier, B. Tests of significance for structural correlations in the linear model of coregionalization. *Mathematical Geosciences* (in press).

Dutilleul, P., Pelletier, B., and Alpargu, G. (2008). Modified *F* tests for assessing the multiple correlation between one spatial process and several others. *Journal of Statistical Planning and Inference*, **138**, 1402–1415.

Dutilleul, P., Stockwell, J.D., Frigon, D., and Legendre, P. (2000b). The Mantel test *versus* Pearson's correlation analysis: assessment of the differences for biological and environmental studies. *Journal of Agricultural, Biological and Environmental Statistics*, **5**, 131–150.

Filion, L. and Cournoyer, L. (1995). Variation in wood structure of eastern larch defoliated by the larch sawfly in subarctic Quebec, Canada. *Canadian Journal of Forest Research*, **25**, 1263–1268.

Filion, M., Dutilleul, P., and Potvin, C. (2000). Optimum experimental design for Free-Air Carbon dioxide Enrichment (FACE) studies. *Global Change Biology*, **6**, 843–854.

Finkenstädt, B., Held, L., and Isham, V., eds. (2007). *Statistical Methods for Spatio-Temporal Systems*. Boca Raton: Chapman and Hall/CRC.

Flandrin, P. (1999). *Time-Frequency/Time-Scale Analysis*. San Diego: Academic Press.

Foroutan-pour, K., Dutilleul, P., and Smith, D.L. (1999). Advances in the implementation of the box-counting method of fractal dimension estimation. *Applied Mathematics and Computation*, **105**, 195–210.

Geary, R.C. (1954). The contiguity ratio and statistical mapping. *The Incorporated Statistician*, **5**, 115–145.

Gneiting, T. (2002). Nonseparable, stationary covariance functions for space-time data. *Journal of the American Statistical Association*, **97**, 590–600.

Golub, G.H. and Pereyra, V. (1973). The differentiation of pseudo-inverses and nonlinear least squares problems whose variables separate. *SIAM Journal of Numerical Analysis*, **10**, 413–432.

Goovaerts, P. (1992). Factorial kriging analysis: A useful tool for exploring the structure of multivariate spatial soil information. *Journal of Soil Science*, **43**, 597–619.

Goovaerts, P. (1997). *Geostatistics for Natural Resources Evaluation*. New York: Oxford University Press.

Graybill, F.A. (1976). *Theory and Application of the Linear Model*. Pacific Grove: Wadsworth and Brooks/Cole.

Graybill, F.A. (1983). *Matrices with Applications in Statistics*, Second Edition. Pacific Grove: Wadsworth and Brooks/Cole.

Greenhouse, S.W. and Geisser, S. (1959). On the methods in the analysis of profile data. *Psychometrika*, **24**, 95–112.

Griffith, D.A. (1987). *Spatial Autocorrelation: A Primer.* Washington: Association of American Geographers.

Haining, R.P. (1990). *Spatial Data Analysis in the Social and Environmental Sciences.* Cambridge: Cambridge University Press.

Helterbrand, J.D. and Cressie, N.A.C. (1994). Universal cokriging under intrinsic coregionalization. *Mathematical Geology*, **26**, 205–226.

Herman, M., Dutilleul, P., and Avella-Shaw, T. (1998). Growth rate effects on temporal trajectories of ring width, wood density and mean tracheid length in Norway spruce (*Picea abies* (L.) Karst). *Wood and Fiber Science*, **30**, 6–17.

Herman, M., Dutilleul, P., and Avella-Shaw, T. (1999). Growth rate effects on intra-ring and inter-ring trajectories of microfibril angle in Norway spruce (*Picea abies*). *IAWA Journal*, **20**, 3–21.

Hughes, G. and Madden, L.V. (1993). Using the beta-binomial distribution to describe aggregated patterns of disease incidence. *Phytopathology*, **83**, 759–763.

Hurlbert, S.H. (1990). Spatial distribution of the montane unicorn. *Oikos*, **58**, 257–271.

Huynh, H. and Feldt, S. (1970). Conditions under which mean square ratios in repeated measurements designs have exact *F*-distributions. *Journal of the American Statistical Association*, **65**, 1582–1589.

Huynh, H. and Feldt, S. (1976). Estimation of the Box correction for degrees of freedom for sample data in randomised block and split-plot designs. *Journal of Educational Statistics*, **1**, 69–82.

Isaaks, E.H. and Srivastava, R.M. (1989). *Applied Geostatistics.* New York: Oxford University Press.

Jardon, Y., Morin, H., and Dutilleul, P. (2003). Périodicité et synchronisme des épidémies de la tordeuse des bourgeons de l'épinette au Québec. *Canadian Journal of Forest Research*, **33**, 1947–1961.

Jenkins, G.M. and Watts, D.G. (1968). *Spectral Analysis and its Applications.* San Francisco: Holden-Day.

Journel, A.G. and Huijbregts, C.J. (1978). *Mining Geostatistics.* London: Academic Press.

Keeling, C.D. and Whorf, T.P. (2005). Atmospheric CO_2 records from sites in the SIO air sampling network (CDIAC, Oak Ridge, TN). http://cdiac.ornl.gov/trends/co2/sio-keel.htm.

Kolasa, J. and Rollo, C.D. (1991). The heterogeneity of heterogeneity: a glossary. *Ecological Heterogeneity*, eds. Kolasa, J. and Pickett, S.T.A., pp. 1–23. New York: Springer-Verlag.

Kotz, S. and Johnson, N.L. eds. (1982). *Encyclopedia of Statistical Sciences, Volume 1: A to Circular Probable Error.* New York: Wiley.

Larocque, G. (2008). *Towards a Coherent Framework for the Multi-scale Analysis of Spatial Observational Data: Linking Concepts, Statistical Tools and Ecological Understanding.* Ph.D. thesis, McGill University, Montréal, Canada.

Larocque, G., Dutilleul, P., Pelletier, B., and Fyles, J.W. (2006). Conditional Gaussian co-simulation of regionalized components of soil variation. *Geoderma*, **134**, 1–16.

Larocque, G., Dutilleul, P., Pelletier, B., and Fyles, J.W. (2007). Characterization and quantification of uncertainty in coregionalization analysis. *Mathematical Geology*, **39**, 263–288.

Legendre, P. and Dutilleul, P. (1991). Comments on Boyle's Acidity and organic carbon in lake water: variability and estimation of means. *Journal of Paleolimnology*, **6**, 94–101.

Legendre, P. and Legendre, L. (1998). *Numerical Ecology*, 2nd English Edition. Amsterdam: Elsevier.

Madden, L.V. and Hughes, G. (1994). BBD – Computer software for fitting the beta-binomial distribution to disease incidence data. *Plant Disease*, **78**, 536–540.

Madden, L.V., Hughes, G., and Ellis, M.A. (1995). Spatial heterogeneity of the incidence of grape downy mildew. *Phytopathology*, **85**, 269–275.

Mantel, N. (1967). The detection of disease clustering and a generalized regression approach. *Cancer Research*, **27**, 209–220.

Marchant, B.P. and Lark, R.M. (2007). Robust estimation of the variogram by residual maximum likelihood. *Geoderma*, **140**, 62–72.

Marcoux, M., Larocque, G., Auger-Méthé, M., Dutilleul, P., and Humphries, M.M. (2010). Statistical analysis of animal observations and associated marks distributed in time using Ripley's functions. *Animal Behaviour*, **80**, 329–337.

Matérn, B. (1972). Poisson processes in the plane and related models for clumping and heterogeneity. *NATO Advanced Study Institute in Statistical Ecology*. College Town: Pennsylvania State University Press.

Matheron, G. (1962). *Traité de Géostatistique Appliquée, Tome 1. Mémoires du Bureau de Recherches Géologiques et Minières, 14*. Paris: Editions Technip.

Matheron, G. (1978). *Estimer et Choisir: Un Essai sur la Pratique des Probabilités*. Les Cahiers du Centre de Morphologie Mathématique No. 7. Fontainebleau: Centre de Géostatistique.

Matheron, G. (1989). *Estimating and Choosing: An Essay on Probability in Practice*. Berlin: Springer.

McElroy, F.W. (1967). A necessary and sufficient condition that ordinary least squares estimators be best linear unbiased. *Journal of the American Statistical Association*, **62**, 1302–1304.

McIntosh, R.P. (1991). Concept and terminology of homogeneity and heterogeneity in ecology. *Ecological Heterogeneity*, eds. Kolasa, J. and Pickett, S.T.A., pp. 24–46. New York: Springer-Verlag.

Mead, R. (1988). *The Design of Experiments: Statistical Principles for Practical Application*. Cambridge: Cambridge University Press.

Milliken, G.A. and Johnson, D.E. (1992). *Analysis of Messy Data, Volume 1: Designed Experiments*. New York: Chapman and Hall/CRC Press.

Moran, P.A.P. (1950). Notes on continuous stochastic phenomena. *Biometrika*, **37**, 17–23.

Nguyen, D., Dutilleul, P., and Rau, M.E. (2002). The impact of nutrition and exposure to the parasite *Plagiorchis elegans* (Trematoda: Plagiorchiidae) on the development of *Aedes aegypti* (Diptera: Culicidae): Analysis by time-dependent transition probabilities. *Environmental Entomology*, **31**, 54–64.

Oden, N.L. (1977). Partitioning dependence in nonstationary behavioural sequences. *Quantitative Methods in the Study of Animal Behaviour*, ed. Hazlett, B.A., pp. 203–220. New York: Academic Press.

Oden, N.L. and Sokal, R.R. (1986). Directional autocorrelation: An extension of spatial correlograms to two dimensions. *Systematic Zoology*, **35**, 608–617.

Pardo-Igúzquiza, E. (1997). MLREML: a computer program for the inference of spatial covariance parameters by maximum likelihood and restricted maximum likelihood. *Computers and Geosciences*, **23**, 153–162.

Patterson, H.D. and Thompson, R. (1971). Recovery of interblock information when block sizes are unequal. *Biometrika*, **58**, 545–554.

Pelletier, B., Dutilleul, P., Larocque, G., and Fyles, J.W. (2004). Fitting the linear model of coregionalization by generalized least squares. *Mathematical Geology*, **36**, 323–343.

Pelletier, B., Dutilleul, P., Larocque, G., and Fyles, J.W. (2009a). Coregionalization analysis with a drift for multi-scale assessment of spatial relationships between ecological variables, 1. Estimation of drift and random components. *Environmental and Ecological Statistics*, **16**, 439–466.

Pelletier, B., Dutilleul, P., Larocque, G., and Fyles, J.W. (2009b). Coregionalization analysis with a drift for multi-scale assessment of spatial relationships between ecological variables, 2. Estimation of correlations and coefficients of determination. *Environmental and Ecological Statistics*, **16**, 467–494.

Pinel-Alloul, B., Guay, C., Angeli, N., *et al.* (1999). Large-scale spatial heterogeneity of macrozooplankton in Lake Geneva. *Canadian Journal of Fisheries and Aquatic Sciences*, **56**, 1437–1451.

Potvin, C. and Dutilleul, P. (2009). Neighborhood effects and size-asymmetric competition in a tree plantation varying in diversity. *Ecology*, **90**, 321–327.

Potvin, C., Lechowicz, M.J., and Tardif, S. (1990). The statistical analysis of ecophysiological response curves obtained from experiments involving repeated measures. *Ecology*, **71**, 1389–1400.

Priestley, M.B. (1981). *Spectral Analysis and Time Series*. London: Academic Press.

Ripley, B.D. (1981). *Spatial Statistics*. New York: Wiley.

Rook, A.J. and Penning, P.D. (1991). Stochastic models of grazing behaviour in sheep. *Applied Animal Behaviour Science*, **31**, 237–250.

Rouanet, H. and Lépine, D. (1970). Comparison between treatments in a repeated-measures design: ANOVA and multivariate methods. *British Journal of Mathematical and Statistical Psychology*, **23**, 147–163.

SAS Institute Inc. (2009). *SAS for Windows, Release 9.2*. Cary: SAS Institute.

Schabenberger, O. and Gotway, C.A. (2005). *Statistical Methods for Spatial Data Analysis*. Boca Raton: Chapman and Hall/CRC.

Scheffé, H. (1959). *The Analysis of Variance*. New York: Wiley.

Schelstraete, I., Dutilleul, P., and Weyers, M.-H. (1997). Neonatal exposure to atypical *Zeitgeber* in the Wistar rat: Effects on development of circadian rhythms of drinking and motor activities. *Biological Rhythm Research*, **28**, 69–86.

Schelstraete, I., Knaepen, É., Dutilleul, P., and Weyers, M.-H. (1992). Maternal behavior in the Wistar rat under atypical *Zeitgeber. Physiology and Behavior*, **52**, 189–193.

Schelstraete, I., Knaepen, É., Dutilleul, P., and Weyers, M.-H. (1993). Simulation chez le rat des caractéristiques chronobiologiques et ontogénétiques des troubles de l'humeur. *Rythmes Biologiques: De la Molécule à l'Homme – Actes du Congrès GERB92*, pp. 169–178. Paris: Polytechnica.

Schuster, A. (1898). On the investigation of hidden periodicities with application to a supposed 26-day period of meteorological phenomena. *Terrestrial Magnetism and Atmospheric Electricity*, **3**, 13–41.

Schwarz, G. (1978). Estimating the dimension of a model. *Annals of Statistics*, **6**, 461–464.

Searle, S.R. (1971). *Linear Models*. New York: Wiley.

Shachak, M. and Brand, S. (1991). Relations among spatiotemporal heterogeneity, population abundance, and variability in a desert. *Ecological Heterogeneity*, eds. Kolasa, J. and Pickett, S.T.A., pp. 202–223. New York: Springer-Verlag.

Shannon, C.E. (1948). A mathematical theory of communication. *Bell System Technical Journal*, **27**, 379–423.

Sklar, L. (1984). *Space, Time, and Spacetime*. Berkeley: University of California Press.

Smith, D.M. (1983). Maximum likelihood estimation of the parameters of the beta binomial distribution. *Applied Statistics*, **32**, 192–204.

Sokal, R.R. (1986). Spatial data analysis and historical processes. *Proceedings of the Fourth International Symposium on Data Analysis and Informatics*, eds. Diday, E., Escoufier, Y., Lebart, L., Pages, J., Schektman, Y., and Tomassone, R., pp. 29–43. Amsterdam: North-Holland.

Sokal, R.R. and Rohlf, F.J. (2003). *Biometry: The Principles and Practice of Statistics in Biological Research*. New York: Freeman.

Stockwell, J.D. and Sprules, W.G. (1995). Spatial and temporal patterns of zooplankton biomass in Lake Erie. *ICES Journal of Marine Science*, **52**, 557–564.

Thomas, J.J. and Wallis, K.F. (1971). Seasonal variation in regression analysis. *Journal of the Royal Statistical Society, A* **134**, 57–72.

Tosic, M., Bonnell, R.B., Dutilleul, P., and Oxenford, H.A. (2009). Runoff water quality, landuse and environmental impacts on the Bellairs fringing reef, Barbados. *Remote Sensing and Geospatial Technologies for Coastal Ecosystem Assessment and Management*, ed. Yang, X., pp. 521–553. Berlin: Springer-Verlag.

Turchin, P. (1998). *Quantitative Analysis of Movement: Measuring and modeling population redistribution in plants and animals*. Sunderland: Sinauer Associates.

Upton, G.J.G. and Fingleton, B. (1985). *Spatial Data Analysis by Example. Volume I: Point Pattern and Quantitative Data*. Chichester: Wiley.

van den Wollenberg, A.L. (1977). Redundancy analysis: An alternative for canonical correlation analysis. *Psychometrika*, **42**, 207–219.

Wallis, K.F. (1974). Seasonal adjustment and relations between variables. *Journal of the American Statistical Association*, **69**, 18–31.

Webster, R. and Oliver, M.A. (1992). Sample adequately to estimate variograms of soil properties. *Journal of Soil Science*, **43**, 177–192.

Whittaker, J. (1984). Model interpretation from the additive elements of the likelihood function. *Applied Statistics*, **33**, 52–64.

Winer, B.J., Brown, D.R., and Michels, K.M. (1991). *Statistical Principles in Experimental Design, Third Edition.* New York: McGraw-Hill.

Zhang, X.F., Van Eijkeren, J.C.H., and Heemink, A.W. (1995). On the weighted least-squares method for fitting a semivariogram model. *Computers and Geosciences*, **21**, 605–608.

Author index

Subject index